Economics of ecosystem management

Tasks for vegetation science 14

Series Editor

HELMUT LIETH

University of Osnabrück, F.R.G.

Economics of ecosystem management

edited by

D.O. HALL, N. MYERS and N.S. MARGARIS

1985 DR W. JUNK PUBLISHERS
A MEMBER OF THE KLUWER ACADEMIC PUBLISHERS GROUP
DORDRECHT / BOSTON / LANCASTER

Distributors

for the United States and Canada: Kluwer Boston, Inc., 190 Old Derby Street, Hingham, MA 02043, USA
for the UK and Ireland: Kluwer Academic Publishers, MTP Press Limited, Falcon House, Queen Square, Lancaster LA1 1RN, UK
for all other countries: Kluwer Academic Publishers Group, Distribution Center, P.O. Box 322, 3300 AH Dordrecht, The Netherlands

Library of Congress Cataloging in Publication Data

International Symposium on Economics of Ecosystem
Management (1983 : Chalkidikē, Greece)
Economics of ecosystem management.

(Tasks for Vegetation science ; 14)
1. Environmental policy--Congresses. 2. Environmental protection--Congresses. I. Hall, D. O. (David Oakly) II. Myers, Norman. III. Margaris, N. S. IV. Title. V. Series.
HC79.E51594 1983 363.7'0068'1 84-15507

ISBN-13: 978-94-010-8928-9 e-ISBN-13: 978-94-009-5502-8
DOI: 10.1007/978-94-009-5502-8

Cover design: Max Velthuijs

Preface

This volume contains selected papers presented at the International Symposium on 'Economics of Ecosystem Management' held in Halkidiki, Greece from 26 to 30 September 1983. The meeting was designed to consider as much as possible the interaction between the economy and ecology, and in parallel to provide a forum for the presentation of the ecological and developmental initiatives of the Ministry of Youth of Greece.

As chairman of the organizing committee I want to express our thanks to OECD, EEC and UNESCO for sponsoring the Symposium. From Greece the Ministries of Culture and Environment are gratefully acknowledged. Last but not least, I want to express our sincere thanks for the continuous support to the Ministry of Youth and Sports of Greece, and especially to the Minister Mr. Kostas Laliotis because he not only adopted the idea for sponsoring events like this, but also because of his interest which became obvious with his presence and talk during the meeting.

The staff of the Ecology Division, namely Drs. M. Arianoutsou and D. Vokou, Messrs. J. Pandis, Th.A. Mardiris, S. Paraskevopoulos and Miss A. Papagianni provided great help concerning the organization, as well as Mrs. A. Karamanli-Vlahopoulou, my secretary, who is always behind these events.

N.S. Margaris

Contents

PART THREE: MANAGEMENT OF SPECIES

PART FOUR: MANAGEMENT OF LAND-BASED ECOSYSTEMS

PART FIVE: BIOMASS

List of contributors

R.A. Beatty
Department of Landscape Architecture
202 Wurster Hall
University of California, Berkeley
Berkeley, CA 94720 USA

J.K. Bowers
School of Economic Studies
The University of Leeds
Leeds LS2 9JT England

F. Brouwer
Institute for Environmental Studies
Free University
P.O. Box 7161
1007 MC Amsterdam, The Netherlands

G.M. Brown Jr.
Department of Economics
University of Washington
Seattle, WA 98195 USA

H. Coccossis
Secretariat of the National Council
for Physical Planning and the Environment
Ministry of National Economy
Athens, Greece

A. Davaki
Ministry of Physical Planning,
Housing and the Environment
17 Amaliados St.
Athens, Greece

S. Fabrewitz
Arbeitsgruppe Systemforschung
Universität Osnabrück
Post Box 4469
D-4500 Osnabrück, FRG

A.C. Fisher
Energy and Resources Group
University of California
Berkeley, CA 94720 USA

N. Garabaghi
Division of Human Settlements &
Socio-cultural Environment
UNESCO
7 Place de Fontenoy
75700-Paris, France

D.O. Hall
Department of Plant Sciences
King's College London
68 Half Moon Lane
London SE24 9JF England

W.M. Hanemann
Energy and Resources Group
University of California
Berkeley, CA 94720 USA

L. Hordijk
Department of Economics
Free University
P.O. Box 7161
1007 MC Amsterdam, The Netherlands

A.M. Jansson
Askö Laboratory
Institute of Marine Ecology
University of Stockholm
10691 Stockholm, Sweden

E.G. Koukios
Laboratory of Organic Chemical Technology
Department of Chemical Engineering
National Technical University
Athens, Greece

M. Lawrence
School of Landscape Architecture
Leeds Polytechnic
Brunswick Terrace
Leeds LS2 8BU England

H. Lieth
Arbeitsgruppe Systemforschung
Universität Osnabrück
Post Box 4469
D-4500 Osnabrück, FRG

M. Lodenius
Department of Environmental Conservation
University of Helsinki
Viikki
SF-00710 Helsinki 71, Finland

L. Louloudis
Department of Comparative Agriculture
Iera Odos 75
Athens, Greece

B.G. Lundholm
Commission for Research on Natural Resources
Box 6710
S-113 85 Stockholm, Sweden

M. Madders
School of Landscape Architecture
Leeds Polytechnic
Brunswick Terrace
Leeds LS2 8BU England

R. Manidaki
Gr. Xenopoulou 5
Athens, Greece

N.S. Margaris
Laboratory of Ecology
Department of Biology
School of Sciences
University of Thessaloniki
Post Box 119
Thessaloniki, Greece

J. Martin
Centre of Environmental Management and Planning
Department of Geography
University of Aberdeen
Old Aberdeen AB9 2UF England

N. Myers
Upper Meadow, Old Road
Headington, Oxford OX3 8SZ England

P. Nijkamp
Department of Economics
Free University
P.O. Box 7161
1007 MC Amsterdam, The Netherlands

M. Pyrovetsi
Division of Zoology
Biology Department
University of Thessaloniki
Thessaloniki, Greece

P. Söderbaum
Department of Economics and Statistics
The Swedish University of
Agricultural Sciences
S-750 07 Uppsala, Sweden

U. Svedin
Wenner Gren Center
Sveavagen 166
S-11346 Stockholm, Sweden

J. Swierzbinski
Department of Economics
University of Washington
Seattle, WA 98195 USA

B. Thober
Arbeitsgruppe Systemforschung
Universität Osnabrück
Post Box 4469
D-4500 Osnabrück, FRG

E. Titus
Department of Regional Sciences
University of Pennsylvania
Philadelphia, PA 19106 USA

A. Tsangridis
University of Athens
Biology Department
Section of Ecology and Systematics
Panepistimiopolis
Athens, Greece

N.K. Vernicos
Akti Miaouli 35-39
Piraeus 18535, Greece

CHAPTER 1

Opening address

GUNTER SCHNEIDER

Mr. Chairman, Ministers, Ladies and Gentlemen,

Allow me, first of all, to convey to you the good wishes of Dr. Narjes, the Member of the Commission who is particulary responsible for environmental protection as well as of Mr. Andreopoulos, Director general of the Department of Environment, Consumer Protection and Nuclear Safety.

Those amongst you who organised this International Symposium know with what unreserved approval and support Dr. Narjes took up the initiative of Minister Tritsis and Prof. Margaris of the University of Thessaloniki to discuss, in the context of an international exchange of experience, the 'Economics of Ecosystems Management for the Mediterranean Region'.

All the more reason, therefore, for Dr. Narjes and Mr. Andreopoulos to regret that they cannot be here today and to assure you hat the recommandations and conclusions of this International Symposium will be duly taken into account by the Commission in its day-to-day work.

You will, of course, not expect me in an opening speech to go into all the various aspects of EEC environment policy which could be of interest to the work of this International Symposium. However, I do think it is necessary to take a brief look at the overall strategy of the European Environment Programme from its beginning to the present day in order for you to benefit from the process by which we have arrived at the necessity of integrated management. After this short survey, Mr. Chairman, I would then like to highlight two or three points about the Community's environmental policy which I think are particularly important and relevant to the present conference.

Finally, I would like to say something in particular about tourism in the Mediterranean area and the relevance of integrated management in such a fragile area. For, indeed, with the accession of Spain and Portugal to the Community, probably in the near future, the 'Mediterranean dimension' of the Community is clearly growing in importance.

The overall EEC strategy in relation to the environment

Mr. Chairman, it comes as a surprise to some that the European Community has a programme of action in the field of environment at all. Yet, in reality, the Community was one of the first bodies to recognise – in the year of the Stockholm Conference on the Human Environment – the need for a vigorous environmental protection policy.

At a summit meeting (the meeting of Community Heads of State and Government) held in 1972 a declaration was made to the effect that 'economic expansion is not an end in itself' and that attention should be given to the 'intangible values and to protecting the environment so that progress', as they put it, 'may readily be put at the service of

D.O. Hall, N. Myers and N.S. Margaris (eds.), Economics of ecosystems management.
© 1985, Dr W. Junk Publishers, Dordrecht. ISBN-13: 978-94-010-8928-9

mankind'. Grand words; but they were followed by a demand to the institutions of the Community to produce an environmental action programme – which was done. It was adopted in 1973; it was renewed in 1977; and it has recently been renewed again, earlier this year.

Mr. Chairman, this latest Action programme for the period 1982–86 particularly stresses the close interrelationship that is essential between environmental and other policies; and underlines right at the beginning that: 'the resources of the environment are the basis of – but also constitute the limits to – further economic and social development'. So the Community is saying that the environment is fundamental and cannot be dissociated, therefore, from measures designed to achieve the basic objectives of the Community.

Consequently, the socio-economic development of the E.E.C. must in future increasingly aim at an adequate protection of the natural habitat, at a rational use of natural resources and a steady improvement of the quality of life. An environment policy, thus understood, must secure its own equal place in relation to other policy sectors within the EEC. Thereby, it is certainly not one policy amongst others, but it is more generally necessary to the overall aims of the Common Market. In other words, environment policy must no longer be seen as a minor policy which restricts the possibilities for development. Its fundamental task is now to serve as the qualitative base without which future economic growth, however it be defined, will come up against limitations.

Fortunately, in all our Community countries it is more and more agreed that 'cutting corners' where the environment is concerned makes no sense at any stage in the economic process.

On the other hand it is, of course, decisive if the objectives of environment policy are to be achieved that the instruments and measures employed be as far as possible economically and ecologically efficient. There is no doubt that in the past this has not been done with all the vigour that was necessary. Perhaps the reason being that environment policy in the sixties and above all in the early seventies, was conceived at a time of general prosperity; at a time therefore when the cost factor of this policy did not automatically – as is necessarily the case now – have priority.

We have now, therefore, to achieve more protection with less money. This is even more pressing in view of the bleak forecast of socio-economic development trends. The forecasts for the period up to 1990 indeed show that the Regions of Europe will achieve only moderate, if any, economic growth, but will still be faced with a worrying increase in unemployment.

The economic position of peripheral rural regions will worsen whilst that of the central densely populated areas will show a relative improvement. The trend towards an increase in regional disparities will thus be reinforced. Given the fact that the Southern Regions of Europe will probably be faced with an increase in the working population and continued growth of urban concentrations, there is likely to be considerable population movement in the search for jobs. These migrations will be concentrated on the tourist areas as well as the industrial urban areas in the South of the Community.

A number of potential socio-economic conflicts could thus arise throughout Europe, which correspondingly could seriously disturb important natural resources – above all the ecologically sensitive coastal and mountain areas, both of which are typical of the Mediterranean environment.

It is clear that there will be increasing conflicts between the protection of the environment and ecological resources and the short-term pressures arising from regional unemployment. However, we believe EEC environmental policy can contribute positively to resolutions to the Community's socio-economic problems, and, in particular to unemployment.

On the one hand, Community economic policy is an investment-orientated policy and one that fosters innovation. In encouraging new capital investment in the private sector it can ensure that less polluting processes are built into the new capital stock. In fostering innovation there are indeed many opportunities for new technologies that are 'clean' technologies. In fostering public sector investment there are many opportunities for 'environmental works' (e.g. those concerned with water resources, or land reclamation, etc., etc.). All of these will make it easier to pursue environmental improvement in the medium term.

On the other hand, environmental improvement and protection, the rational use of resources and the minimisation of waste is a 'sine qua non' of future socio-economic development – both quantitative and qualitative. This is so for three main reasons:

Firstly, despite some slackening of oil prices and stability of raw material costs, many such non-renewable resources remain relatively expensive in the short term and in the medium to the long term will retain their strategic importance. Hence the relevance of rational use of resources, waste minimisation and recycling.

Secondly, new physical developments in towns and rural areas require basic infrastructure – much of which is job-creating environmental work. Examples are the provision of water supplies, drainage, water and effluent treatment, waste disposal facilities, combating coastal erosion, etc.

Thirdly, the social and political demand for the maintenance and the improvement of environmental quality is increasing rather than diminishing. Hence, there is a continued demand for environmental 'goods': clean air, minimisation of environmental health risks, protection and conservation of the countryside and historic towns, etc. Small Labour-intensive projects could promote employment and improve the environment.

Mr. Charmain, there is no doubt that independently of these positive contributions, there will be, as we have seen, increasing socio-economic demands on the extremely fragile Mediterranean environment and its resources. This situation has given considerable impetus to the Commission to develop a series of measures which aim at complementing a policy of integrated environmental planning especially in this area. I will only mention two of these measures: the Information System on the State of the Environment in Europe and – in many ways a complementary project – a compulsory Environmental Impact Assessment Procedure.

As far as the first of these instruments is concerned, there is a proposal to be made to the Council in the near future for the adoption of a work programme for the implementation of the information system on the state of the environment and the natural resources in the Community, the methodological basis of which is now operational after a long and fruitful cooperation with all competent institutions and authorities in the Community. It is of some interest to mention that this proposal expressly makes clear that the Mediterranean is one of the priority areas for immediate implementation.

It is clear that if it is to meet its objectives the Information System must go beyond merely storing and mapping primary data on individual environmental variables; it must also provide a system for analysing and integrating these data to give information on more complex and more 'derived' aspects of the environment. Thus, although the original specification talked of a classification of the territory of the Community, 'ecological mapping' is more strictly a means of evaluation. It should also be pointed out that the aim of the Information System is to map and evaluate environmental conditions throughout Europe *not* as a basis for encouraging further development of the environmental resources at any price, but as a means of facilitating their protection and conservation. Finally, particular attention will be paid to the identification and protection of biotopes of European significance.

With regard to the second of these measures, Mr. Chairman, allow me to say that, if a preventive approach is to be undertaken in the Mediterranean, it is clear that new administrative procedures will have to be put into practice, bearing in mind, of course the specificities of the Mediterranean region. One of the central tools in this preventive approach is going to have to be environmental impact assessment, a proposal which hopefully will be adopted before the end of this year with the Greek Presidency of the Council of Ministers. It should be gradually introduced into the planning and preparation of all forms of human acitivity likely to have a significant effect on the environment. From the Commission's viewpoint this proposal for a directive on environmental impact assessment is seen as being extremely important in relation to the develoment of a preventive approach. The Commission has indicated that, in the light of experience, it will put forward proposals for the gradual extension of the field of application of environmental impact assessments; and will further take the necessary initiatives to ensure that all Community measures which are expected to affect the environment significantly are gradually subjected to prior assessment. To be really effective, a procedure for the environmental impact assessment may require the collaboration of a range of specialists. It will be the role of the ecologist to have a broad understanding of living systems and therefore to be able to bring together and integrate an appropriate range of studies in order to anticipate the effects of development activities in relation to the total environment. Mathematical precision is not required.

However, the application of ecology purely as a descriptive science is not sufficient. Where possible, techniques should be applied to classify, quantify and interrelate the various elements of ecosystems in order to make predictions of ecological change as accurate as possible.

I would therefore like to express our firm desire that during this conference, you should consider how your specialist knowledge can be made *operational* as quickly as possible within an environmental impact procedure.

Mr. Chairman, let me make a last few remarks about tourism which, as I know, will be an essential topic of this symposium and about which the Community will, I am sure, take increased and essential action to protect the environment in the Mediterranean basin.

The Mediterranean, as you are all aware, is the world's foremost tourist area. It is not surprising to find therefore that the Commission has addressed a communication to the Council on 'Initial Guidelines for a Community policy on tourism'. The European Parliament has discussed this document and recommends in particular that environmental considerations be given their due weight in newly discovered tourist development areas which are usually in less favoured areas where tourism is developing in a chaotic manner.

Uncontrolled tourism has harmed the environmental heritage – beaches, sanddunes, adjacent woods – in addition to the damage caused in more traditional, well-established areas of marine tourism by construction work. In addition to the environmental impact procedure, specific rules need to be introduced to regulate the phenomenon of mass tourism:

- the rapid introduction of rules to protect, at the very least, those sectors most vulnerable to damage, that is, coastlines and mountain areas
- protection of the environment should be a precondition for all Community support for the development of tourism
- projects to renew, restore and improve tourist areas should be supported financially, and finally
- environmental and cultural heritage protection should be included in all tourist management training courses.

It is fundamental that tourism should be prevented from destroying the bases of its own existence.

I would like to conclude by saying that whatever the priorities of EEC policy may turn out to be, the high importance of a sound forward-looking and preventive environmental policy, as a key element in the whole body of the Community's socio-economic policies, can no longer be neglected. Whatever the priorities, the central message remains the same: it really does not pay to ignore or set aside the needs of the environment. They will catch up with us sooner or later; and when they do so, the bill will be much heavier than if they had been properly dealt with in the first place.

The message is clear. We must aim at reaching a situation in which adequate protection of the environment is universally seen and accepted as a necessary part of good management of our scarce natural resources.

I am sure that this Symposium will give us new arguments to further the transformation of our increasing knowledge into practical action.

CHAPTER 2

Economics, evaluation and environment

PETER SÖDERBAUM

Introduction

Environmental or ecosystem-related problems are in most cases highly complex in character. They are multi-dimensional and interdisciplinary and often involve a number of uncertainties and risks. In addition to monetary variables, such as financial costs, non-monetary variables are generally involved. Especially for the latter, the phenomena of irreversibilities, commitments and other forms of inertia are important features. Conservation of ecosystems or resources that are judged to be unique or otherwise of great value, exemplify this concern about irreversible non-monetary processes.

Environmental problems are in most cases not only environmental in the sense that they are limited to certain ecosystems such as forests, wetlands or lakes: in many cases, problems that are observed at the level of specific ecosystems are related to the development over time of socio-cultural systems, i.e. systems of human beings. For instance, tourism may have ecological impacts and there may be concern about the conservation not only of ecosystems but also of indigenous 'human structures' as suggested in subtitles from the scientific programme of this conference.

Another typical feature of environmental problems in a broad sense (involving social as well as ecological systems) seems to be that they involve conflicts of interest between different parties and that these parties often refer to different philosophical perspectives, ideologies and values. This existence of conflicting philosophical perspectives and ideologies raises important questions for many of us who identify with the role of 'economist' and who have become accustomed to regard ourselves as experts in evaluation. Do we possess abilities that allow us to point to a philosophical perspective as the 'correct' one, which in turn enables us to evaluate alternative courses of action 'correctly' and point to the 'optimal' solution? Or, should we reconsider our role as experts?

On competition between paradigms

Some months ago, I participated in an international symposium on organic, biodynamic and other forms of non-conventional agriculture (Dlouhý, 1983). It was agreed that the philosophy behind organic or biodynamic farming differs considerably from the philosophy behind conventional farming. To this one may add that the philosophy behind conventional farming as regards efficiency criteria is equivalent to that of mainstream economics with its simplistic ideas about allocative efficiency, profits and gross margin in monetary terms. At the same conference I suggested that if there is non-conventional farming as opposed to conventional farming, then there may also be non-conventional economics as opposed to conventional economics. And ideas on evaluation and efficiency may be different in some specific kind of

D.O. Hall, N. Myers and N.S. Margaris (eds.), Economics of ecosystems management.

non-conventional economics from those of conventional economics.

This existence of competing perspectives or paradigms seems to be the rule rather than the exception within fields which are judged important in practical public policy. Agriculture is one example but forestry and energy systems are other areas where perspectives differ. Among social sciences outside economics in which competing paradigms exist, psychology, political science and educational science are perhaps the best examples.

As already indicated, a specific perspective or paradigm within one field may have characteristics in common with specific paradigms in other fields. The philosophy behind organic farming may be highly compatible with the philosophy behind institutional economics as a non-conventional paradigm within economics. And some observers argue that institutional economics in turn is highly compatible with the mainstream of political science (Lindberg, 1982). One may even identify some interdisciplinary paradigm which predominates within the social sciences and in my optimistic moments I feel that institutional economics is more compatible with such an interdisciplinary paradigm than is conventional economics. If this is so, then institutional economists belong to the majority at this interdisciplinary level, while being members of a minority group among economists.

One reason for the existence of competing paradigms within many social sciences is that each paradigm (with its conceptual framework, theories and methods) has a specific ideological content. The choice between paradigms is therefore not a purely scientific matter. Another reason has to do with the philosophy of science and the emergence of alternatives to positivism or the Newtonian worldview, for instance the Heisenbergian perspective (see Weisskopf, 1979). In short, such competition between paradigms can be regarded as part of the pluralism of Western societies and perhaps as a fundamental element of a democratic society.

Competition between paradigms may also be a fundamental prerequisite to progress within a discipline. One may question whether a specific paradigm can be properly understood if no reference is made to alternatives. And a dialogue between advocates of different schools may be constructive and influence the particular development of each paradigm over time. Advocates of paradigm A may internalize into their own thinking elements that they regard as positive in paradigm B, and so on.

On the meaning of institutional economics in relation to environmental management

It is not possible here to go into detail on the meaning of institutional economics (see, for instance Kapp, 1968, 1976; Myrdal, 1978) but some hints may be given. According to Gunnar Myrdal in Europe and Allan Gruchy in the US, the ambition to be holistic (rather than atomistic or reductionistic) is the main characteristic of contemporary institutionalism. Where mainstream, neoclassical economics concentrates on the market and market-related phenomena, institutionalists point out that the market represents only one specific kind of institution and that in addition other institutions as well as actors and their attitudes must be studied as elements in a process of economic development.

Conventional (mainstream) economists tend to limit the territory of economics to market characteristics with the motive that such aspects are 'more accessible to economists' (Mishan, 1982, p. 30) and sometimes make distinctions between 'economic' and 'non-economic' factors. Institutional economists seem to be more problem-oriented and according to them the problems facing our societies seldom respect administrative borderlines between disciplines. Rather, there is a considerable overlapping between disciplines. If economics is a moral science, as has been argued by Keynes and Kenneth Boulding (1970) among others, then ethics is certainly in a sense a part of economics. In a similar way psychology and sociology enter into economics when economists make assumptions about the behaviour of individuals or organizations. All this means that there is a considerable interdisciplinary area of social science where economists should be ready to learn from other disciplines rather than remain self-sufficient.

The fact that other disciplines enter into economics does not, however, make it meaningless to attempt to define economics as a science. A definition in terms of the management of resources, where resources are means towards human ends or means to satisfy human wants – certainly does not cover all aspects of economics, but it is useful for our present purpose of dealing with choice and evaluation. But *in practice* definitions of this kind have been interpreted by conventional economists as resource allocation in monetary terms, where financial or money resources *and the monetary aspects of other resources* (the price) have been emphasized. For instance, a piece of land is certainly a *natural* resource but the 'economic' meaning of this land tends to be associated with its price (or the discounted value of future money earnings connected with that piece of land). Scarcity tends to be measured by the *present* price, which in turn is a matter of the knowledge, information, values and perceptions of present actors, whether potential buyers or sellers.

As an alternative to such a perspective exclusively (or almost exclusively) in monetary terms, I will once more point in the direction of holism as opposed to reductionism. Economics refers to management of resources (or housekeeping) in a holistic perspective where non-monetary and monetary courses of events (or processes) are kept apart. According to this perspective, which is compatible with institutional economics, any tendency to *reduce* complex health, socio-cultural, physical or ecological processes or impacts to their alleged monetary equivalent should be opposed. The belief among conventional economists that monetary equivalents can meaningfully be estimated for all kinds of non-monetary processes or impacts will be referred to as *monetary reductionism*.[1]

This appeal to avoid monetary reductionism is not limited to the concept of economics itself, but also holds for 'resource', 'capital', 'wealth', 'cost', 'benefit'. All these concepts have to be reconsidered to allow for a serious treatment of non-monetary aspects.

The concept of resource is understood in its qualitative as well as its quantitative sense. Resources may degrade qualitatively from one point in time to another (acid rain which deteriorates the soil which again negatively influences the growth of forests) or improve qualitatively (human resources as a result of education).

Similarly 'cost' ('benefit') refers to negative (positive) consequences from the point of view of a particular interested party and may involve non-monetary as well as monetary aspects.

When Barry Commoner talks about 'biological capital' in connection with 'planetary housekeeping' and the importance of 'keeping the planet's skin', the biosphere, in good condition, he is not speaking of some monetary value attaching to the biosphere or to ecosystems: 'This machine is our biological capital, the basic apparatus on which our total productivity depends. If we destroy it, our most advanced technology will become useless and any economic and political system that depends on it will founder' (Commoner, 1971, p. 13).

Instead Commoner has broadened the content of the term 'capital'. Similarly 'capital degradation' and 'depletion of resources' may be understood in this broad sense. At the time of the referendum on nuclear power in Sweden, one conventional economist after another stood up to argue that it would obviously mean serious capital degradation not to start a reactor that had been built at an investment cost of around three billion Swedish crowns. If 'capital degradation' is taken also to include contamination of the reactor, radioactive pollution of nearby waters and the atmosphere, aspects of work

[1] It is true that conventional economists sometimes talk about 'real' resources as opposed to monetary resources. There is always some 'real' counterpart to any observed monetary event. What is challenged here is not a lack of awareness about the existence of 'real' processes but the simplistic treatment of these processes. In addition, they are better named 'non-monetary' than 'real', in order to emphasize their multi-dimensional and qualitative as well as their quantitative character.

It may be added that my questioning of conventional economics in this respect is not limited to third party effects, so called externalities, but also holds for the parties directly involved in a market transaction. Producing and selling or buying and using a commodity is normally accompanied by a number of social, cultural, health or environmental effects. The fact that there is a mutual trade agreement between buyer and seller does not justify any reduction of the economic significance of this process to certain quantities being sold at a particular price.

environment, successive positional changes in the stock of nuclear waste, risks of minor and major accidents, feelings of insecurity among individuals living in the neighbourhood of the reactor, and so on, then it is no longer obvious which of the two alternatives – starting or not starting the reactor – involves more (or less) capital degradation.

This question cannot be settled with reference to economic science alone, but is rather a matter of doing without the energy-producing capacity of the reactor, of the effects of using non-nuclear energy sources as alternatives and, in the end, a matter of subjective value judgements. Conventional economists seem to be largely unaware of the values and ideologies built into their concepts, theories, methods and hence arguments.

This way of defining capital, resources (where 'capital' and 'resources' can actually be used interchangeably), costs, benefits, wealth, debts, savings, etc. clears the way for multi-dimensional thinking where conventional definitions tend to limit our attention to monetary aspects. We may therefore conclude that institutional economics is holistic in three respects (all of them certainly interrelated), i.e. focus of attention, relationships to other disciplines, and the role of monetary in relation to non-monetary aspects. In all these three ways conventional economists, in their theories, and even more in their practical approaches, tend to be reductionistic.

Another area where views may differ between mainstream, neo-classical economists and institutionalists relates to the role of values in economics and within the social science generally. How is the work of an economist influenced by his own values and ideology or by values in his socio-cultural environment and what rules of behaviour for the economist as a scientist or expert are appropriate considering this problem of values?

The tendency to regard concepts, theories and methods as neutral with respect to values or 'value-free' and thereby 'objective' is a major weakness of the conventional approach. By avoiding a discussion of value issues the normative elements inherent in conventional analysis (as indeed in all analysis) will remain hidden from the reader or user of the analysis.

According to the institutionalists, all stages of the research process may be influenced by valuations. This is true of the choice of issue or problem to be addressed, the way the problem is perceived and formulated, the choice of contacts with interested parties as part of the research process, the choice of terminology, concepts, theories, methods, the choice of observations, presentation of conclusions, etc. The following often quoted formulation by Myrdal is significant to the institutionalist position:

> Valuations are always with us. Disinterested research there has never been and can never be. Prior to answers, there must be questions. There can be no view except from a viewpoint. In the questions raised and the viewpoint chosen valuations are implied. (Myrdal, 1978, p. 6)

Each conceptual framework, postulate about the behaviour of the consumer, business firm, theory, or method, has a specific valuational or ideological content. This can be seen by comparing the approach of conventional economics with some non-conventional approach, such as the institutional. There is for instance a considerable difference between conventional cost–benefit analysis and the 'positional analysis' that I have been advocating for some time (1973, 1980, 1982a).

Myrdal's main recommendation is that we should aim for consciousness of the kind of values that influence our work and that we should try to be open and explicit about values. In his study of the Negro problem in the USA (1944), Myrdal pointed to the democratic ideals of equality and justice formulated in the American constitution. When studying development problems in India and other Asian countries Myrdal (1968) referred to 'modernization ideals' defined in a specific way. Such values may be questioned by individuals who believe in other ideals and therefore conclusions should be regarded as conditional upon the declared viewpoint.

According to this view, the fact that it is impossible to be conscious of one's own values in any complete sense is no excuse for not attempting to be explicit.

In addition to aspiring to be conscious and explicit, I believe that we should sometimes be many-sided with respect to valuational viewpoints. As far as studies of public policy issues like roads, airports, pipelines, dams, housing, etc. are concerned, where conflicting interests can be expected, it does not seem fruitful to refer to 'correct' values invented by the economists themselves and point to the 'optimal' solution that follows from these values. Instead many-sidedness with respect to the alternatives, impacts and interests considered and also with respect to different possible valuational points of departure should be stressed and conditional conclusions suggested. If we assume two value profiles and three alternatives, the conclusion may be that the total effect profile of alternative 3 is the one which is most compatible with value profile 1, while the excepted impacts of alternative 1 are largely consistent with value profile 2.

Working with conditional conclusions in this way is quite different from the traditional expert role of economists which includes an expertise in 'correct' values according of conventional cost–benefit analysis and an ability to point to unanimous conclusions derived from such values. Peter Self at the London School of Economics uses the term 'econocracy' when referring to cost–benefit analysts who believe in 'correct' prices:

> Econocracy, as I use the word, is the belief that there exist fundamental economic tests or yardsticks according to which policy decisions can and should be made. Thus stated, econocracy is much more ambitious, and consequently more dangerous to the public than any kind of technocracy. (Self, 1975, p. 5)

So, the idea of 'scientifically correct' values is incompatible with democracy and should therefore be rejected. But of course this does not rule out the possibility that there may be consensus among politicians about objectives on other grounds concerning specific groups of decisions.

From textbooks on the history of economic ideas we have all learnt that Adam Smith was a professor or moral philosophy and many of us have tended to regard this as indicative of the backwardness of economics at that time. Considering the present situation at our departments of economics it would today, in my opinion, represent a large step forward if we all realized that the attemps made to assume away issues of ethics and morals have been failures. A clear declaration that economics is a moral science certainly does not make economics less 'scientific'. Rather the opposite is true. Unawareness of values and ideologies is a sign of a poor standard.

Some observations from the literature

The idea of pointing to institutional economics as a more promising paradigm in relation to environmental problems is not a new one. One of the main European advocates of institutionalism in this century, K. William Kapp, has repeatedly questioned the usefulness of the conceptual framework of neoclassical economics as an instrument for understanding, evaluating and otherwise dealing with environmental issues. Kapp regarded concepts such as 'externalities' as inadequate and preferred himself to refer to 'social costs' where the term 'costs' was used in the broad sense that I have previously recommended. About evaluation in terms of conventional cost–benefit analysis, Kapp made the following statement:

> In fact, the more we admit that *all* benefits (secondary, indirect, intangible etc.) of control measures ultimately have to be included in benefit–cost calculations, the more problematical becomes any evaluation in terms of one single monetary standard. In short I fail to see that cost–benefit analyses as they stand today have a solution of the problem of evaluation either the social costs of the disruption, or the social benefits of the improvement of our environment by control measures. (Kapp, 1977, p. 26)

It is interesting to note that Kapp with his institutionalist perspective made a comprehensive study of the nature and magnitude of environmental problems as long ago as 1950 in his book 'The Social Costs of Private Enterprise'. His work anticipates that of other early environmental economists such as Ezra Mishan and Allen Kneese by around 15 years and is quite impressive in its enumeration and analysis of a large number of the environmental problems that most of us only became aware of in the 1970s. Conventional environmental economists have not always been quick to refer to Kapp in their history writing, but I am glad to announce that I have found one economist, Roefie Hueting, who considers himself a mainstream economist and who devotes a large number of pages to discussing Kapp's contribution (1980, pp. 67–72). This discussion would probably have been even more interesting had Hueting been acquainted with the paradigm of institutional economics. Judgements of 'right' or 'wrong' should somehow be related to the conceptual perspective chosen. What is 'right' within the institutional paradigm may be 'wrong' from the point of view of neoclassical economics.

It should also be emphasized that mainstream thinking does not give unqualified support to evaluation in terms of conventional BCA (benefit–cost analysis). On the one hand, the point of departure in the literature of environmental economics is generally the same. Referring to Krutilla (1961) and Mishan (1971), Coddington, Opschor and Pearce indicate this position as follows:

> As 'normally' practiced BCA is an application of neoclassical welfare economics. Using the appropriate shadow prices, to say that benefits are in excess of costs is formally equivalent to saying that gainers can overcompensate losers. In selecting projects where net social benefits are positive, or in selecting the scale of output that maximizes net social benefit, BCA provides a mechanism for choosing policies on the basis of a potential Pareto improvement, (Coddington et al., 1972, p. 119)

On the other hand some of the well-known mainstream environmental economists point to the limits of the approach they are suggesting. Coddington et al., for instance, point to the principle of consumer sovereignty as a basis for BCA and then go on to question the ability of consumers to evaluate in terms of willingness to pay possible losses of future welfare related to such things as 'pesticides, radiation, pollution'. Another point raised is '*whose* preferences are to count'. BCA points to the willingness to pay of the present generation, while – because of the often irreversible and accumulative character of environmental impacts – those who will bear the possible welfare losses often belong to future generations.

I regard this article by Coddington et al. as a step forward but the length of this step should not be exaggerated. At the end of their paper the authors are eager to assure their readers that 'our concern here has not been to suggest that BCA has no part to play in the formulation of environmental policy' (p. 122).

Another sign of an awakening among mainstream economists is an article by Ezra Mishan (1980) 'How Valid Are Economic Evaluations of Allocative Changes?', which was published in the *Journal of Economic Issues*. In this article Mishan, himself a well-known textbook writer on BCA, admits that conventional BCA has a specific valuational content and that the practical usefulness of the approach is dependent on the existence of political consensus within society concerning these values. If such consensus does not exist, then conventional BCA should be withdrawn from the scene until some possible future situation when such consensus is to be found.

Other conventional economists have warned against the application of BCA within specific fields, such as energy systems or more precisely nuclear power, where discounting procedures in monetary terms are incompatible with some of the long-term impacts of these systems. It is clearly difficult or impossible to find meaningful estimates of willingness to pay for physical impacts of radioactive pollution several hundred years into the future. Additional problems of evaluation are connected with the presence of minor and major risks of accidents (Kneese, 1977; Pearce, 1979).

This awareness among mainstream economists about the limits of conventional BCA, and even about the possible dangers of using this approach in specific situations, seems to be growing over time. At some stage these forces may contribute to a shift in paradigm and to economics textbooks whose content is more in line with the institutional position. The process of change in thinking habits is, however, a very slow one and I share Hazel Henderson's disappointment about the recent development of mainstream environmental economics as it appears, for instance, in books from Resources for the Future, Washington (Henderson, pp. 230–231). In my pessimistic moments, I feel that most mainstream economists are more eager to save neo-classical economic theory than to save the environment. In any case, our vested interests in one or other of the different schools of thought should not be neglected as a force shaping the future of economics as a science and also the future of our societies.

Categories of approaches to evaluation

What alternatives to conventional BCA are available for evaluating projects from a societal point of view? An answer is provided by Table 1 in which possible approaches have been categorized with respect to aggregation/disaggregation of (1) effects of different kinds (social, cultural, health, ecological, monetary, etc.); (2) effects related to different points in time or periods of time; (3) effects related to different individuals or interested parties.

Conventional benefit–cost analysis, BCA, belongs to the aggregated category, the idea being to estimate a present value in monetary terms for each project considered, by discounting future monetary costs and benefits. Many economists stick to this conventional approach and do not seem to listen to any criticism. And those who do listen tend to move as little as possible, for instance, by excluding some sectors such as nuclear power plants and related investments from the domain in which BCA is applicable.

There are also revised forms of BCA in which the monetary discounting procedure is the same but prices reflect other values than those of the willingness-to-pay approach. Such forms of revised BCA have been suggested for the appraisal of projects in developing countries. (See Ezra Mishan, 1982, who describes these attempts and recommends a return to the original version of BCA. See also my comment (1983) to Mishan.)

Environmental economics is an example of a field in which the complexity of impacts and the difficulty in determining correct prices has made some conventional economists recommend 'cost-effectiveness' analyses, i.e. the intermediate category of Table 1. The aim may be to choose the alternative which minimizes the monetary cost of achieving some nonmonetary environmental goal, expressed as an aspiration level.

But as Table 1 illustrates, there are also a number of approaches of a more disaggregated kind. Many of these were developed in the 1970s and reflect a disbelief in the meaningfulness of far-reaching aggregation. Social, cultural, physical, and environmental effects can be described in their own terms of each period or point in time rather than being monetized and added together. Effects related to one interested party or group of individuals can be kept separate from effects related to other groups.[2] Some limited aggregation may still be meaningful, such as summary statements of the monetary (or financial) effects related to one interested party or group of individuals.

It needs no pointing out that the disaggregated approaches have their origins outside conventional economics. They are more related to interdisciplinary approaches like cybernetics, systems philosophy, systems theory or decision theory as a branch of business administration. As an example of systems analysis of the disaggregated kind I would refer to Quade (1974). A number of books on Policy Analysis have been produced in recent years, for instance Dror (1971) and Dunn (1981). Other interdisciplinary approaches include Technology Assessment (an approach that has become institutionalized through the Office of Technology Asssessment, Washington) and Environmental Impact Statements, EIS (see for instance Bendix & Graham, 1978). Reference is sometimes made to

[2] It is true that distributive effects can be considered as part of conventional cost–benefit analysis, This may be done as a restriction that discriminates between projects that will and those that will not be considered for further analysis. The usual way to handle distributive effects, however, is to summarize effects related to all groups as part of a present value or benefit–cost ratio and then as extra information describe what happens to different groups. This extra information will probably be regarded as merely a modification of the main impression embodied in the present value.

Table 1. A classification of approaches to evaluation in public decision situations.

Categories of approach		Examples
I	Highly aggregated	
	a) Objective function in monetary terms	Conventional benefit–cost analysis, revised versions of BCA
	b) Objective function in other terms e.g. utility terms	Some of the approaches referred to as multiple-objectives, multi-criteria or multi-attribute analysis
II	Intermediate	Cost-effectiveness analysis
III	Highly disaggregated	Some kinds of systems analysis (or policy analysis) such as trade-off analysis, environmental impact statements, positional analysis

Social Impact Statements and Employment Impact Statements (Henderson, 1981, p. xvii). Positional Analysis is an approach I have suggested (1973, 1982a) which is characterized by a reconsideration of the concept of economics along the lines outlined in this paper and by positional thinking, systems thinking and an approach to the analysis of interests related to an issue. Just as with positional analysis, Nicholas Ashford's trade-off analysis can be seen as a reaction against conventional BCA. Ashford (1980, 1981) is mainly concerned with the issue of regulating chemicals in the work environment.

Conventional economists tend to regard such disaggregated approaches as comparable to the first step of a conventional cost–benefit analysis. Initially all kinds of effects can be described within the scope of conventional BCA, it is argued. As I see it, in many cases such a description of the disaggregated approaches is not correct. The latter approaches, moreover, often differ from conventional BCA in their conceptual framework. For instance, the connotations or meanings attached to words like economics, cost, benefit, value, resource, capital, and efficiency may not be the same. These concepts may be components of a different conceptual perspective or paradigm.

Institutionalists point to the need for holism as opposed to reductionism. Conventional BCA can be regarded as a good example of reductionism. More precisely, it is a case of monetary reductionism. A very complex set of effects of different kinds referring to different periods of time and different categories of individuals are *reduced* to some alleged monetary equivalent. Simplification is certainly necessary and permitted, but this is over-simplification. The result is not only a simplified view but also a distorted one.

Institutional economics is also associated with interdisciplinarity. Institutionalists are ready and eager to learn from other social science disciplines and regard compatibility with contemporary developments in political science, psychology, sociology, anthropology, and business economics as important when judging the value of their studies. Benefit–cost analysis in its conventional form is largely limited to one discipline, with the declared purpose of remaining 'in occupation of a clearly marked and defensible territory' (Mishan, 1982, p. 42).

Institutional economists often prefer open-ended rather than closed models. Monetary aggregation clearly reflects an *aspiration* to capture everything in closed models. The use of more open-ended models means that many aspects and effects treated as 'intangibles' within the scope of conventional BCA can be regarded as tangible. 'Tangible effects' are defined by Mishan as 'effects whose magnitudes are derivable from the market data which are more accessible to economists' (Mishan, 1982, p. 30). Market data are certainly important, but is there any good reason to make such data more central to economic analysis than, say, health or environmental data?

Institutional economists often exhibit sensitivity to the problem of values in relation to social and public issues. As argued before, all kinds of approaches to public decision making have a valuational content. The set of rules applying 'correct' prices and a correct rate of discount in conventional

BCA involves a recommendation to apply a specific kind of ethics. If consensus cannot be expected, which I regard as the normal case in many important fields of public decision making (concerning land use, transportation, energy systems, pollution, etc.), then the *choice* of one specific objective function involves a dubious political stand. Those who share the values built into the objective function will benefit at the expense of all those who do not. In such a situation the economist becomes a 'devoted analyst' in relation to those parties who share the values built into conventional benefit–cost analysis.

It is only at a higher level in the value hierarchy that we can look for a consensus of values. Marc Tool (1979) points to the imperatives of democracy as an important set of rules when dealing with public policy making, and as I see it, the ideals of democracy point to a role for the economist or policy analyst which is partly different from that entailed by conventional benefit–cost analysis. He/she must avoid devotedness and submissiveness to specific parties. Analysis should clarify the many sides of an issue in order to serve citizens and politicians who differ on questions of values. The *many-sidedness* of alternatives considered, resource dimensions, interests affected, and possible ideologies or values seems to be a key concept. This idea of 'non-manipulation', that is, of not concealing certain alternatives, effects, or interests, will secure some independence for the analyst.

Such a role for the economist is more compatible with the disaggregated approaches of Table 1 than with aggregated approaches. Aggregation in situations where the parties involved have conflicting interests can be criticized by some parties as 'manipulation' or submissiveness to certain interests.

Another imperative of democracy relates to the accumulation of knowledge in society. What will the different actors involved (the politicians and the analyst included) learn from the application of a specific disaggregated approach compared to what they will learn from the conventional BCA approach?

Will a disaggregated approach be preferable to an aggregated one in illuminating possible discrimination against certain parties? Will the disaggregated approach provide a better picture of possible environmental and health effects? In what way will qualitative and irreversible effects (Georgescu-Roegen, 1979) be considered according to different approaches? Will decisions be regarded as 'once-and-for-all' decisions or only as the first step(s) in a step-by-step adjustment process or means–ends continuum?

The choice of one approach over another implies the application of values. In recommending the more disaggregated approaches, I have referred mainly to values at higher levels in the value hierarchies, that is to values on which one can expect a high degree of consensus. I have also tried to indicate that such disaggregated approaches are more compatible with institutional economics.

Relying only on higher level values, we cannot expect to 'resolve' all conflicts of interest or point to optimal conclusions. Instead our ambition will be to clarify the conflicts of interest involved and to point to conditional conclusions.

Environmental impact statements and positional analysis as two disaggregated approaches

In relation to the present discussion of paradigms in economics, I would like to make some further comments on two of the disaggregated approaches, namely Environmental Impact Statements (EIS) and positional analysis. Of the two EIS is certainly the better known. Although similar developments have occurred in other countries the EIS approach is mostly connected with the requirements of the US National Environmental Policy Act (NEPA) which came into effect in 1970.

The idea behind EIS is to illuminate all kinds of possible, significant, environmental impacts of proposed actions, to compare environmental and other impacts of the proposed action with two kinds of alternatives i.e. (1) a 'no-action' or 'no-build' alternative, and (2) other action alternatives than the proposed one. In addition, possible mitigation measures to reduce negative environmental impacts should be explicitly considered and negative irreversible environmental effects (judged to be unavoidable) clearly stated. The idea behind all this is to reduce the negative environmental impacts as much as possible and 'to know what one is doing' if a decision is taken to adopt to the proposed alternative.

It is interesting to note that NEPA points explicitly in the direction of a 'systematic interdisciplinary approach which will insure the integrated use of the natural and social sciences and the environmental design arts in planning and decision making'. Another feature of NEPA is that it contains important procedural stipulations, for instance relating to the stage in the planning process at which public hearings should be held (see for instance Caldwell, 1983).

One possible disadvantage with EIS is that it starts with an alternative that is already at the stage of being proposed, whereas conventional BCA and positional analysis enter into the planning process at an earlier stage, at which each of a set of mutually exclusive alternatives has a more equal chance and hopefully fewer commitments have been made. The fact that EIS concentrates to a high degree upon environmental impacts is in a way a strength of the approach. The law and the procedure were designed to counteract the previous emphasis upon more narrow technical and monetary efficiency criteria. Interpreted in this way, EIS is something that is added to the other kinds of information which will be brought forward under established procedures.

Positional analysis (PA) is more comprehensive in scope in that it attempts to illuminate all kinds of impacts and differences between alternatives. One distinction is made between non-monetary and monetary effects and another between flows and positions, where flows refer to periods of time and positions to points in time. It is argued that analysis in monetary terms is insufficient and that our techniques to describe changes in non-monetary terms have to be improved. Non-monetary flows (discharges of mercury into a nearby lake from a pulp factory) are important, but in order to get an idea of changes in the state of the environment we also have to estimate changes in positions (parts per million of mercury in fish caught at a specific place in the lake) following as a result of a specific sequence of flows. Decision-trees in positional terms are used to illustrate relationships between present options and future positions with their associated options. Irreversibilities of non-monetary processes, whether physical or social, are at the heart of this analysis.

At an early stage of the analysis, differences between actors or interested parties with respect to perceptions of the problem (including views on relevant alternatives) and the identification of alternatives to be considered are emphasized. Many-sidedness is an important criterion in this selection of alternatives for further study. In spite of this 'many-sidedness' the choice of alternatives is always a stage of potential manipulation.

Systems thinking and analysis of interests related to an issue are two other elements of PA. Against the background of a specific set of alternatives to be considered, an attempt is made to identify all the systems of different kinds (social, physical, ecological, etc.) which will be influenced differently depending upon the alternative chosen. This system's structure may then be used as a basis for describing differences between the impacts of alternatives.

In a similar way the analyst attempts to identify all human activities (e.g. recreational activity, transportation activity, agricultural activity) which are expected to be affected differently depending upon the alternative chosen. For each activity an *assumption* is made regarding the goal-direction of those participating. Activity plus assumed objective defines an interest and a matrix indicating orders of preference in relation to the alternatives from the point of view of each interest can be constructed. In this way conflicts or trade-offs can be presented at the level of effects and/or at the level of interests. Only rarely will it be possible to point to one alternative as the best in all dimensions and in relation to all interests. Instead conditional conclusions which relate to possible value profiles and also to possible uncertainties are suggested.

What have been the reactions of conventional economists to these disaggregated approaches? In the case of EIS, my experience is certainly incomplete but my impression is that conventional economists have been rather indifferent and that little has been written about the relationships between EIS and BCA in journals and textbooks on environmental economics. One hypothesis that can explain this is that EIS does not fit well into the paradigm of conventional economics. Comprehensive lists of environmental effects make ideas of shadow prices, revealed preferences and the like appear unrealistic. And in many cases the conception of the environment as valuable in itself, non-tradable or non-reductible to monetary terms, does not go well with established thinking habits.

There are some recent cases, however, where conventional economists are to be credited for taking up EIS in their writings. In his book 'Resource Economics' Alan Randall (1981) does this in a way which might be expected of a conventional economist. His cognitive fixation on the conventional paradigm makes him interpret NEPA and EIS in BCA terms: 'Evaluate effects. Place values on the significant beneficial and adverse effects in monetary terms where applicable, quantitatively where possible, and qualitatively in any event' (p. 312). 'The national economic development impacts of proposed projects and programs should be assessed and estimated, following established procedures of benefit-cost analysis' (p. 313). Pointing to the 'Flood Control Act of 1936 which requires benefit cost analysis of federal water projects' (p. 314), Randall cannot see why NEPA should be different. But he adds that he knows that his view is controversial.

In my view the ideas and principles behind EIS and BCA are fundamentally different. The ideological content of the two approaches differs and any attempt to press the reductionistic BCA philosophy upon the EIS methodology is therefore essentially a political act. The advocacy of BCA as the correct approach to evaluation and of the use of this approach in all public affairs is certainly not based upon scientific principles alone. And the result in terms of future states (or positions) of the environment will be quite different if BCA is given a central role rather than a more disaggregated approach.

One thing is a bit confusing, however, in Randall's book. He talks about BCA as a narrow approach 'focusing on economic efficiency (or natioal economic development)' (p. 313) and argues in favour of a more comprehensive view of *economic* (!) impacts including regional economic development, state and local government finance and national defence. This acceptance of some of the recommendations of the original EIS procedure means that Randall (in the name of a compromise?) departs from the conventional view based upon aggregation (and avoidance of 'double-counting').

The other approach, positional analysis, has only recently been described in international journals. It has been practised in Sweden and Norway mainly for studies in the fields of transportation and land-use planning. For a long time conventional economists remained indifferent, relying on BCA as the one and only 'correct' approach, insensitive to the argument that BCA, compared to other approaches, has a specific ideological content. In the last few years at least one textbook writer in Sweden has given up the conventional idea that there is only one correct theory for each type of economic agent or level or analysis and included PA along with other alternatives such as Lichfield's planning balance sheet (Lichfield et al., 1975).

Concluding comments

In this paper I have pointed to competing paradigms within economics and also commented upon 'paradigmatic' relationships to other disciplines. I have argued that these issues of conceptual framework or 'spectacles' are extremely important when we formulate and try to deal with environmental problems. These discussions about perspectives should therefore continue rather than be avoided.

I have also pointed in the direction of institutional economics as my personal preference. I believe that this perspective will gain ground in the future and that it offers a kind of response to the criticism of conventional economics that has been expressed by interdisciplinary thinkers like Johan Galtung (1984) and Hazel Henderson. But institutional economics is not a perspective which is different in every respect from present-day mainstream economics. In some areas institutionalists rely on the same concepts and approaches as conventional economists, in other areas they call for additions and supplements or a more holistic picture of reality. As far as certain fields are concerned, however, institutionalists believe that we should forget about essential elements of the conventional approach as quickly as possible. Conventional benefit–cost analysis with its simplistic idea of allocative efficiency is one of the parts which in my judgement should be replaced.[3]

Having said this I should remind myself and others of the fact that few resources in terms of researchers and students are now engaged in the development and application of non-conventional paradigms. In my judgement the 'marginal utility' (or rather expected non-monetary positional changes with respect to knowledge, health, social systems, environment) of reallocating scholars from conventional to non-conventional research tasks is high. As far as environmental economics is concerned, conventional economists are very close to a monopoly position and monopolies of this kind are not entirely different in their adverse effects from monopolies in commodity markets.

Considering the inertia characterizing institutions and systems of social scientists (and I include myself in this appraisal) the best one can do is to emphasize tolerance and dialogue as elements contributing to the long-term progress of science.

Acknowledgement

Work on this paper was financially supported by the Swedish Council for Building Research, Stockholm.

References

Ashford, Nicholas A. et al., 1980. Benefits of Environmental, Health, and Safety Regulation. Prepared for the Committee on Governmental Affairs, United States Senate, by the Center for Policy Alternatives at the Massachusetts Institute of Technology, Boston. Mimeographed.

Ashford, Nicholas A., 1981. Alternatives to Cost–Benefit Analysis in Regulatory Decisions. Ann. N.Y. Acad. Sci. 363: 129–137.

Bendix, Selina & Graham, Herbert, 1978. Environmental Assessment, Approaching Maturity. Ann Arbor Science Publ., Ann Arbor.

Boulding, Kenneth, 1970. Economics as a Science. McGraw-Hill, New York.

Caldwell, Lynton K., 1983. EIA Procedures in the USA 12 years on. A Personal View. In: Symposium Papers: Environmental Impact Assessment Symposium, Crete April 1983, P.A.D.C. EIA-Unit, Dept. of Geography, University of Aberdeen, Scotland.

Coddington, Alan, Opschoor, Hans & Pearce, David, 1972. Some Limitations of Benefit–Cost Analysis in respect of Programmes with Environmental Consequences. In: Problems of Environmental Economics, OECD, Paris.

Commoner, Barry, 1971. The Closing Circle. Nature, Man and Technology. Alfred A. Knopf, New York.

Dlouhý, Josef (ed.), 1983. Proceedings. International Scientific Colloquium on Comparisons between Farming Systems. Swedish University of Agricultural Sciences, Department of Plant Husbandry, Report 124, Uppsala.

Dror, Y., 1971. Design for Policy Sciences. Elsevier, New York.

Dunn, William, 1981. Public Policy Analysis. Prentice Hall, Englewood Cliffs, N.J.

Galtung, Johan, 1984. Economic Theory as Carrier of Cosmology. Some Causes and Consequences. In: Bergström, Sören (ed.), Economic Growth and the Role of Science. Proceedings from a symposium in Stockholm, August 9–11, 1983, Stockholm University, Stockholm.

Georgescu-Roegen, Nicholas, 1979. Methods in Economic Science. J. Econ. Issues 13: 317–28.

Henderson, Hazel, 1981. The Politics of the Solar Age. Alternatives to Economics. Anchor, New York.

Hueting, Roefie, 1980. New Scarcity and Economic Growth. More Welfare Through Less Production? North-Holland, Amsterdam.

Kapp, William, 1950, 1971. The Social Costs of Private Enterprise. Shocken New York.

Kapp, K. William, 1968. In Defense of Institutional Economics. Swed. J. Econ. 70: 1–18.

Kapp, K. William, 1971. Environmental Disruption: General Issues and Methodological Problems. Soc. Sci. Inf. 9: 15–32.

Kapp, K. William, 1976. The Nature and Significance of Institutional Economics. Kyklos 29: 200–232.

Kneese, Allen W., 1977. Benefit–Cost Analysis and the Atom. In: R. Steppacher (ed.), Economics in Institutional Perspective. Heath, Lexington.

Krutilla, J., 1961. The Welfare Aspects of Benefit–Cost Analysis. J. Pol. Econ. 69: 226–235.

Lichfield, Nathaniel, Kettle, Peter & Whitbread, Michael, 1975. Evaluation in the Planning Process. Pergamon, Oxford.

[3] In this paper I have concentrated my discussion on evaluation at the public level. For a more comprehensive study of normative institutional economics in relation to assumptions about objectives, evaluation, and accounting practices at the three levels of society, organization and individual, see Söderbaum (1984, 1980).

Lindberg, Leon, 1982. The Problems of Economic Theory in Explaining Economic Performance. Ann. Am. Assoc. Pol. Soc. Sci. 459: 14–27.
Mishan, Ezra, 1971. Cost–Benefit Analysis. Allen & Unwin, London.
Mishan, Ezra J., 1980. How Valid Are Economic Evaluations of Allocative Changes? J. Econ. Issues 14: 29–47.
Mishan, Ezra J., 1982. The New Controversy about the Rationale of Economic Evaluation. J. Econ. Issues 16: 29–47.
Myrdal, Gunnar, 1944. An American Dilemma: The Negro Problem and Modern Democracy. Harper & Row, New York.
Myrdal, Gunnar, 1968. Asian Drama: An Inquiry into the Poverty of Nations. Pantheon, New York.
Myrdal, Gunnar, 1978. Institutional Economics. J. Econ. Issues 12: 771–783.
Pearce, D.W., 1979. Social Cost–Benefit Analysis and Nuclear Futures. In: G.T. Goodman (ed.), Energy Risk and Management. Academic Press, London.
Quade, E.G., 1974. Analysis for Public Decisions. Elsevier, New York.
Randall, Alan, 1981. Resource Economics. An Economic Approach to Natural Resource and Environmental Policy. Grid, Columbus, Ohio.
Self, Peter, 1971. Econocrats and the policy process. The politics and philosophy of cost–benefit analysis. Macmillan, London.
Söderbaum, Peter, 1973. Positional analysis for decision making and planning. An interdisciplinary approach to economic analysis. Esselte Studium, Stockholm (in Swedish).
Söderbaum, Peter, 1980. Towards a reconciliation of economics and ecology. Eur. Rev. Agric. Econ. 7: 055–077.
Söderbaum, Peter, 1982a. Positional Analysis and Public Decision Making. J. Econ. Issues 16: 391–400.
Söderbaum, Peter, 1982b. Ecological Imperatives for public policy. Ceres (FAO Review on Agriculture and Development) 15: 28–32.
Söderbaum, Peter, 1983. Ezra Mishan on Economic Evaluation. A Comment. J. Econ. Issues 17: 206–213.
Söderbaum, Peter, 1984. The GNP-Growth Issue in Relation to Conventional and Institutional Economics. In: Bergström, Sören (ed.), Economic Growth and the Role of Science. Proceedings from a symposium in Stockholm, August 9–11, 1983, Stockholm University.
Tool, Marc R., 1979. The Discretionary Economy: A Normative Theory of Political Economy. Goodyear, Santa Monica.
Weisskopf, Walter, A., 1979. The Method is the Ideology: From a Newtonian to a Heisenbergian Paradigm in Economics. J. Econ. Issues 13: 869–883.

CHAPTER 3

Integrated regional economic-environmental modeling

FLOOR BROUWER, LEEN HORDIJK and PETER NIJKAMP

1. Introduction

Methodologies for analyzing multidisciplinary phenomena in an integrated way are becoming increasingly necessary due to a rise in the complexity of societies.

In the present paper, a multi-dimensional modeling approach is chosen as a frame of reference for analyzing economic-ecological systems characterized by many interrelationships between the variables. The main focus of this paper will be on problems inherent in designing and estimating an integrated economic-ecological model.

Economic models include such variables as production, consumption, investment and employment, while ecological models include variables reflecting the association, stability and diversity of ecosystems such as aquatic and terrestrial vegetation. The variables from these different disciplines interact with each other when they are analyzed in an integrated modeling approach. The size and composition of terrestrial vegetation, for example, have a close relationship with the level of recreational use; SO_2 emissions are related to the nature of employment, and so forth.

There are a number of conceptual problems associated with integrated regional economic-ecological systems, dealing with the dimensions of space and time and the level of measurement of the data (see also Nijkamp, 1983).

(1) Economic variables are expressed in units of money, quantities or index numbers while many ecological variables are measured in a categorical sense.

(2) Economic variables are measured at a regional or a national level, while the spatial dimension of ecological variables is often much smaller (e.g. m^2).

(3) The time horizon of economic and ecological variables is quite different.

In Section 2 we will concentrate on methodological issues in integrated regional economic-environmental systems with systems analysis as a frame of reference for analyzing the structure of strongly interdependent components.

The subsequent section will deal with conceptual problems in economic-environmental modeling, with a specific view of space and time dimensions (Section 3).

The next issue to be dealt with concerns a structure analysis of complex systems, based on graph-theoretical methods like qualitative structure analysis (Section 4).

Then specific econometric and statistical problems will be discussed, with particular emphasis on non-metric data analysis (Section 5).

Finally, the above mentioned concepts will be illustrated by an empirical application to an integrated land use planning model in the Netherlands.

D.O. Hall, N. Myers and N.S. Margaris (eds.), Economics of ecosystems management.

 ISBN-13: 978-94-010-8928-9

2. The analysis of complex economic-ecological systems

In this section we deal with a modeling approach known as *systems analysis*, to analyze the structure of strongly interrelated multidisciplinary components (see also among others Bennett & Chorley, 1978; Caswell et al., 1972; Wilson, 1981, for a wide variety of applications of systems analysis in many disciplines). A system is simply characterized by three elements, viz. an *input*, say x, and *output*, say y, related to each other by a *transfer function* f, or: $y = f(x)$. The characteristics of some input element is determined by the existing interrelationships. We can distinguish between a closed system and an open system in the following way. The characteristics which do not belong to some defined system are called the environment. A closed system has, unlike an open system, no input effects from the environment.

An example of a closed system is a simultaneous model of endogenous variables, without exogenous variables that influence the model. On the other hand, when exogenous variables are part of a simultaneous model, we have an open system. A schematic representation of a closed system, which consists of only two components A and B interacting with each other, is given in Figure 1.

The components A and B are related to each other in a mutual way, represented in analytical terms by $B = f(A)$ and $A = g(B)$. The operators f and g are called feedbacks, and the output of B becomes input into component A which determines the characteristics of B, etc.

The structure properties of a system like Figure 1 can be summarized as follows (see also Bennett & Chorley, 1978).

(1) The properties of each component influence the system as a whole.
(2) The properties of each component influence at least one other component.

The structure of a system is determined by the interrelationships. We can conclude from (1) and (2) that a system cannot be partitioned into sets of mutual independent components. Such system has properties quite different from the separate components so that 'the behavior of the 'whole' system is usually much richer than the sum of the parts' (Wilson, 1981, p. 3).

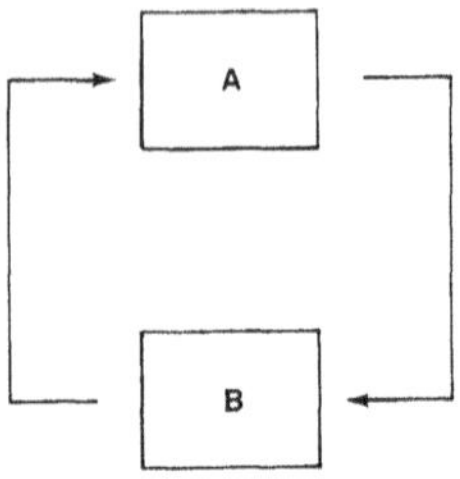

Fig. 1. Schematic representation of a closed system.

There are four conceptual and methodological problems in dealing with the analysis of complex economic-ecological systems, viz:

(a) A *spatial* aspect which has to do with the implications to model results of the aggregation level and scale of variables, either economic or ecological.
(b) A *time* aspect of variables measured at either a discrete or a continuous level.
(c) The *structure* of relationships in either a binary, a qualitative or a quantitative sense.
(d) *Data* measured at a ratio scale, interval scale, ordinal scale or a nominal scale.

3. Concepts of space and time in integrated economic-ecological modeling

In a systems analytic approach for a spatial system we can distinguish between homogeneous and functional regions.

The classification of homogeneous regions is determined by their attributes, while a functional classification is determined by multiple relationships which exist between regions (see also Fischer, 1978). A graph-theoretical approach is a useful tool to operationalize the classification procedure. A homogeneous or natural region can be characterized by, for example, geographical, physical, climatological or economic factors with special emphasis on 'human behavioral processes, such as perception, learning, attitude formation and decision making' (Gale & Golledge, 1982, p. 60). A functional classification is determined by binary relationships between spatial units (see also Paelinck & Nijkamp, 1976).

We will now stress the importance of the aggregation level as well as the scale level to modeling results. The subjective choice of areal units is often described as the problem of ecological fallacy in spatial analysis; it has a great importance for modeling results: 'we must emphasize the necessity, in this type of work, of not losing sight of the fact that our results depend on our units' (Yule & Kendall, 1950, p. 313). The modeling results in regional analysis 'measure, as it were, not only the variation of the quantities under consideration, but the properties of the units which we have imposed on the system in order to measure it' (Yule & Kendall, 1950, p. 312).

The problem of the choice of *areal unit* and the consequences for spatial analysis can be subdivided into two: a *scale* problem and an *aggregation* problem both of which depend upon each other (see also Openshaw & Taylor, 1979, 1981).

The scale problem is defined by the different modeling results when areal data are defined by increasingly larger areal units; the aggregation problem deals with the alternative possibilities of partitioning an area into a fixed number of units. The scale problem in spatial analyses is comparable with micro- and macro-analysis in economics. An illustration of the wide range of modeling results dependent on different aggregation and scale levels can be found in Openshaw & Taylor (1979).

We can conclude from the above that the areal unit problem is a very complex one in regional modeling, especially when economic and ecological variables, with different regional scales, are analyzed in an integrated way (see also Grunfeld & Griliches, 1960). The fundamental question is at what regional scale we wish to investigate both the economic and ecological variables.

The spatial aspects of an integrated model has multi-dimensional components and this will also occur with the temporal aspect of a model. Events in time generate an ordering because time distinguishes between past, present and future.

An integrated analysis of individual activities in space and time is developed by, among others, Hägerstrand at the Lund School in Sweden, called time-geography (see also Thrift, 1977a).

An individual time–space path can be represented in a four-dimensional space which consists of a three-dimensional space component and a one-dimensional time component. An isomorphism or one-to-one correspondence is assumed to exist between the elements of time and space.

The dynamic components in different disciplines express divergent temporal levels (see also Thrift, 1977b). We distinguish between the superstructure level, built environment level, activity level and the individual attitudes level.

A schematic representation of these four temporal levels in, for example, psychological models and socio-ecological models, is given in Figure 2 (see also Thrift, 1977b, p. 69).

'Psychological time is most associated with explanation of observed behaviour at the activity system and attitude level' (Thrift, 1977b, p. 71), while the opposite occurs with socio-ecological models.

4. Structure analysis in integrated economic-ecological modeling

4.1 *Introduction*

The analysis of the causal structure of economic-ecological models is discussed in this section, with special emphasis on the level of measurement of the variables included in such models. In terms of the analysis of the structure of a model, the following measurement levels may be distinguished (see also Brouwer & Nijkamp, 1984a).

(1) A *binary* relationship. This indicates whether or not a certain variable has an impact on another variable. If the direction of impacts is the

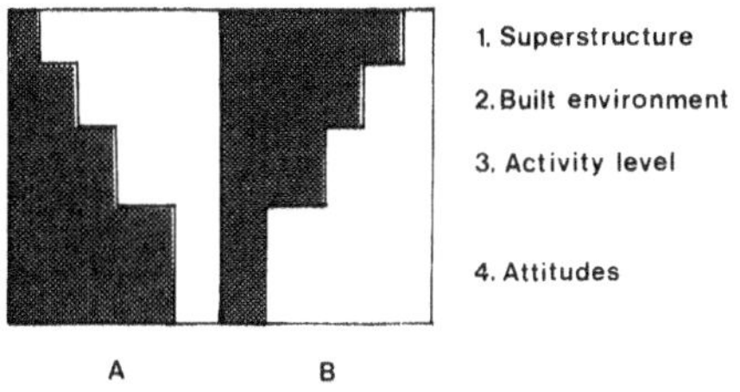

Fig. 2. The relationship between temporal levels and discipline, where A = psychological time and B = socio-ecological time.

only information available, graph theoretical methods are a useful tool to analyze the causal structure of such models (see also Roberts, 1978).

(2) A *qualitative* relationship. When the sign of the binary relation (viz. positive, negative or zero) is known, indicating the qualitative direction of the impact of the one variable upon the other one, qualitative calculus can be applied to operationalize such models (see also Greenberg & Maybee, 1981). We will discuss hereafter the conditions to solve a linear model with qualitative information as well as conditions for qualitative stability of a system of linear differential equations.

(3) An *ordinal* relationship. When the order of magnitude of impacts is indicated in terms of rank numbers, the relationship can be studied by means of ordinal econometrics.

(4) A *cardinal* relationship. When the causal relationship between variables is quantified by path coefficients, path analysis can be employed, based on an ordinary least squares estimation procedure to estimate model parameters (see also Leitner & Wohlschlägl, 1980).

In this paper, the attention will mainly be focused on graph-theoretical methods to operationalize the analysis of complex structures of integrated economic-ecological models.

4.2 *Model structure and binary relationships*

In this section we deal with graph-theoretical methods, especially directed-graphs, when a binary relationship is the only information about the structure of integrated economic-ecological models.

A directed graph or digraph G is defined by a finite, non-empty, irreflexive and binary relationship with a set of vertices V. The set of relations are called the edges E, and G is represented by $G = (V, E)$ (see also Roberts, 1978).

The information about the model structure which follows from the digraph G can be expressed in matrix terms when Boolean algebra is used (see also Tinkler, 1977).

Consider a graph G with a set of vertices $V = (V_1, \ldots, V_I)$ and a set of directed edges $E = (E_1, \ldots, E_J)$. A matrix of adjacency R, with elements r_{ij}; $i, j = 1, \ldots, I$, is defined in the following way:

$$r_{ij} = \begin{cases} 1 \text{ if there exists an edge from vertex } i \text{ to vertex } j \\ 0 \text{ in all other cases} \end{cases}$$

Matrix multiplication simply shows that the matrix R^n gives the total number of possible paths between each pair of points with length n $(n \geq 1)$.

The information about the stimulus–response relations (represented by the matrix of adjacency) can also be expressed by the matrix of incidence T and the cyclomatic matrix U. All possible cycles are compared with the number of cycles which really exist. A cycle is defined by a closed chain from some vertex into itself.

The matrix of incidence T, with elements t_{ij}, i=1, $\ldots$, I; j=1, $\ldots$, J is defined by

$$t_{ij} = \begin{cases} 1 & V_i \text{ starting-point from } E_j \\ -1 \text{ if} & V_i \text{ end-point from } E_j \\ 0 & \text{in all other cases} \end{cases}$$

Matrix T represents the structure of a graph, just like the matrix R. The rank number of the matrix T equals I-x, with x the total of unconnected subgraphs of the graph G.

The cyclomatic matrix U, with elements u_{jk}, j= 1, $\ldots$, J; k= 1, $\ldots$, K, is determined by the total number of possible cycles (K in total), and defined as

$$u_{jk} = \begin{cases} 1 & j \text{ belongs to cycle } C_k \text{ with the same direction as } G \\ -1 \text{ if} & j \text{ belongs to cycle } C_k \text{ with the opposite direction as } G \\ 0 & \text{in all other cases} \end{cases}$$

The rank number of U equals the total of independent cycles of the graph G and is called μ (see also Gallopin, 1972)

$$\begin{aligned} \mu &= \text{rank } (U) = J - I + x \\ &= J - \text{rank } (T) \,. \end{aligned} \tag{1}$$

The number μ, also called the cyclomatic number, represents the number of fundamental circuits from a graph, because the total number of vertices and edges is corrected for the number of unconnected subgraphs.

A number of indices have been developed to represent the complexity of a graph as well as the level of association within a graph, viz. the alpha, beta and gamma index. The index alpha is developed as an index of association within a graph by means of the observed fundamental circuits and is defined as

$$\alpha = \frac{\mu}{2I - 5} \tag{2}$$

A very simple measure of complexity within a graph is the index beta, which relates the total number of edges to the number of vertices

$$\beta = \frac{J}{I} \tag{3}$$

A high value of β indicates a complex graph structure of a model. The index gamma is defined as

$$\gamma = \frac{J}{3\,(I - 2)} \tag{4}$$

for planar graphs and is a measure of association (just like the α-index).

In the above we discussed structure analysis of complex systems, e.g. with an economic-ecological modeling approach where the binary relationship is the only information about the impact between variables. In the next subsection the presence of qualitative information will be discussed.

4.3 *Qualitative structure analysis*

The analysis of qualitative relations, or qualitative calculus, has been developed to analyze the relationships between variables in a qualitative way, i.e., when the sign of the impact on some dependent variable is obtained from prior knowledge concerning the signs of the structural parameters. This type of information is especially relevant in economics, ecology and chemistry (see also among others Jeffries, 1974; Lancaster, 1962).

One main topic of qualitative calculus is the sign-solvability problem. Sufficient and necessary conditions are developed to solve a set of linear equations, expressed by

$$Ax = b \tag{5}$$

for variable x. The vector b and matrix A contain qualitative (in terms of positive, zero or negative impact parameters) information and the solution of (5) is (provided A is a non-singular matrix):

$$x = A^{-1}b \tag{6}$$

The inverse operation of A with qualitative information is computed in an analogous way as the inverse operation of a matrix with quantitative information. Consider A is a matrix of order $n \times n$ with elements a_{ij}, $i, j=1, \ldots, n$.

Sufficient and necessary conditions for sign-solvability of (5) in a unique way are (see also Bassett et al., 1968).

(1) $a_{ii} < 0$, $i=1, \ldots, n$

(2) all cycles in A of length larger than one are non-positive

(3) $b \leqslant 0$

(4) if $b_k \neq 0$, *then every path* $i \rightarrow k$ is non-negative for every i; $i = 1, \ldots, n$; $i \neq k$

While quantitative models encompass information about the direction as well as the size of the changes, qualitative calculus only gives information about the direction of the changes of variables. When a qualitative model is not fully sign-solvable, i.e. solvable in a unique way, we need additional information (at an ordinal or cardinal measurement level) to obtain solvable results.

We now deal with another topic which concerns a structure analysis of complex systems, viz. stability in a qualitative way from a set of linear differential equations. Qualitative stability or sign stability can easily be interpreted in the following way. Consider a set of linear differential equations with a qualitative information matrix A.

$$\dot{x} = \frac{dx}{dt} = Ax \tag{7}$$

Suppose the magnitudes, but not the sign, of the non-zero cell elements of the matrix A are changed at random. If all of these matrices are asymptotically stable, matrix A which only contains qualitative information, will be called sign stable or qualitatively stable.

Necessary conditions for qualitative stability are formulated by Quirk and Ruppert (1965) as follows

(1) all diagonal elements of A to be non-positive, i.e.

$a_{ii} \leq 0, i=1, \ldots, n$

(2) $a_{ii} < 0$ for at least one i

(3) all cycles of length two are non-positive, i.e. $a_{ij}a_{ji} \leq 0, i \neq j$

(4) there are no cycles of length m ($m \geq 3$) in the graph which corresponds to the matrix A

(5) the matrix A is non-singular.

These conditions are necessary but by no means sufficient. A reformulation from these conditions so that necessary and sufficient conditions for qualitative stability are obtained, is discussed in Brouwer & Nijkamp (1984a).

Stability analysis in a qualitative way is especially relevant for the analysis of growth patterns in either ecological or economic systems with only limited information.

4.4 *Model structure and path analysis*

In Sections 4.2 and 4.3 we have analyzed two types of model structures, represented by either digraphs or signed digraphs. Such modeling approaches become especially relevant where there is limited information about the impacts between variables, e.g., when knowledge about the level of impact between variables is not available.

In this section, the analysis of path models is discussed when the relations between variables are represented by path coefficients. Path coefficients indicate the causal relationship between variables, the so-called path-diagram (see among others Hall & Taylor, 1983; Leitner & Wohlschlägl, 1980; Macdonald, 1977).

The starting point for path analysis, after a definition of the relevant variables, is the specification of the causal ordering between the variables. The set of variables assumed to be irrelevant for the causal relationships, are called exogenous variables. The path coefficients p_{ij} reflect the path from variable X_j to variable X_i, and are estimated with an ordinary least squares approach when the following assumptions hold

(1) the causal relationships are correctly specified

(2) the error terms E_i are independent from the causal variables X_j

(3) homoscedasticity or constant variance, from the error terms

Model evaluation procedures then take place by means of specific statistical tests, e.g. whether one or more paths in an equation are significant by different from zero or to what extent one or more paths from a model can be deleted.

The significance of some coefficient is tested by means of a two-sided t-test. If we want to test whether two or more coefficients in an equation are different from zero, we use a test-statistic which follows a Fisher distribution. The causality between two variables consists of a direct effect as well as an indirect effect. The difference between the direct effect and the indirect effect is tested by a statistical procedure in an analogous way to the parameter tests. A practical application of path-analysis modeling using health facilities as an example is given by Hall & Taylor (1983).

5. Exploratory and explanatory data analysis

5.1 *Introduction*

Various metric and non-metric methods for spatial data analysis will be discussed in this section, with particular emphasis on recent developments in the field of soft econometrics. We have already noted in Section 2 that one can distinguish between variables measured at a nominal, ordinal, interval or a ratio scale.

In Section 5.2 we deal with some correlation and association measures and the main disadvantages of widely used correlation measures in regional analysis. The analysis of contingency tables with nominal data is briefly discussed in Section 5.3 by means of log-linear models. The log-linear model is part of a broad family of generalized linear models. Such methods are especially suitable for disaggregate data.

Many individual decisions are based on a limited set of discrete choice alternatives, e.g., the choice of a house, or a decision to move. A number of statistical analyses with discrete choice models have been developed recently. In Section 5.3 we also discuss a model classification which distinguishes between explanatory variables and response variables.

5.2 *Analysis of correlation and association*

A number of correlation and association measures have been developed in the literature to quantify the association for pairs of variables, measured at either a cardinal, ordinal or a nominal scale.

The product-moment correlation coefficient developed by Pearson for pairs of continuous variables x and y relates the covariance between variables to the separate variance parts, viz.

$$r^{xy} = \frac{\text{cov}\,(x,y)}{\sqrt{\text{var}(x)\ \text{var}\,(y)}} \tag{8}$$

When the variables are measured at an ordinal scale with rank numbers, due to, for example, a great unreliability of the observations, the rank correlation coefficient from Spearman is an ordinal analogue of Pearson's correlation coefficient.

A main disadvantage of the rank correlation coefficients of Spearman and Kendall in spatial analysis is that they do not take into account the spatial structure and the spatial relationships between variables. A measure of association for spatial variables has been developed by Tjøstheim (see also Tjøstheim, 1978).

A recently developed measure of association with nominal data which makes use of entropy measures and information theory is called zèta (ζ). Here, a decomposition of association is possible when three or more variables are analyzed in a simultaneous way (see also Lammerts van Bueren, 1982). A number of entropy measures in ecology and regional living preferences can be found in Brouwer & Nijkamp (1984b) and Colwell (1974).

We can conclude from various empirical applications with the association measure ζ that the level of association between variables is sometimes rather weak. This can be explained by a specific empirical situation, but a more relevant reason for a low value of the association measure may be the nominal measurement level from the observations due to a particular classification procedure.

A general conclusion drawn from this section is that correlation and association analysis is a useful tool in an exploratory research phase using spatial data from different disciplines.

5.3 *The log-linear model in contingency table analysis*

We have already remarked that in spatial analysis, much information is frequently non-cardinal in nature. For instance, surveys in economic and ecological analyses often contain categorical data measured in qualitative terms. Categorical data on surveys may be represented in contingency tables (see also Bishop et al., 1975). The log-linear model for spatial analyses may be regarded as a member of a broad family of generalized linear models and is discussed in Brouwer et al. (1983). An application of the log-linear model to an ecological system is given by Whittam & Siegel-Causey (1981).

Consider a three-way $I \times J \times K$ contingency table with elements n_{ijk} and sample size N. There are a number of statistical tests which can be analyzed with this table, e.g. (see also Payne, 1977)

$$H_0 \ : \ n^*_{ijk} = \frac{N}{IJK} \qquad \text{for all } i, j \text{ and } k \tag{9}$$

$$H_0 \ : \ n^*_{ijk} = \frac{n_{i\cdot\cdot}\, n_{\cdot j\cdot}\, n_{\cdot\cdot k}}{N^2} \qquad \text{for all } i, j \text{ and } k$$

$$H_0 \ : \ n^*_{ijk} = \frac{n_{ij\cdot}\, n_{\cdot\cdot k}}{n_{i\cdot\cdot}} \qquad \text{for all } i, j \text{ and } k$$

$$H_0 \ : \ n^*_{ijk} = \frac{n_{ij\cdot}\, n_{i\cdot k}}{n_{i\cdot\cdot}} \qquad \text{for all } i, j \text{ and } k$$

$$H_0 \ : \ \begin{array}{l} n^*_{ij\cdot} = n_{ij\cdot} \\ n^*_{\cdot jk} = n_{\cdot jk} \\ n^*_{i\cdot k} = n_{i\cdot k} \end{array} \qquad \text{for all } i, j \text{ and } k$$

This gives the following hierarchical log-linear models

$$\begin{aligned}
\ln n^*_{ijk} &= \mu \qquad\qquad (10)\\
\ln n^*_{ijk} &= \mu + \mu_1(i) + \mu_2(j) + \mu_3(k)\\
\ln n^*_{ijk} &= \mu + \mu_1(i) + \mu_2(j) + \mu_3(k) + \mu_{12}\,(ij)\\
\ln n^*_{ijk} &= \mu + \mu_1(i) + \mu_2(j) + \mu_3(k) + \mu_{12}\,(ij) + \\
&\quad \mu_{13}(ik)\\
\ln n^*_{ijk} &= \mu + \mu_1(i) + \mu_2(j) + \mu_3(k) + \mu_{12}(ij) + \\
&\quad \mu_{13}(ik) + \mu_{23}(jk)
\end{aligned}$$

μ is the overall mean effect; $\mu_1(i)$, $\mu_2(j)$ and $\mu_3(k)$

are the main effects for variables 1, 2 and 3. The first order interaction effects are $\mu_{12}(ij)$, $\mu_{13}(ik)$ and $\mu_{23}(jk)$, while $\mu_{123}(ijk)$ is the second-order interaction effect between variable 1, 2 and 3.

The five statistical hypotheses mentioned above are tested by means of a test statistic which follows a chi-squared distribution asymptotically.

Maximum likelihood parameter estimates are obtained for the interaction between variables. The log-linear modeling approach is not a real explanatory analysis because no distinction is made between response variables and explanatory variables.

Wrigley (1979) developed a model classification which distinguishes between response variables and explanatory variables. See also Figure 3.
Cell elements (*a*), (*b*) and (*c*) give the traditional regression models with quantitative variables, while categorical data analysis methods are given in cell elements (*d*), (*e*) and (*f*). The cells (*d*) and (*e*)consist of among others, logistic/logit and probit regression models while the weighted least squares regression model is part of cell (*f*). Some statistical analyses with discrete choice models, developed in transportation research, have been recently applied in regional research. Many decisions have only a limited set of discrete alternatives, e.g. choice of a house, traffic route choice, decision to move, etc. An interpersonal utility interpretation may then consist of a classification of a population based on maximization of individual preferences.

6. Illustration of an integrated economic-ecological model

In this section is given an illustration of a recently

		Explanatory variables Continuous	Mixed	Categorical
Response variables	Continuous	(*a*)	(*b*)	(*c*)
	Categorical	(*d*)	(*e*)	(*f*)

Fig. 3. Model classification with measurement level.

developed integrated economic-ecological model for regional planning purposes (see also Arntzen & Braat, 1982). This model was developed as an instrument for spatial planning on a regional level for the medium-long term and economic, geographical and spatial elements are analyzed in relation to the natural environment.

The case study is modeled after the Dutch region West-Brabant where a regional planning extension scheme is being developed. A plan situation and a reference situation (development without the extension scheme) are compared with each other.

Figure 4 shows the Markizaat area which is part of the West-Brabant region.
The integrated model structure consists of 5 components, viz. a demographic module A, a facilities module B, a (socio)economic module C, an environmental module D and an intermediate module E. A simplified structure representation of such an integrated model is given in Figure 5.
The model structure in Figure 5 is interpreted as a closed system, without input effects from the outside (see also Section 2). The different modules are really open systems because of the modelized interrelationships between modules. In Figure 6 we present an example of an open system with mutual relationships between ecological variables and the level of recreation. The mutual associations between six ecological variables are given in graph-theoretical terms by signed digraphs.

We present the following variables in Figure 6:
1 = quantity of algae
2 = quantity of herbivorous birds
3 = quantity of aquatic macrofauna
4 = quantity of oxygen
5 = quantity of phosphate
6 = quantity of aquatic vegetation

The subsystem of ecological variables is interpreted as a closed system, when the recreation component and its mutual influence on the quantity of algae are excluded from the analysis. In such a case the ecological module is a simultaneous system of endogenous variables (see also Section 2). An increased number of people engaged in recreation has a positive effect on the growth pattern of

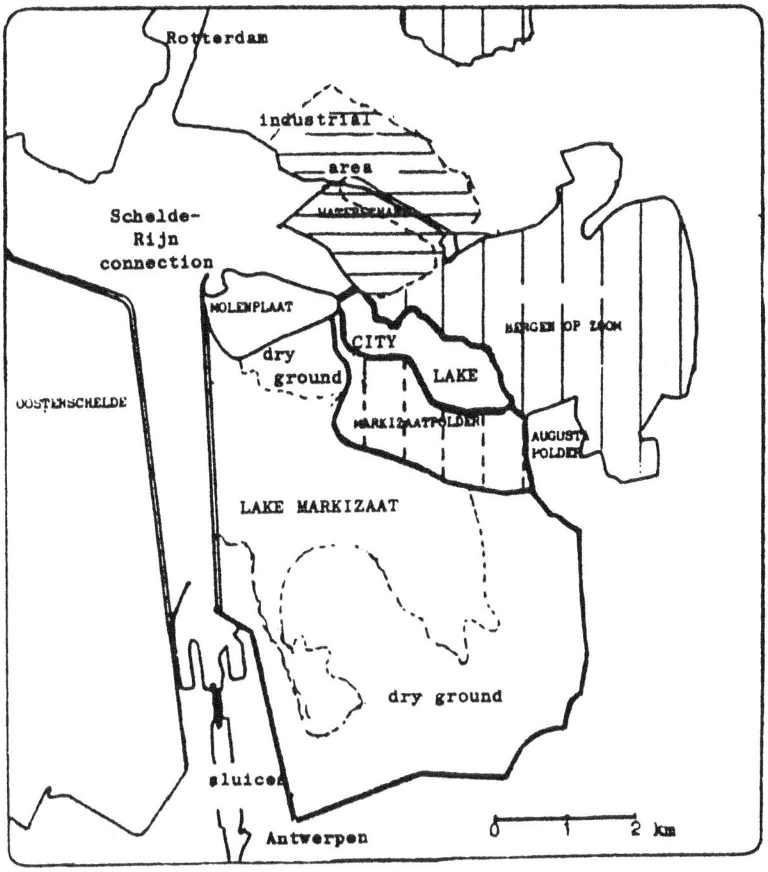

Fig. 4. The Markizaat area.

algae and the growth pattern of algae has a negative effect on the level of recreation. The recreation variable has no effect on the conditions of stability in a qualitative way but it influences the equilibrium value of the ecological variables.

In a set of linear differential equations

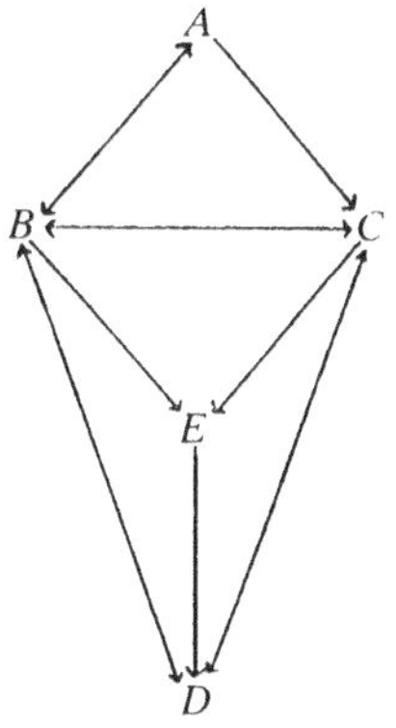

Fig. 5. An integrated economic-ecological model.

$$\dot{x} = \frac{dx}{dt} = Ax \tag{11}$$

the matrix sign (A), containing qualitative information about the impacts between the variables represented in Figure 6 becomes

$$\text{sign}\ (A) = \begin{bmatrix} - & + & + & 0 & - & + \\ - & - & 0 & 0 & 0 & - \\ - & 0 & - & - & 0 & 0 \\ 0 & 0 & + & - & - & + \\ + & 0 & 0 & + & - & + \\ - & + & 0 & - & - & - \end{bmatrix}$$

The necessary and sufficient conditions for stability in a qualitative way hold for the matrix sign (A) (see also Section 4.3, and Brouwer & Nijkamp, 1984a). Every matrix of the type sign (A) has eigenvalues with negative real parts. If some equilibrium situation is disturbed, e.g. an exogenous change, a

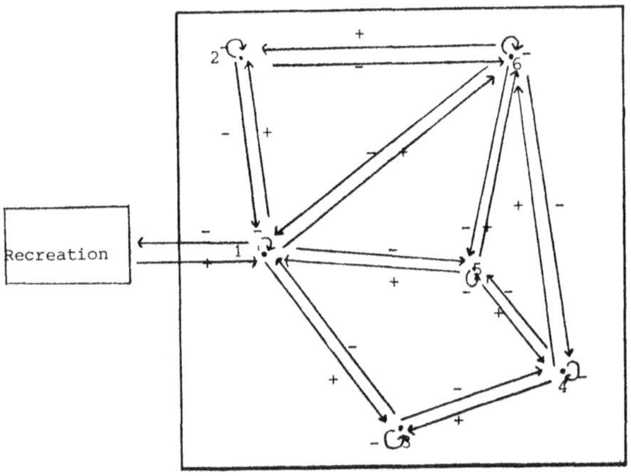

Fig. 6. Graph representation of ecological variables.

new equilibrium situation will be attained independent from the level of the impact variables a_{ij}.

Figure 5 shows that each module is related – either directly or indirectly – to all other modules. In addition to this simplified representation of an integrated economic-environmental model one may also distinguish between intra-relationships (within each module) and inter-relationships (between the modules).

The various specified relationships between the elements, consisting of both the intra- and the inter-relationships can be represented as a Boolean matrix to indicate the directions of the influence in a zero-one sense.

Lack of data and the complex way in which variables from different disciplines interact, made a combination of qualitative and quantitative analyses indispensible for a proper analysis of the variables and the structures they fit in.

7. Conclusion

A number of methodological issues have been discussed in this paper in the context of a multi-dimensional modeling approach for an economic-ecological system. Systems-analytical methods have been used to analyze the structure of strongly interrelated multidisciplinary components, with special emphasis on four conceptual and methodological problems which deal with complex economic-ecological systems. These four problems are related with the *spatial* and *temporal* aspects when integrating these disciplines, the data measured at *different levels of measurement* as well as the *structure of relationships* in either a binary or a qualitative sense.

The choice of the spatial aggregation level as well as of the spatial scale is relevant for modeling results, because in economic-ecological models, the socio-economic variables are usually measured at a regional level (particularly in a spatial planning context), while ecological variables are often measured on a more disaggregated spatial scale. In this regard, time-geography has been discussed in this paper as a methodological tool for analyzing individual observations in a multi-dimensional approach between time and space. Furthermore, mathematical methods like graph-theory have been developed to analyze fundamental relationships between variables when only information about the structure of relations is available. These methods are applied in, among others, the fields of energy, transportation, economics and pollution (see also Roberts, 1978). The above mentioned methodological issues are relevant to the analysis of integrated economic-ecological models in a spatial context.

Acknowledgement

This study is supported by the Netherlands Organization for the Advancement of Pure Research (ZWO), project number 46–079.

References

Arntzen, J.W. & Braat, L.C., 1982. An integrated environmental model for regional policy analysis. In: T.R. Lakshmanan & P. Nijkamp (eds.), Systems and Models for Energy and Environmental Analysis, Gower, Aldershot. pp. 50–63.

Bassett, L., Maybee, J. & Quirk, J., 1968. Qualitative economics and the scope of the correspondence principle. Econometrica 36: 544–563.

Bennett, R.J. & Chorley, R.J., 1978. Environmental Systems: Philosophy, Analysis and Control. Methuen, London.

Bishop, Y.M., Fienberg, S.E. & Holland P.W., 1975. Discrete Multivariate Analysis: Theory and Practice. MIT Press, Cambridge, Massachusetts.

Brouwer, F. & Nijkamp, P., 1984a. Qualitative structure analysis of complex systems. In: P. Nijkamp, H. Leitner & N. Wrigley (eds.), Measuring the Unmeasurable, pp. 509–532. Martinus Nijhoff, The Hague.

Brouwer, F. & Nijkamp, P., 1984b. Multiple rank correlation and association analysis. In: G. Bahrenberg, M.M. Fischer & P. Nijkamp (eds.), Recent Developments in Spatial Analysis: Methodology, Measurement Models, pp. 317–334. Gower, Aldershot.

Brouwer, F., Nijkamp, P. & Scholten, H.J., 1983. Contingency table analysis for categorical data with an emphasis on temporal stability and structural zeros, paper presented at the 23rd European Conference of the Regional Science Association, Poitiers.

Caswell, H., Koenig, H.E., Resh, J.A. & Ross, Q.E., 1972. An introduction to systems science for ecologists. In: B.C. Patten (ed.), Systems Analysis and Simulation in Ecology, Vol. 2, pp. 3–78. Academic Press, New York.

Colwell, R.K., 1974, Predictability, constancy and contingency of periodic phenomena. Ecology 55: 1148–1153.

Fischer, M.M., 1978. Zur Lösung funktionaler regionaltaxonomischer Probleme auf der Basis von Interaktionsmatrizen: ein neuer graphentheoretischer Ansatz. Karlsruher Manuskripte zur Geographie, Geographisches Institut, Heft 25, Karlsruhe.

Gale, N. & Golledge, R.G., 1982. On the subjective partitioning of space. Ann. Assoc. Am. Geogr. 72: 60–67.

Gallopin, G.C., 1972. Structural properties of food webs. In: B.C. Patten (ed.) Systems Analysis and Simulation in Ecology, Academic Press, New York. Vol. 2: 241–282.

Greenberg, H.J. & Maybee, J.S. (eds.), 1981. Computer-assisted Analysis and Model Simplification. Academic Press, New York.

Grunfeld, Y. & Grilliches, Z., 1960. Is aggregation necessarily bad? Rev. Econ. Statist. 42: 1–13.

Hall, G.B. & Taylor, S.M., 1983. A causal model of attitudes toward mental health facilities. Environm. Plann. 15: 525–542.

Jeffries, C., 1974. Qualitative stability and digraphs in model ecosystems. Ecology 55: 1415–1419.

Lammerts van Bueren, W.M., 1982. Measuring Association in Nominal Data. Ph.D. thesis, Rotterdam.

Lancaster, K., 1962. The scope of qualitative economics. Rev. Econ. 29: 99–123.

Leitner, H. & Wohlschlägl, H., 1980. Metrische und ordinale Pfadanalyse: ein Verfahren zur Testung komplexer Kausalmodelle in der Geographie. Geogr. Z. 68: 81–106.

Macdonald, K.I., 1977. Path analysis. In: C.A. O'Muircheartaigh & C. Payne (eds.), The Analysis of Survey Data: Model Fitting, Vol. 2 pp. 81–104. John Wiley and Sons, Chicester.

Nijkamp, P., 1983. Economic and ecological models: a qualitative multidimensional review. Series Diskussienota's 1983-2, Free University, Department of Economics, Amsterdam.

Openshaw, S. & Taylor, P.J, 1979. A million or so correlation coefficients: three experiments on the modifiable areal unit problem. In: N. Wrigley (ed.), Statistical Applications in the Spatial Sciences, pp. 127–144. Pion, London.

Openshaw, S. & Taylor, P.J., 1981. The modifiable areal unit problem. In: N. Wrigley & R.J. Bennet (eds.), Quantitative Geography: a British View, pp. 60–69. Routledge and Kegan Paul, London.

Paelinck, J.H. & Nijkamp, P., 1976 Operational Theory and Method in Regional Economics. Saxon House, Farnborough.

Payne, C., 1977. The log-linear model for contingency tables. In: C.A. O'Muircheartaigh & C. Payne (eds.). Vol. 2, pp. 105–144. The Analysis of Survey Data: Model Fitting, John Wiley and Sons, Chicester.

Quirk, J. & Ruppert, R., 1965. Qualitative economics and the stability of equilibrium. Rev. Econ. Stud. 32: 311–326.

Roberts, F.S., 1978. Graph theory and its applications to problems of society. Series: Society for Industrial and Applied Mathematics, Vol. 29. Philadelphia, Pennsylvania.

Thrift, N., 1977a. An Introduction to Time Geography. Series: Concepts and Techniques in Modern Geography, Vol. 13. Institute of British Geographers, London.

Thrift, N., 1977b. Time and theory in human geography (part 1). Progr. Human Geogr. 1: 65–101.

Tinkler, K.J., 1977. Introduction to Graph Theoretical Methods in Geography. Series: Concepts and techniques in Modern Geography. Vol. 14. Institute of British Geographers, London.

Tjøstheim, D., 1978. A measure of association for spatial variables. Biometrika 65: 109–114.

Whittam, T.S. & Siegel-Causey, D., 1981. Species interactions and community structure in Alaskan seabird colonies. Ecology 62: 1515–1524.

Wilson, A.G., 1981. Geography and the Environment. Wiley, New York.

Wrigley, N., 1979. Developments in the statistical analysis of categorical data. Progr. Human Geogr. 3: 315–355.

Yule, G.U. & Kendall, M.G., 1950. An Introduction to the Theory of Statistics. Charles Griffin, London.

CHAPTER 4

Economic and ecological theory: differences and similarities

UNO SVEDIN

Personal introduction

If you want to address such an ambitious question as that on the differences and similarities between economic and ecological theory it might be necessary to start with a statement on personal perspectives. Of course there is nothing like a homogeneous body of theory, in either economics or ecology, that can be compared. There is also no such thing as a single objective outsiders elevated judgement.

So what are my credentials for indulging in this type of comparison? I am neither an economist nor an ecologist. But with a background in systems analysis, technology assessment and related fields I have been involved in futures studies of natural resources issues (Bertelman et al., 1980) and I am also interested in the question of methodology within the realm of futures studies (Schwartz et al., 1982). So I have seen close collaborators of mine working hard with their input–output matrixes in economics and have spent lots of time listening to my friends in the ecology camp explains the latest problems in energy mapping of ecosystem dynamics. Indeed I also have had the privilege of listening to a very fine presentation of this Conference's topic last year at the Wallenberg Symposium at Saltsjöbaden, Sweden, where I also had the task of summing up the last day's discussions. Indeed, this paper draws heavily on what I learnt on that occasion, and to some degree on my attempt, successful or not, to synthesize a few observations.

Monopolistic claims – reductionistic approaches

To talk about the relationship between economic and ecological theory is to examine two very broad structures of thought, both of which by some proponents have been associated with totalistic aspirations. Thus we are discussing approaches of knowledge which, like early Marxist theory and scholastic theology of the Medieval times, claim to provide the final explanation of phenomena relating nature and man.

However, it is important to observe that as in any 'culture' there are 'subcultures'. The culture of economic theory as well as the culture of ecology, thus consist of many different approaches. And not all of these have totalistic aspirations. This emerging pluralism is an important feature of the current situation. In addition there are 'new' theory structures emerging which claim to explain phenomena within the overlap region between economic and ecological theory. One of these – energy analysis – partially a child of ecology, has emerged with strong claims of its own.

When dealing with questions concerning systems of thought with totalistic claims, the key question of *reductionism* immediately comes into focus. In fact both sides – economy as well as ecology – have accused the other side of this fallacy. As ecology has been the less established scientific discipline, ecologically oriented circles have attacked economic theory for its, it is claimed, unjustified ways of reducing every fact of life to monetary units.

D.O. Hall, N. Myers and N.S. Margaris (eds.), Economics of ecosystems management.

The mutual challenges

The challenge for economy has been:

(a) Many things which might in principle be included in a strictly monetarized theoretical structure have not been included so far. (Examples are the value of the existence of the life support system; the value of genetic diversity; or that energy conservation laws have not been taken into account when formulating the laws of economy.) The answer to this challenge from conventional economic theory has been to improve, through the inclusion of more and more externalities into the calculus.

Against this strategy the critics claim:

(b) that it is not possible to convert everything into these categories.

Criticism from other quarters than from ecology, i.e. from social science, add to these points a third attack:

(c) that within the realm of social phenomena basic ideas on 'economic man' and social rationality used within the structure of economic theory are not only dubious but in many cases obviously wrong.
Closely associated with this position is a claim for a rejection of the reductionist model of man, which reduces him to an agent acting in a simplified manner on the micro level.

The critique on the other hand against ecological reductionism has other features:

(a) that ecology in its present shape has a very simplified notion of man as social phenomenon, including the role of man's creation of 'technology';
(b) that ecology, on a subtler level, has too strong an emphasis on direct causality explanations, well known in natural science, but not at all cherished in all areas of social science.

Indeed what the conflict is about is not simply a question of different 'language', but a clash between different underlying patterns of thought, involving different views on what is to be regarded as assumed and what is not. Differences are also there regarding what the key issues are, as well as the means to mirror or explain relationships. We shall in the following paragraphs deal one by one with some important comparative issues.

System problems

Within a systems analysis approach you have to consider among other things the

- systems definition and restrictions
- ways to break down the system into subsystems
- selection of key phenomena
- ways to select and treat key phenomena and relationships
- ways to view the dynamic factor or time in general

In all these cases there might be differences between economic and ecological approaches. Let us start to discuss the various approaches to a system's definition and limitation. In the two cases of ecological and economic analysis we have seemingly a rather similar object – that of society embedded in and dependent on a natural habitat. But in the two types of analysis *spatial boundaries* selected might be utterly different. In an ecological approach the need is to choose an area within which the main streams of materials and energy in the system circulate. Of course there might be subsidiary flows, but they must not be sufficiently important in regard to the explanations of the phenomena you want to address your attention to. This means that the borders of the chosen system are set by natural phenomena and their dynamics.

In an economic analysis the important boundary conditions might be of a political kind or might try to incorporate key facts of a web of geographically distributed economic relationships (e.g. trade patterns). In any case it is rather obvious that the spatially selected boundaries in the two cases will not generally be the same. This has led to problems in studies where the task is to use ecologically based information on energy flows and to relate these to social and economic features. The system's boundary chosen out of ecological concerns is, as an example, the *island* of Gotland in the Baltic and its surrounding waters (Jansson & Zucchetto, 1978). But that system's boundary is not the optimal one in regard to regional economics. The strong economic ties between a region of a country and the overall economic development of the nation (and indeed of world economy) makes such a decoupling of a geographical part more difficult. In this case perhaps other part systems might be more rewarding objects of study, such as dealing with different branches of industry. In addition to this problem we might have administrative system's boundaries to take into account, which do not follow ecologically rewarding system's limitations.

Also in the *selection of time span* there might be obvious differences. Not infrequently the ecological approach claims a longer time perspective than in economic analysis. This might or might not be true in specific cases. We do have ecological studies where the time boundary assigned is the sequence of the seasons of *one* year only. And we do have economic studies with a substantially longer time perspective. What can be stated is that the phenomena of interest are often different regarding their time depth in the two cases, resulting in different selection choices.

The different weights various phenomena are given in different forms of analysis are reflected in the choice of the units chosen for the analysis. The choice between dollars, kilograms and kilowatt hours as basic units with which to express the selected phenomenas could give amusing differences of emphasis. If we take the example of input for a pillar 1 meter high, and designed to support a load of 1000 tons, the answer could be put in terms of the cost in dollars for the alternatives made of steel, brick or concrete. However, the impression given by the comparison is quite different if it is given in terms of energy incorporated in the materials (expressed in 'liters of oil'). The result might be looked upon still differently if it were presented as the investment in work hours for the construction of the pillars. Or why not the combined cost of the initial investment and the continuous maintenance of the structure. Indeed what our example shows is that the resource has to be expressed *in such a denomination that is pertinent to the question* which is being studied. And that rule, simple as it is, holds true not only for resources but for entire analytical approaches to complex systems. The rule hits hard against any simple-minded reductionist mode of analysis where a single type of denominator is taken for granted.

Emphasis on limits

The role of and emphasis on limitations as compared with interest in various expansion-oriented mechanisms tells quite a deal about fundamental attitudes in various analytical schools of thought. The former approach works from the outside inwards, from a judgement on boundary conditions towards partial mechanisms contained within the limits. The second approach works from the inside outwards, discussing generating internal mechanisms as the key starting point. Ecology very much seems to be oriented towards the first type of approach emphasizing, for example, conservation laws for energy, carrying capacity concepts and possible threshold breaking points for ecosystems under stress. Economic theory has so far been little concerned with these types of constraining factors. If constraining factors have been of interest, they have had an institutional character, rather than factors relating to physical limits.

The proper choice of principal approach of course depends on the phenomenon involved. If the limits are rather remote in relation to the phenomena under study then the role of borderline effects might not be that interesting. If on the other hand material and energy limits are close by, then the importance of dealing with these effects increases. Let us take as an example the difference in approach between Malthus and Mill. For the former scholar, the limits of available usable cropland were imminent. In that case the expanding population had to be seen in a tragic light. If on the other hand the limits of natural possibilities are seen as adjustable, at least to some extent, and if more phenomena than material conditions alone could change the rules of the game (e.g. the development of technology improving the efficiency of the use of material conditions) then the outlook might change drastically. Indeed this was the position taken by Mill. He recognized the role of improved infrastructure, social organization, etc. In such a case other forms of 'limits' might come into play – such as 'the social limits to growth' (Hirsch, 1977) – or institutional limits of various sorts with entirely different features than the material or energy ones.

In any case the examples show the varying roles of an interest in boundaries of any sort might have, versus the role of generating mechanisms inside the boundaries themselves, or in adjusting and expanding these.

In a way this problem is related to the question of emphasis on micro- or on macro-phenomena. Generally it seems as if macro-oriented studies are more keen on the role of limits than are micro ones. However the situation is not clearcut. What is obvious though is the need for studies, both in economy and ecology which try to link the macro and the micro levels.

Man as a systems agent

When comparing different approaches it is sometimes useful to dig into the set of hidden assumptions underlying these patterns of thought. Upsetting differences between sciences and even direct difficulties in the communications between them are often found, not only in differences in the choice of object of research ('systems definition'), but also in different underlying conceptual perspectives.

One such item in the very most philosophical end of the spectrum of problems deals with the view on *'man as a system's agent'*.

In economic theory many basic assumptions have been connected to the notion of 'economic man'. Much has been done to diversify the classical and rude view of an atomistic actor continuously making self-oriented choices of optimization. In ecology on the other hand, not infrequently another type of reductionism can be found explicitly or implicitly. Here man is seen as a biological input/output organism only, with intake of food and delivering of excreta as output as dominant features. Biologically inclined population studies have thus been more oriented towards the use of population dynamic models relevant also for bacteria and mice, than penetrating into the realm of social dynamics and cultural control. These aspects have, in short, not been regarded as sufficiently strong and determining factors for descriptions of long-term development. Some proponents have even made the analogy more strictly biological: man has been seen as a cancer cell on the 'body of the planet earth'.

Man's goal setting and innovative faculties have sometimes, but not always, been incorporated in what might be termed 'the engineer's view'. In such cases the description of dynamics of the system has absorbed the historical record of man's potential for creating changed conditions. Not seldom however this has been done without going into the social and psychological details of how this innovating capacity came into play. Let's call this view 'man the creative but *closed box*'.

It is interesting to see how a philosophical debate of the medieval times is relevant in this context. The followers of Aristotle claimed that man is a passive recipient of actions from outside. The opposition, through e.g. 15th century scholar Nicolaus of Cusa (Cusanus), claimed that man has been given a genuine creative capacity from God. This capacity, for good or worse, has a potentially transforming power on the environment (Lyttkens, 1979).

In all these cases the focus has been directed towards different facets of man's existence: the role of man as a self-oriented optimizing agent in the market place, a change creating actor in the technology arena or as a passive biological mechanism.

Views on information

These facets are closely linked to different views on information. In economic theory information is of key interest, but not always explicitly. It is a free flow of information which is at the heart of the 'perfect market idea'. Without the free flow of information no 'juste' and independent judgement could be performed by the actors who are the atoms of the system. In the case of monopolistic information control the free market is distorted and the theory has to deal with market failure perturbance analysis. During the last decade an increased interest has been shown within the economic research community towards the effects of distortions in the information system, that is the problem of 'information asymmetry'. Thus information transfer (here meaning roughly 'knowledge of the state of the art') is of key importance in economic theory either directly as an object of study or as a prerequisite for further analysis.

In the case of 'the engineer's view' the mental map of information transfer is similar to an electronics circuit chart. Any connection of information has to be drawn explicitly in a chart symbolizing interconnections. In the systems script of Jay W. Forrester (1971) the information connections are drawn explicitly as dotted lines (Fig. 1).

Physical flows ↓ run from sources to sinks. A knob can regulate the flow. However dotted lines show connecting paths of influence from various auxiliary variables ○.

In such a way full charts of connections could be built. Figure 1 shows a part of a 'world model' chart. Our key point here is not to dwell on specifics of the chart, but to emphasize once more that the lines of influence – what might be termed 'control information' – are drawn explicitly and are separate from other flows.

In ecological theory information is never distinguished from its material (or energy) carrier. The counting of the substrate also takes care of the information content. H.T. Odum's energy analysis script provides a nice example in this respect. Here no specific links of information, separate from flows of energy and materials, have been found necessary. Figure 2 gives an example (Jansson, 1972).

Flows go from (energy) sources ○ to sinks through various 'work gates' as multipliers, retarding processes etc. and through intermediate storage. If we compare Figure 2 with Figure 1 we see that explicit reference to flows of influence *other than* flows of material and energy are absent. Indeed they are probably not only regarded as unnecessary but impossible.

In this perspective *all* influence is exerted through its energy content. This is reasonable in an analysis of ecosystems with biological links only. However, in society information content (meaning 'knowledge of the state of art') is something more than the energy substrate carrying the message.

The lack of making this distinction limits the approach to such systems where complex control is not exerted through the exchange of highly structured messages.

The way such ecological energy analysis regards

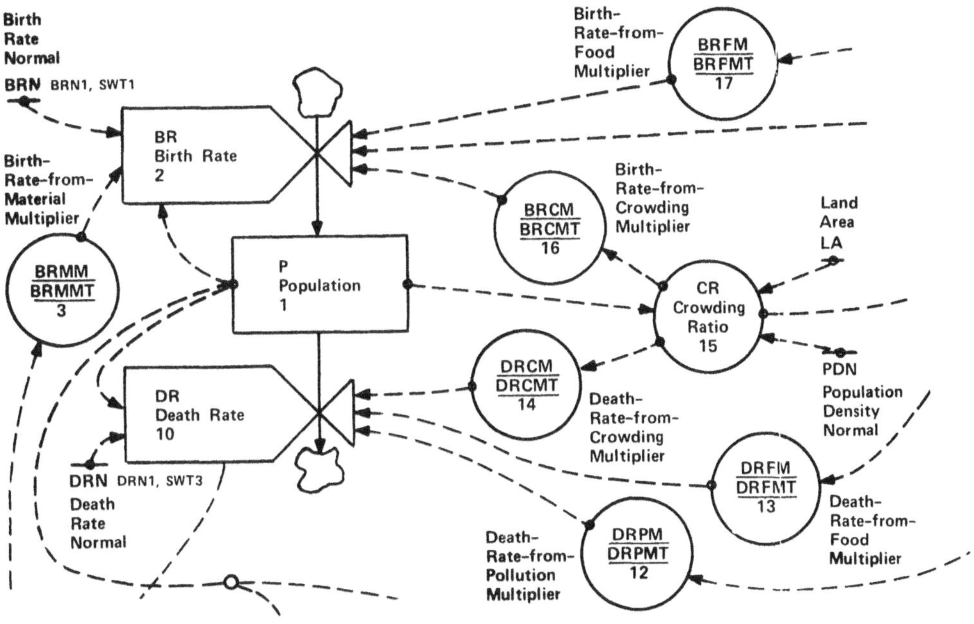

Fig. 1. Part of systems chart by J.W. Forrester (world model).

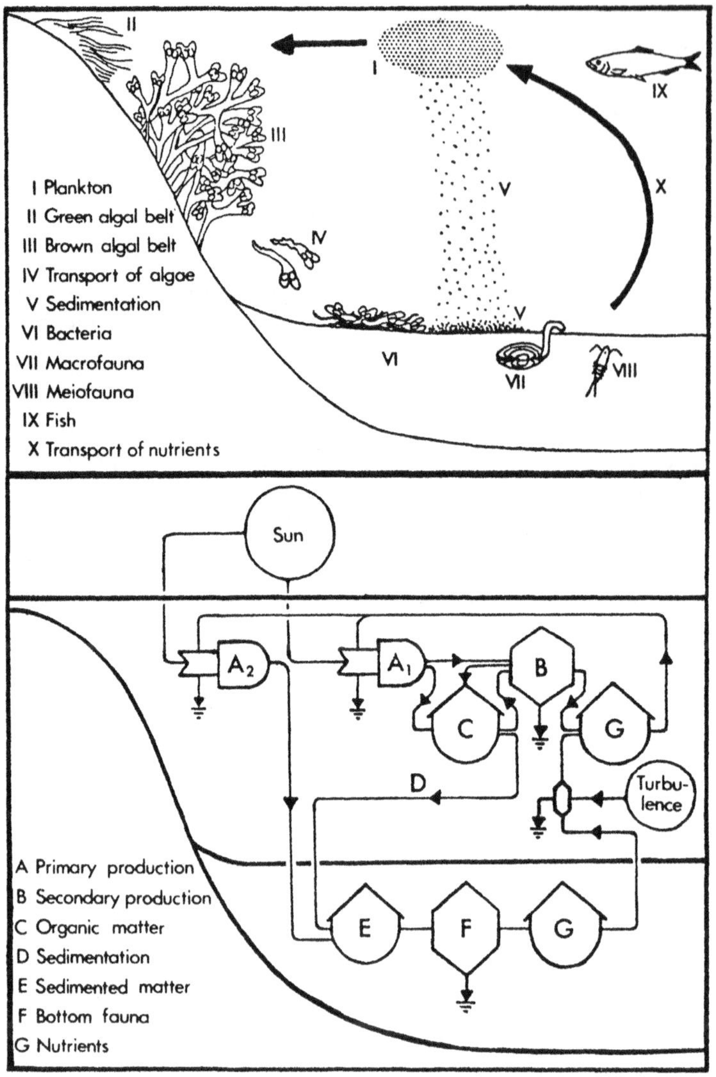

Fig. 2. Flow of energy and matter within a coastal area. Above: a naturalistic picture. Below: the same system expressed in H.T. Odum's energy language.

information reflects, if driven to its extreme, the reductionist view that everything could be measured in the same units – energy. In one way this is true, in another it is not. Of course any information must be associated with a flow of energy. But it is the pattern of the often tiny amount of energy necessary to perform the control function which matters when we are talking about the information role in general.

In some quarters the 'theory of information' related to names like Shannon has had a confusing effect in this respect. Hopes have been raised that the structural content also in 'normal messages' could be expressed in a single type of unit borrowed from thermodynamics in physics. Here the entropy concept expresses the level of structural order. Thus some scholars have taken this idea as a starting point for a separation of budgets of different kinds of energy depending on its structural quality. The measure of this quality has been associated with different levels of 'exergy' or 'negentropy', the entropy concept developed within the physics of thermodynamics.

This is all well and constitutes an advance in the

accounting systems as separate forms of energy have distinctive, different inherent technical performance capacities embodied. However, by so doing, and incorporating a two-dimensional character of energy in the ecological theory framework, which is based on physical unit counting, another type of reductionism is often inserted. The belief is that having incorporated exergy measures in the theory structure (and thereby 'information structure') a complete system of reference is given of phenomena in nature as well as in society. But this approach still leaves out many other aspects, not the least within the realm of information transfer.

Not infrequently, the exergy calculations are in practice associated with claims of understanding systems performance in general, but are based on understanding the behaviour of inanimate matter. The features of such systems are often generalized and the results are used to draw farreaching conclusions about systems of higher order, as those with self-organizing capacity (e.g. living systems). In dealing with resources questions in general, such overgeneralizations might prove erroneous. For example the dispersion of matter into lower forms of organization as time goes by (and if no energy input is used to maintain the forms) is on a nonliving level of systems rightly regarded as an inevitable process of 'decay'. This statement is based on fundamental laws in thermodynamics. However for a partial system this one-directional flow of events could be counteracted locally by biological processes which convert 'simpler' structures to ones of higher order as in the familiar process of photosynthesis of green plants. Such biological units could be said technically to act locally against the overall structural degradation indicated by thermodynamically based time. The problem of exergy reductionism arises as in other cases of reductionism not through the tool of investigation itself, which is beneficial within its own realm, but through the claim of *total* power of explanation of the phenomena involved.

Technology and the issue of dynamics

Linked to the information issue are varying views on technology. Traditional scientific ecology does not deal with technology at all (but might be involved with the effects of technology – as pollution). Its object is natural phenomena. Technology in the form of man made artifacts and mechanisms is by definition not included. However ecological theory has expanded and in some cases aspirations have been raised to apply 'ecological tools' to objects surpassing the domain of 'natural phenomena'. Some authors thus discuss ecological approaches to nature and society and thereby try, most legitimately, to expand the domain of the ecological approach. In that case the existence of technology within the boundary of the newly defined system creates problems, as technology not only generates products but also generates changes in the previous rules of the game.

This feature has not only been an obstacle to ecological theory, but also to several classical economic analysis approaches e.g. traditional input–output analysis which is basicly static in its design. There have been attempts to overcome these obstacles. However the potential of change connected to technology most often has had to be inserted from outside. This has in practice meant, for example, that historical records have been used and less an understanding of underlying phenomena. That the task is formidable is demonstrated by the meager results produced by e.g. technological forecasting in order to predict technological development. Indeed the key mechanisms of technological development are still not understood in their full cultural and societal contexts.

Obviously the historical record shows that technological development has had a strong transforming capacity on material conditions. Earlier perceived resource limits have been transcended through the means of new organization and of technological innovation. Barnett & Morse (1963) showed in their now classical work that costs per unit of output in extractive industry declined in the period 1870–1957. This means that during that period technological development more than compensated for the potential scarcity the extraction of minerals constantly invites.

We do not need here to go into detail of later

criticism of these findings, on the grounds that environmental degradation had not been internalized as a cost. The point here is to hint at the *contextual* quality of this capacity of change. The overall effect to keep scarcity away at one level or the other is not to be associated with any particular event or invention. Instead it is an expression of a combined societal, economic and technological system's effect over many years, covering such a varied set of issues as changed trade patterns, new financial structures for the mobilization of capital, international and domestic political control, new technical devices and changed habits of the population at large.

The importance and difficulty of context

What we have seen in the case of technology holds true also for many of the other approaches we are comparing. Many phenomena have the character of being contextually sensitive. By that it meant that the interpretation of their behaviour cannot be drawn from local comparisons only. The 'entirety' influences the particular in such a decisive way that this influence cannot be neglected.

It is probably fair to say that both economic theory based on *atomistic* assumptions on economic agents as well as ecological theory based on a scrutinizing of *elements* of flows building a map of an ecosystem, despite their holistic claims are basicly built on a non-contextual approach. It would be interesting to see how a contextual sensitive economic theory could be built e.g. a covering theory of economic events in Sweden which would in some way reflect general social and cultural tendencies at the macro level into the description of phenomena at the micro level.

It would be equally interesting to see how an ecological theory could be built based on an understanding of overall systems characteristics in an interplay with discrete local events.

In general that would imply the development of a theoretical structure in both cases where non-local phenomena are given importance, i.e. that something which happens at a far distance to a particular event has direct influence on that very event. But we already understand the tremendous problems of such an endeavour. We already know some of the problems with instabilities through small fluctuations in large meteorological model systems. And in that case we have still not included such non-local influences as was indicated above.

Optimization, the use of analogies and the role of values

In many cases people argue as if optimization is something unconditional and clear and not needing specification. But which type of unit is used in the optimization ($, KWh or tons) and the selected boundary condition depend on the purpose of the study. All these types of specifications have to be added to the question before it can be answered. And the reflection of different foci of interest, different system selections and choice of limitations, different views of mechanisms all will result in different views on what optimization is about.

In general you could argue that the problem of how to pose the questions is embedded in the value structures inherent in all sciences. Here the use of analogies in the various approaches to the formation of conceptual patterns is essential. We have for example seen the use of the model of the machine, the entropy constrained physical system, various images of balance, the prototype idea of the biological cell, as well as that of creative man. All of these mental pictures flavour the design of a certain scientific approach. And they convey in a subtle way a whole array of implicit assumptions. It has been the aim of this paper to put a finger on some of these underlying patterns of thought.

If we want a real and genuine integration in the future between economical theory and ecological theory it is necessary for us to understand these assumptions, to build upon their complementary elements and find out the extent to which irreducible differences could be coped with through selection of systems limits. It is through the deliberate choice of a specific analytical approach to a given problem that we can gain success in the individual case. The integration aimed at should thus not be seen as a hybrid marriage intended for general use under all circumstances. Rather it should be seen as a continuous design of selective tools for specific problems.

References

Barnett, H.J. & Morse, C., 1963. Scarcity and Growth. Baltimore.

Bertelman, T., Hollander, E., Olsson, C.-A., Parmsund, M., Sohlman, S. & Svedin, U., 1980. Resources, Society and the Future. Pergamon.

Forrester, J.W., 1971. Counterintuitive Behaviour of Social Systems. MIT.

Hirsch, F., 1977. Social limits to growth. Routledge and Kegan Paul, London.

Jansson, A.-M. & Zucchetto, J., 1978. Man, Nature and Energy Flow on the Island of Gotland. Ambio 7: 140–149.

Jansson, B.-O., 1972. Ecosystem Approach to the Baltic Problem. Bulletins from the Ecological Research Committee No. 16. Swedish Natural Science Research Council.

Lyttkens, H., 1979. Gud, människan, samhället och naturen (God, Man, Society and Nature). In: G. Agrell (ed.), Människan i miljön – Vårdare eller Vandal?. Skeab.

Schwarz, B., Svedin, U. & Wittrock, B., 1982. Methods in Future Studies – Problems and Applications. Westview.

CHAPTER 5

Three basic concepts: man as part of the environment, carrying capacity, conservation. Some further consideration

NICOLAS VERNICOS

Introduction

Ten years of research in the fields of economic and sociocultural ecology, mainly conducted in relation to a number of projects developed by UNESCO, have offered a unique opportunity to familiarize ourselves with several theoretical and applied issues of an emerging science dealing with the interplay of mankind's economic, social and cultural activities, and his natural environment.

During this time, Greek society (to which this paper is more particularly addressed) became aware of the environmental problems: a majority of people come to the conclusion that the national political bodies would have to deal with these problems in a coherent, consistent and viable manner.[1] This goal is, however, far from being achieved because of the tremendous dearth of basic information and the pervasive lack of training, both of decision makers and of field staff.

In this respect it should be noted that recent studies, sponsored by international organizations and relating to ecosystem management, have proved that interdisciplinary work towards collecting and interpreting data must be constantly encouraged; specialists find it particularly difficult to overcome the embedded attitudes of the lone academic and are generally belittled by different scientific outlooks. Yet, only interdisciplinary and system approaches may provide us with sufficient knowledge of what the issues involved in ecosystem management are.

Participation in research aimed at formulating guidelines for the 'eco-management' of minor Mediterranean islands and in identifying major issues relating to the management of land-based natural systems of the Mediterranean coastal zones have shown that several basic concepts or ideas are not yet fully assimilated. Hence the need of repeating them continuously and illustrating them through applications and field training projects.

With the view of contributing to the sound economic and social management of the numerous island ecosystems of Greece we will endeavour to reexamine three of these concepts, accompanying them with reflections drawn from our own field observations or from mundane facts that are too often overlooked. In our view these concepts that follow apply to all ecosystem management efforts.

The 'man in the environment' concept focuses on the manager of an environmental setting and implicitly deals with the reason for which there is a need for ecosystem management. 'Carrying capacity' is obviously related to the substance to be managed: the resources and the human population. Finally 'conservation' refers to the goals of management, it will be adressed by the question: 'are we to preserve or to improve?'

[1] The above consideration delineates the emergence of 'ecopolitics', see Deutch, Karl (ed.), 'Ecosocial Systems and Ecopolitics', UNESCO, Paris (1977).

D.O. Hall, N. Myers and N.S. Margaris (eds.), Economics of ecosystems management.

 ISBN-13: 978-94-010-8928-9

Man is part of the environment

In the beginning of 1974 a task force was convened by UNESCO to evaluate the scope, form and content of the contribution by social sciences to the 'Man and the Biosphere Programme' which is the major international program for ecosystem studies and management. Along with other participants we formulated the following statement of fact:

> 'Man is simultaneously part of the environment and – with his awareness and capacity for deliberate action – responsible for its stewardship. Thus, man is not merely an actor on an environmental stage but has partly created his own setting.'[2]

We consider that this statement governs all environmental situations, especially those found in areas, such as the Mediterranean, where man has been intensively active for over ten thousand years. Nevertheless we still tend to think in terms of balanced, self-regulating 'natural' systems, disregarding the numerous disruptive processes that past and present human activities have introduced. Lack of information is one of the reasons that we fail to take into account the history of resilience processes of human origin.

There is however an important ideological bias[3] that prevents several ecological studies from taking into account, in a consistent manner, the fact that deliberate human actions have shaped nature and to clearly perceive how man contributes to the creation of the natural settings we observe in the Greek landscape. This bias has its origin in the Aristotelian view of the relationship between man and environment, each occupying a distinct sphere that never meet or mingle. The concept of 'design in nature' formulated by Aristotle was further reinforced by Christian philosophy that made God the basis for the design. Needless to say that in such a physico-theology the seesaw processes of survival upon which resilience studies concentrate do not fit in very well. Even in the secular version of this world view, where 'natural laws' and scientific rationality replace the divine purpose, the 'catastrophy' theory explanation of ecosystem change is ridden with negative connotations. This is amply displayed by the unfortunate choice of the word 'catastrophy'; as a result of which we are forced to consider the various disruptive effects of the activities of a society upon nature in terms of 'catastrophy'. Yet it is these very activities that cause evolutionary changes in both the human race and in ecology and these 'spurts' and 'crashes' have provided selective advantages over millenia.

If we fully accept the fact that man is part of the environment, and that he has partly created the settings, this creates a number of important implications.

In the first instance we may stress the concept of the ecosystem by taking into consideration that both flora and fauna and the physical characteristics of the environment, have been interfered with many times throughout history with varying degrees of intensity by humans. As a result a proper investigation of every Greek landscape may inform us of the impacts of past and present 'human use systems'.

We must therefore define a set of ecologically pertinent criteria to judge and identify successes and failures in man's stewardship of the natural environment. As an example we need scientific criteria in order to be able to determine whether a modern maquis landscape is 'good' or 'bad' or whether a reforested area will be ecologically 'better' or 'worse'. In a similar way, is a small islet where rodents have proliferated a 'natural' setting simply because humans rarely visit? Furthermore, is greenery and the current esthetic criteria of beauty attached to it sufficient to sanction the establishment or the preservation of ecological communities composed of exotic plants and introduced animals?

[2] UNESCO, Programme on Man and the Biosphere (MAB), 'Task Force on: the Contribution of the Social Sciences to the MAB Programme' (Paris, 28 February–2 March 1974), UNESCO, MAB Report Series No. 17 (SC. 74/Conf. 672/1).

[3] An important document came to our knowledge during the Symposium: The (Swedish) Committee for Future Oriented Research, 'Natural Resources in a Cultural Perspective', Report No. 37-S May 1981, Uppsala (1981). We stronggly recommend this volume as it takes up several of the questions dealt with in the present paper.

Thirdly, we also need to spell out our management targets relating to ecological management, always taking into consideration existing criteria of success and failure, as well as the more general needs of the society in terms of economy, sociology, culture and individual psychological well-being.

Man's stewardship of the environment further implies a thorough knowledge of the resilience processes that are associated with 'human use systems' which are the organizations and activities through which resources are managed. This means that we must be conscious of a human influence on the functioning of 'niches'.

Indeed, since we define a 'niche' as a distinctive life style of an organism, usually described through a unique use of resources, or as the manner in which a population makes a living in its habitat, human influence may be traced by his tampering with the resources, in his effect on the habitat or even in determining the composition of the organisms making up these eco-communities.

Finally, we note that much to our regret there is no possibility of restoring the ancient Greek landscape. The modern Greek countryside has been modified beyond return, and probably beyond recognition, by the generations that followed Perikles and Alexander of Macedon.

Carrying capacity

A second basic concept that requires some further consideration is that of 'carrying capacity'. It is defined as the 'theoretical limit to which a population can grow and still be supported permanently by the environment'. This concept involves the calculations of the maximum number of individuals forming the stable population.

Studies to measure an ecosystem's carrying capacity usually provide this figure by dividing the estimate of the total available strategic resource or mix of resources, considered to be the limiting factor, by the individual resource need over a certain time period. The divisor (individual need) is normally weighted by a coefficient that reflects effectiveness of resource use and possible energy maximizing (e.g. the feeding strategy in which organisms maximize food yields). One can immediately understand that it makes a lot of difference whether the resources are renewable or not.

We have previously noted that an ideological bias may interfere with the study of the impact of human actions on environment and with full acceptance of the resilience approach in ecological studies. We will now caution against some logical pitfalls that surround the definition and the applications of the 'carrying capacity' concept.

'Carrying capacity' is derived from the idea that an organism can exist only within a limited range of physical conditions. This immediately reminds us of the terminology used, and work done by, cyberneticians who also start from the idea that every species has a number of interlinked variables which are closely related to 'survival'. And indeed the analytical tools of cybernetics (variables, states of a system, lines of behaviour, fields or phase-spaces of a system, etc.) can be readily applied to ecological systems. In such a context 'survival' of a cybernetic organism will also refer to essential variables which are to be maintained within given numerical limits. From there it is within our capabilities to engineer cybernetic analogs and models of ecosystems and to apply a sufficiently rigorous logic and technology for handling the constructed complex mechanism. But then we should start worrying about the logical pitfall involved in postulating an identity or an equivalence between ecological concepts and cybernetics' terminology. For one thing we should remember that cybernetic reasoning presupposes a chain of derived definitions that amount to tautologies in ecology: in cybernetics adaptation may be limited to mean a system's behaviour that maintains the essential variables within physiological limits. In other words adaptation is a form of 'stability' and simultaneously adaptation is 'survival'.

Let us therefore be cautious in our technical treatment of concepts such as the 'carrying capacity', especially if convenience tends to dictate the search for single limiting factors which may not be acceptable in describing the complexity which is paramount in ecological studies.

It is our contention that further research in 'carrying capacity' should seek to demonstrate how it varies from one generation to another, integrating thus the seesaw processes of resilience. It is true carrying capacity is not conceptualized as being constant when applied to humans, as technology and environment induce variation. Yet, research into island ecosystems has shown that the 'dynamics' of carrying capacity are far from developed as an analytical tool.[4]

Further studies are therefore necessary to provide insights into how carrying capacity may be modified by changes in the effectiveness of resource use or by substitution in the resource mix. Modification of niche life styles and significant shifts in the interactions within the ecological community resulting from resilience processes are also bound to affect the carrying capacity.

Obviously those who manage ecosystems will need to know how the carrying capacity of an environmental unit will vary for each of the significant populations, prior to taking any action.

The study of the minor Mediterranean islands has demonstrated another major shortcoming: the enumeration of the local human population is particularly difficult, thus rendering problematic calculations of a carrying capacity for humans. This is due to the large number of people travelling from one place to another and to commercial trading networks extending far afield. Agreement between local groups and the boundaries of their islands or communes was never achieved in the past. Today intensive human migrations that include emigration, immigration, seasonal tourism and weekend visitors render the very idea of such an agreement unimaginable. Local groups in all areas of the Mediterranean produce a number of specialized crops or artifacts which they trade with other groups to obtain whatever else is needed. (According to archeological discoveries, this pattern may date back to the seventh millennium following the discovery of artifacts made of obsidian brought to the dwellers of the Franchthi cave from Melos and from the island of Yali.)

In all these cases the 'ecological population' needed for the calculation of a human 'carrying capacity' is not the local group but all the groups participating in the trading network. Similarly the resource mix locally available has to take into account all commercial and technological inputs and outputs.

A similar problem is encountered with the concept of the genetic population, as defined by means of a closed mating sustem. As marriages occur with persons from groups that may be located as far away as overseas countries, the effective genetic population must also be defined as the entire network of groups that enter into such a mating system.

Let us therefore dispense with the idea that there exist local groups isolated in cosy little habitats, or island niches, leading protected lives that are not exposed to disruptive external influences.

To preserve or to improve?

The constant encroachment of modern activities upon the physical environment, accompanied by intense population pressure on the limited spatial resources of urban areas has created a desire for the preservation of wildlife and of natural surroundings. The call for preservation is particularly strong in urban areas that are still attracting people while daily expanding and radiating all sorts of communication networks.

The migration from cities to suburban areas, recently observed in the United States, has not yet been identified in the countries bordering the Mediterranean. This new settlement process is bound to alleviate the predicaments of the modern metropolis, but we doubt that important changes in current environmental issues will result within the next ten or twenty years. In the meantime with the accumulation of all sorts of waste products, the inevitable products of the economic process, severe health hazards are locally induced. As a result, the public is increasingly aware of the ecological realities, while experts toll the death bell of animal

[4] A UNESCO-University of Malta workshop held in Malta from 24–27 October 1983 clearly demonstrated this point. A computer model for the study of the island of Gozo is now under development to test the analytical and problem-solving capacity of an adapted Mesarovic-Pestel model that includes a number of 'carrying capacity' functions.

and plant life, not to mention the many images of mankind's doomsday in the media.

There is no doubt that the number of extinct and endangered species call for action to conserve what exists and to preserve species under threat from man's constant modification of their habitat.

Conservation as well as preservation of species and of genetic resources must clearly be included among environmental management targets and balanced against all other needs and priorities.

Generally speaking, a consistent and viable policy for solving ecosystem management problems must necessarily be founded on a dynamic perspective of the development processes. This calls for an understanding of the major developmental trends and also requires a consensus of what is the central point from which decisions for planning and for action may stem.

It is our belief that this central point is the dynamics of mankind's population and its need or effective demands. A global understanding that allows for a coherent management of mankind's environment calls for the assessment of present and future trends of:

(a) the human population, its structure, its socio-cultural variety and the regulatory processes societes command through social organizations;
(b) settlements, built structures, networks and artifacts;
(c) the complex set of demands and needs for resources, amenities, leisure time, quality of life and symbolism.

The estimate of the growth in demand for resources, land and improved environmental quality can then be set against available resources. It is at this stage that we need a proper assessment of environmental resources, both terrestrial and marine, in which we must clearly include measures of biomass, carrying capacity and resilience limits.

Coherent, consistent and viable environmental management strategies and action planning must then be devised to integrate needs, means, availabilities and ecological constraints. Ecological constraints present in relation to preservation policies may sometimes conflict with human needs. Given the present situation prevailing in a large number of countries, environmental concerns should not dictate policies that condemn human groups to poverty or frustrate the quest for development in the Third World, or create further delays for the realization of a New Economic Order.

All sound ecosystem management efforts within global socio-economic policies call for priority actions and data collection. They must above all elicit a general sense of man's responsibility towards nature. Man is part of the environment and human society is now the driving component of the terrestrial biosphere. It is therefore his responsibility to monitor natural equilibria, reduce ecosystem damage, and to enhance through better 'use systems', his life-supporting resources.

Among these resources the entire genetic information stock of terrestrial life is included: hence the necessity to conserve it, in the light of potential future use. Thus there is a growing concern to preserve the species-rich Mediterranean environment by alleviating the extreme fragility that human activities have created.

The Mediterranean experience points to the necessity of ecosystems management aiming not at mere 'conservation' but at upgrading and improving the environment. Upgrading of nature, with a parallel improvement the quality of human life calls for proper social, economic and cultural development.

In this sense improving the environment implies:

- enhancing renewable resources;
- correcting past environmental errors and erasing the scars of degradation, destruction or mismanagement;
- the development of a sharper sense of purpose to allow for the survival of live species in their niches;
- restoration and preservation of genetic variety by recreating habitats and protecting fragile environments;
- self-imposed restriction to preserve resources for future generations and to avoid further mismanagement: the result of insufficient knowledge.

We trust that a majority will agree that mere conservation may be considered to be too passive a goal, when there is so much to improve and upgrade.

CHAPTER 6

Pour une politique d'éco-développement touristique: réalisation pilote de structures polyvalentes d'accueil

NINOU K. GARABAGHI

1. Tourisme de participation active et tourisme balnéaire de masse: opposition ou complémentarité?

1.1 *Le développement du tourisme balnéaire de masse et sa remise en cause: la genèse d'une nouvelle demande*

Avec l'avènement de la société de consommation, le tourisme de masse est devenu un des secteurs essentiels de l'activité économique et des pays industrialisés où le droit aux'vacances commence à être une réalité effective et des pays en voie de développement qui espèrent se développer en tirant profit des avantages du tourisme.

Nous tenons à souligner que nous nous limiterons ici au cas du tourisme balnéaire et côtier qui, de par son importance et ses impacts spécifiques sur des ressources naturelles limitées, a été à l'origine de la remise en cause du tourisme de masse qui a non seulement créé des déséconomies externes mais aussi des déséconomies d'échelle.

Le développement du tourisme balnéaire de masse qui est un tourisme 'concentré', 'enclavé' et 'commercial' a été à l'origine de différents coûts sociaux: déséconomies externes de production et de consommation. Les activités touristiques de masse sont génératrices de coûts externes qui sont d'ordre économique, culturel et écologique. Elles sont à l'origine de la destruction de ressources touristiques qui, à l'heure actuelle, eu égard à la demande, peuvent être estimées comme étant des ressources rares; ceci alors même quel'étude du développement touristique futur, nous oblige à accepter la réalité d'une expansion inévitable de la demande à long terme.

Le problème ici, c'est que cette expansion freinée dans son élan par la récession économique mondiale, ne doit pas nous masquer la réalité d'une demande changeante. Une nouvelle demande en gestation a été identifiée tant par les experts du tourisme que par les diverses autorités nationales et les différents organismes nationaux et internationaux.[1]

[1] Le Conseil de l'Europe, dans son Journal Officiel 1.271/2 (octobre 1980) établit un programme d'aide à certaines régions du Sud de l'Italie et de la France, en particulier dans le domaine de la promotion du tourisme rural, la formation des opérateurs dans ces domaines, ainsi que des animateurs, particulièrement dans les régions où il y a déséquilibre de l'afflux touristique à la faveur des zones côtieres. Des classes vertes et bleues[2] sont envisagées ainsi que des contributions pour la mise en oeuvre d'infrastructures d'un tourisme lié à des activités récréatives et culturelles. Ces contributions pourront aller jusqu'à 70% des investissements.

[2] Le ministre de l'environnement et du cadre de vie, le ministre de l'éducation, le ministre de l'agriculture et le ministre de la jeunesse, des sports et des loisirs français, dans la Circulaire du 6 juin 1980 relative au développement de l'initiation à l'environnement, souhaitent développer 'les centres permanents de classes vertes et de mer, les centres permanents d'initiation à l'environnement[3] les fermes et jardins scolaires' tout en précisant vouloir promouvoir de nouvelles structures telles que 'les ateliers bleus, les centres ethnologiques de l'environnement'.

[3] Les thèmes d'activités cités comme exemple pour un centre permanent d'initiation à l'environnement ou atelier bleu, por-

D.O. Hall, N. Myers and N.S. Margaris (eds.), Economics of ecosystems management.
 ISBN-13: 978-94-010-8928-9

En effet, avec le refus croissant de la société de consommation forcée et orientée, la naissance d'associations de défense de diverses sortes: mouvements de défense des consommateurs, de défense de la nature; avec l'élargissement du temps libre donc des loisirs, la prise de conscience de la nécessité d'une gestion rationnelle du temps, l'aspiration à un mode de vie intégré, rien d'étonnant à voir naître une nouvelle demande touristique qui pourrait être qualifiée de demande de tourisme de participation active.

La question à laquelle nous tenterons ici d'apporter des réponses est de savoir si, compte tenu de la genèse de cette nouvelle demande touristique, il n'est pas possible de prévoir des infrastructures d'accueil qui permettraient de maximiser les bénéfices et de réduire les coûts liés aux activités touristiques traditionnelles.

1.2 *Pour une politique d'éco-développement touristique*

La question que nous nous posons ici est pourquoi une offre touristique écologique?

La première réponse qui viendrait à l'idée consisterait à dire une demande existe, cette demande n'est pas uniquement 'solvable' mais correspond à un 'besoin objectif', il nous faut y répondre.

Non, la raison ne se limite pas dans une réponse à une demande latente fondée, nous sommes en présence d'une stratégie de développement touristique à objectif multiple.

tent sur 'cinq thèmes d'activités avec une orientation nettement économique et contemporaine'. Ces cinq thèmes restent valables pour un atelier bleu:

(1a) ecologie d'estuaire: fonctionnement des éco-systèmes en milieux salés permanents et intermittents et perturbations quantitatives et qualitatives de ce fonctionnement par les pollutions; évolutions des formes de l'estuaire;
(1b) écologie marine (chaînes alimentaires, écosystèmes du plateau continental);
(2) ornithologie;
(3) connaissance des paysages et des gens par randonnées pédestres et cyclistes;
(4) travail des hommes dans la construction de l'espace et activités humaines contemporaines; aquaculture;
(5) energies renouvelables et observation météorologique.

Une politique d'éco-développement touristique opte pour un 'tourisme communautaire intégré' ce qui signifie intégration du tourisme á tous les éléments économiques, sociaux, culturels, écologiques de la zone d'accueil.

L'idée centrale réside dans l' 'éco-développement', le tourisme doit contribuer au développement de l'économie régionale et locale et non à sa colonisation.

Le tourisme doit contribuer à la préservation des ´equilibres éco-systématiques essentiels au maintien de la vie, voire aider à en créer de nouveaux et non pas oeuvrer à les détruire.

L''éco-tourisme' se présente comme un instrument efficace de revitalisation des économies en voie d'extinction. Dans les pays occidentaux, le tourisme en espace rural, est une réponse au problème des prix agricoles qui doivent être artificiellement soutenus pour assurer à juste titre le revenu des agriculteurs.

D'une façon générale, nous allons voir comment une 'politique d'éco-développement touristique' permettrait d'accroître les 'bénéfices privés' de par l'exploitation d'un plus grand éventail de ressources touristiques et par l'étalement du tourisme dans le temps (bénéfice correspondant au 'manque à gagner' engendré par le sous-emploi des équipements touristiques classiques lourds). Elle est également à l'origine de 'bénéfices privés induits' de par son caractère endogène: réinvestissements par la population locale des bénéfices tirés du tourisme dans des activités économiques propres à la région.

Avec la nouvelle modalité du tourisme, nous ne sommes plus en présence d'une situation de concurrence où le secteur touristique s'est avéré être à l'origine de déséconomies externes de production et de consommation; au contraire le tourisme étant intégré à l'economie locale, elle se présente comme un secteur dynamisant et diversificateur.

Une politique d'éco-développement touristique permettrait également d'accroître les 'bénéfices sociaux'. En réduisant le 'manque à gagner' dû aux fuites des recettes touristiques à l'étranger, et en accroissant de par son caractère 'diffus' les possibilités de mise en valeur des ressources touristiques dans l'espace, elle est à l'origine de bénéfices économiques sociaux accrus.

Pour rester dans le domaine de l'économie, étant basée sur le principe de la 'participation communautaire', elle s'avère être un instrument de justice économique accrue.

Pour ce qui concerne les coûts liés aux activités touristiques, la nouvelle modalité du tourisme intégré permettrait de réduire les coûts sociaux directs et indirects engendrés par le développement du secteur touristique de masse. Ainsi qu'il a été précisé avec la modalité du tourisme intégré les coûts externes de la production touristique se trouvent réduits. La problématique du coût d'opportunité des emplois touristiques par exemple est résorbée. Et, d'une façon générale, la modalité du tourisme intégré et diffus tend à réduire les déséconomies externes d'ordre non seulement économique mais aussi culturel et écologique.

Dans le domaine des coûts directs liés aux activités touristiques, les coûts sociaux de la production touristique se trouvent réduits du fait de la dimension réduite des installations touristiques. En effect, l'éco-tourisme est basé sur le principe d'une utilisation rationelle et modérée des ressources locales en main-d'oeuvre, en infrastructure, en énergie, en eau, etc.

Pour ce qui concerne le coût des installations touristiques proprement dit – coûts privés des investissements hôteliers – il est possible d'affirmer que les coûts de ces unités économiques de petite et moyenne dimension en dépit de l'absence d'économie d'échelle, peuvent être inférieurs à ceux des grandes installations touristiques qui ne correspondent plus tellement aux desiderata de la clientèle touristique actuelle et potentielle qui aspire à un mode de vie intégré.

2. Project éco-tourisme: développement du tourisme de participation

L'éco-tourisme signifie une offre touristique endogène, intégrée et diffuse, c'est-à-dire d'abord et surtout participation de la population locale de la zone d'accueil à l'activité touristique, mais aussi des aménagements touristiques à dimension humaine et intégrés aux caractéristiques de l'environnement naturel et culturel de la région d'accueil. 'Diffuse' signifie une offre diversifiée qui se réaliserait par la mise en valeur de l'ensemble de l'espace considéré grâce à l'utilisation d'un nombre adéquat, éventuellement sous forme de réseaux, d'infrastructures d'accueil de petite dimension localisées dans des sites choisis.

Vu sous l'angle de la demande touristique, écotourisme signifie aspiration des touristes à un mode de vie intégré, c'est-à-dire participation active des touristes à la vie sociale, économique et culturelle de la région d'accueil.

Ces formes 'écologiques' du tourisme devraient permettre de préserver les ressources touristiques qui, avec le développement incontrôlé et anarchique du tourisme de masse, sont devenues des ressources rares, tout en contribuant à la mise en valeur des ressources touristiques inexploitées par le tourisme balnéaire de masse.

2.1 *Caractéristiques de la modalité du tourisme proposé: éco-tourisme participatif*

Comme il a été précisé 'éco-tourisme participatif' signifie intégration du tourisme à tous les éléments économiques, sociaux, culturels, écologiques de la zone d'accueil.

Nous sommes en présence d'un 'tourisme de rencontre' fondé sur une double participation: participation de la population de la zone d'accueil à l'activité touristique (c'est-à-dire initiative locale) et participation des touristes à la vie sociale, économique et culturelle de la communauté d'accueil.

2.1.1 *Tourisme de participation active: la participation des touristes à la vie sociale, économique et culturelle de la région d'accueil.* Avec l' 'éco-tourisme participatif', nous sommes en présence d'un nouveau type de touriste, curieux et respectueux des particularités culturelles, sociales et écologiques du lieu d'accueil qu'il découvre grâce à son voyage de loisirs et dont il cherche à mieux connaître en participant à sa vie quotidienne. Nous sommes de ce fait en présence d'un touriste 'actif', c'est-à-dire désireux de participer à la vie sociale, économique et culturelle de la société d'accueil.

Les activités économiques, sociales et culturelles de la communauté d'accueil se présentent comme des 'ressources touristiques', cette mise en valeur des ressources naturelles et culturelles locales en tant que ressources touristiques (activités de loisirs) se présente comme un instrument de préservation des richesses naturelles et culturelles des lieux d'accueil.

L'éventail des ressources touristiques disponibles dépend des caractéristiques naturelles et culturelles de la région d'accueil, on peut mentionner d'une part les ressources relatives à la flore et à la faune, l'artisanat utilitaire ou artistique. Il y a par ailleurs tout un éventail de ressources potentiellement disponibles dans le patrimoine culturel de la communauté d'accueil, exemples: langue locale, musique et danse folkloriques, cuisine et couture locales, traditions populaires, etc.

2.1.2 *Tourisme communautaire intégré: la participation de la population locale à l'activité touristique.* Comme il a été précisé les formes 'écologiques' du tourisme participatif utilisent les lieux tels qu'ils sont et se moulent dans les sociétés locales telles qu'elles sont.

Nous sommes en présence d'un tourisme léger qui utilise l'infrastructure de la zone d'accueil. Ainsi des unités touristiques de petites dimensions devraient permettre d'éviter d'avoir à faire face au phénomène d'engorgement des équipements collectifs. Nous faisons allusion ici aux rendements sociaux décroissants dus à l'encombrement dans l'usage des facteurs fixes qui se sont manifestés sous forme de déséconomies externes dans le secteur touristique traditionnel caractérisé par un développement rapide et concentré.

Il est à noter toutefois que le tourisme balnéaire a été à l'origine d'économies externes d'ordre écologique. En effet, le tourisme balnéaire classique étant essentiellement basé sur l'utilisation de ressources naturelles telles que le climat, les plages, etc., il se présente comme un moyen de contrôle de la 'pollution industrielle'. Mais le problème est qu'avec le développement incontrôlé du tourisme de masse, le tourisme balnéaire s'est avéré être lui-même générateur de diverses catégories de pollutions qui ne préservent pas ses propres ressources.

L' 'éco-industrie touristique' à base de technologies douces devrait permettre d'apporter de nouvelles solutions aux problèmes du tourisme face à la crise de l'énergie et à la dégration de l'environnement.

Les unités touristiques devraient être bâties selon le principe d'une utilisation rationelle des ressources locales, on vise à atteindre une plus grande 'autosuffisance'.

Ainsi dans l'optique de l'éco-énergétique, les régions ensoleillées feraient appel à l'utilisation des 'énergies renouvelables d'origine solaire' ce qui serait à l'origine d'une plus grande indépendance énergétique de la région.

Dans la lignée d'actions anti-pollution, l'écotourisme fait appel à une politique de 'récupération des rejets', d'où l'étude pour la mise en oeuvre de différentes techniques d'épuration d'eau et de recyclage des déchets.

Les constructions destinées aux touristes devraient permettre de préserver l'unité architecturale de la région d'accueil parce que pour les édifier on aura utilisé les matériaux du pays et mis en pratique les mêmes techniques de construction, ce qui a par la même occasion, l'avantage de rabaisser les coûts des constructions.

Ces installations pour touristes seront décorées avec des objets artisanaux fabriqués par la population de la zone d'accueil. Ainsi les produits artisanaux ne sont plus uniquement fabriqués pour être emportés en tant que souvenirs ou objets esthétiques et culturels. Nous sommes d'une part en présence d'une intégration des objets artisanaux dans l'environnement des touristes et par ailleurs comme il a été mentionné les touristes participent aux activités artisanales en tant qu'activité de loisirs.

Le point crucial dans la modalité du tourisme proposé réside dans le terme de 'tourisme communautaire intégré'. Nous sommes en présence d'un tourisme de participation communautaire fondé sur le principe de 'l'initiative locale'. Ici les activités touristiques sont déléguées à l'ensemble de la population locale associée en 'coopérateurs', étant entendu que celle-ci devra bénéficier de l'aide de l'Etat.

Il n'est pas inutile de rappeler ici le rôle joué par les pouvoirs publics dans le développement du secteur touristique privé traditionnel. Le marché touristique n'est pas né d'une façon spontanée mais bien par l'intervention des pouvoirs publics, qui ont soutenu des investissements directs dans l'ensemble du monde, autour desquels se sont agglomérées des entreprises touristiques nationales et transnationales qui ont prospéré dans une atmosphère mixte. L'Etat devra comme toujours apporter son aide au développement de secteur touristique coopératif, aide financière certes mais aussi aide promotionnelle, aide à la formation de gestionnaires, d'animateurs, etc.

Il est inutile de préciser que les coopératives de tourisme doivent nécessairement être 'intégrées' à la structure générale de la vie économie de la région d'accueil. Comme il a été précisé, le tourisme doit se presenter comme un secteur dynamisant et diversificateur et non comme une 'enclave'.

Cette modalité du tourisme 'intégré' et 'endogène' devrait permettre de mettre un frein au dépeuplement des régions rurales, dans la mesure où avec le tourisme, nous sommes en présence d'une diversification des activités économiques de la région d'accueil; et les activités touristiques de par leur caractère novatrices peuvent inciter les jeunes à rester ou à revenir sur leur terre natale.

2.2 *Avantages du tourisme de participation*

Comme il a été précisé, le project éco-tourisme s'incrit dans la lignée d'une politique générale d'éco-développement qui a pour principe: 'penser globalement et agir localement'. L'objectif essentiel consiste à prévoir autant que possible les effets dits 'd'anti-développement', on opte pour une 'régulation par anticipation', et l'on essaye de réduire autant que possible les inévitables 'régulations par rétroaction, par erreur'.

Le développement du tourisme balnéaire de masse a ét à l'origine d'effets 'd'anti-développement' qui consistent dans les effet néfastes que les activités touristiques de masse exercent sur le milieu naturel en altérant ses rythmes ou en rompant ses équilibres.

Les activités touristiques de masse sont génératrices de nuisances, pollution de l'air, de la mer, pollution sonore, pollution architecturale, dégradation du paysage, encombrements, destruction d'espèces animales ou végétales, etc.

Avec une politique d'éco-développement touristique, le développement touristique prend comme postulat le social, l'écologique, l'espace et ses habitants ce qui tend à limiter les effets d'anti-développement.

Avec le développement du tourisme balnéaire de masse, les touristes se trouvaient être à l'origine de 'déséconomies externes de consommation'. Nous ne faisons pas allusion ici au rôle des touristes dans la modification des modes de consommation (effets d'imitation). Les conséquences économiques et socio-culturelles liées à ces 'effets d'imitation' sont bien connues: extraversion de l'économie, fuites des recettes touristiques affectées à l'importation de biens de consommation à l'instar des biens d'equipement, frustration de la population qui reste en marge des nouvelles modes de consommation, prostitution, délinquance, etc.

Les économies externes de consommation auxquelles nous faisons allusion ici sont liées à la situation de 'concurrence au niveau de la consommation locale'. En effet, avec le développement du tourisme de masse les touristes se trouvent être en concurrence directe avec la population locale pour se qui concerne la consommation de biens et services collectifs: concurrence au niveau de l'utilisation des moyens de transport en commun, concurrence au niveau de la consommation d'eau douce, d'électricité d'où des coupures de courants, encombrement des plages publiques, etc. (exemples: Malte, Famagouste à Chypre et ailleurs).

Les touristes 'autosuffisants' sont, par ailleurs, accusés d'être à l'origine de la hausse du coût de la vie. Les touristes de la classe moyenne et du milieu ouvrier sont en concurrence plus directe avec la population des zones d'accueil pour ce qui concerne l'achat de denrées alimentaires, ce qui fait monter le prix des denrées de première nécessité (fruits, légumes, etc.).

Les touristes de la classe moyenne font également concurrence au niveau de l'achat d'articles de l'artisanat local qui pour la population de certains pays d'accueil ont toujours une valeur d'usage et non point une valeur exclusivement esthétique ou artistique.

Ainsi les prétendues raisons culturelles du rejet du tourisme enclavé par la population locale seraient d'ordre économique: lorsque la population locale ne profite pas directement des 'bénéfices' dûs au développement touristique, elle n'est tout naturellement pas prête à payer les 'coûts' liés aux méfaits du tourisme de masse.

Avec la modalité du tourisme endogène, intégré et diffus, les touristes devraient être à l'origine d' 'economies externes de consommation'.

Le tourisme diffus est basé sur le principe du respect de la capacité d'accueil des zones touristiques; on cherce à exclure tout risque de surcharge, les problèmes liés à la surconsommation des biens et services collectifs ne devraient pas se poser.

Avec le tourisme endogène et intégré, les touristes sont à l'origine d'économies externes de consommation, ils permettent d'exploiter des ressources jadis inutilisées. Vecteur de modernisation au sens positif du terme, le tourisme ne se présente pas uniquement comme un facteur de revitalisation d'une économie en voie de dislocation, mais comme secteur dynamisant (et diversificateur).

Ainsi, comme il a été précisé avec l'éco-tourisme, la problématique des 'déséconomies externes de production' ne devrait plus se poser. En effet, avec le développement rapide et concentré du tourisme balnéaire, l'industrie touristique était en concurrence avec les autres secteurs de l'économie au niveau de l'utilisation des ressources productives de la région: concurrence au niveau du capital, de la main-d'oeuvre, de la terre, de l'eau, etc.

Pour la terre, on peut citer comme exemple de concurrence entre économie agricole et économie touristique, le cas des agriculteurs maltais qui ont perdu leur terre à Mellieha.

Pour l'eau, on peut citer comme exemple, le conflit entre les producteurs d'agrumes et les hôteliers dans la zone Hammamet-Nabeul qui a conduit le Gouvernement tunisien à devoir financer en 1973, un vaste projet de pompage d'eau pour sauvegarder les vergers d'agrumes de la Péninsule de Cap Bon. Nous sommes ici en présence d'un exemple typique de socialisation des coûts privés de l'entreprise touristique, alors qu'une régulation par anticipation aurait permis d'internaliser ces coûts externes. Une réglementation adéquate aurait pu obliger l'entreprise hôtelière à prévoir des projets de déssalement de l'eau de mer qu'elle aurait financé elle-même, ce qui aurait évité d'imposer à la collectivité tunisienne le coût des loisirs des touristes étrangers. Il n'est pas inutile d'évoquer ici la judicieuse décision prise par le Gouvernement maltais d'imposer une taxe sur les piscines d'eau douce; cette taxe a été accompagnée par une forte hausse du prix de l'eau douce consommée par les hôtels.

Pour la main-d'oeuvre, les exemples ne manquent pas; il y a le cas fâcheux des Seychelles où le boom touristique a été à l'origine d'un exode de main-d'oeuvre agricole vers les emplois rémunérateurs de l'industrie touristique (bâtiment, hôtellerie).

Il en a été de même à Sousse où l'expansion continue du tourisme est survenue à un moment où le marché du travail était déjà sollicité; il en est rèsulté un exode de travailleurs qui ont quitté l'agriculture et la pêche comme aux Seychelles parce que le travail en usine et dans les hôtels, moins éprouvant physiquement, offrait en outre l'avantage d'un revenue fixe.

Ainsi, à Sousse, mis à part la tendance inflationniste au cours de la période estivale, lorsque le mouvement touristique est à son comble, il s'est produit une hausse du prix des denrées alimentaires du fait de la baisse de la productivité agricole due à l'exode de la main-d'oeuvre agricole locale, qui a été remplacée par une main-d'oeuvre étrangère moins expérimentée.

Les Seychellois aussi ont dû pâtir de la hausse du prix des denrées alimentaires. L'industrie de la pêche par exemple a connu, en 1972, une forte baisse de l'offre due à la baisse du nombre de pêcheurs marins alors même que la demande croissante de poisson du secteur touristique venait s'adjoindre à la demande de la population seychelloise. Ce qui soulève le problème du 'coût d'opportunité social' de l'emploi touristique qui, comme on le voit, ne peut se confondre avec les coûts d'opportunité privés.

La modalité du tourisme proposé de par les caractéristiques décrites devrait permettre de réduire les déséconomies externes d'ordre non seulement économique et écologique, mais aussi culturel.

Il n'est pas inutile de préciser ici qu'avec la modalité du tourisme de rencontre, la société d'accueil se transforme; elle s'enrichit tout en préservant et cultivant sa propre identité.

Les activités touristiques lui permettent de prendre conscience de sa différence, à l'épreuve de l'altérité, la population d'accueil se découvre à elle-même.

Pour ce qui concerne le coût des installations touristiques proprement dites – coûts privés des investissements hôteliers – il n'est pas erroné d'affirmer que les coûts de ces unités économiques de petites et moyennes dimensions peuvent être inférieurs à ceux des grandes installations touristiques classiques.

A Chypre, par exemple, le coût moyen par lit – pour les hôtels de une à cing étoiles, se serait élevé à 5.000 dollars des Etats-Unis, alors que le coût moyen par lit à Casamance, au Sénégal, où une expérience de 'tourisme communautaire intégré' a été menée, a été 'estimé' à 333,3 dollars des Etats-Unis.

Les avantages du tourisme de participation ne se limitent pas à la maximisation des bénéfices par la réduction des coûts.

Une politique d'éco-développement touristique permettrait d'accroître les revenus résultant des activités touristiques de par l'exploitation d'un plus grand éventail des ressources touristiques et par l'étalement du tourisme dans le temps.

Les bénéfices résultant du 'tourisme de week-end' correspondent au 'manque à gagner' dû aux sous-emploi des grandes installations touristiques classiques. Celles-ci ne peuvent fonctionner qu'à partir d'un certain seuil de clientèle qui ne peut être atteint non seulement à cause du problème du trajet lié à la 'distance' des infrastructures d'accueil concentrées, mais également á cause de la 'nature' de l'offre en question.

Une politique d'éco-développement touristique permettrait également d'accroître les 'bénéfices sociaux' en réduisant le 'manque à gagner' dû aux fuites des recettes touristiques dont une partie est dépensée pour acheter à l'étranger des biens et des services destinés au secteur touristique classique.

Pour rester dans le domaine de la réalisation des buts sociaux, le projet éco-tourisme participatif étant basé sur le principe de la participation communautaire est à l'origine d'une redistribution des revenus (zone urbaine/zone rurale, Nord/Sud), il se présente donc comme un instrument de justice économique par le développement de la production.

CHAPTER 7

Regional ecodevelopment: ecological initiatives of youth (opening Pandora's box . . .)

N.S. MARGARIS

1. Decision-making

Traditional forms of economic development followed up to now in Greece were based on administrative procedures where:
- every decision was taken in centrally;
- all information and knowhow relevant to the decision was making importend from outside;
- responsibility for the realization of the decisions taken was held by a directorate at the top of the administrative hierarchy, the members of which had reached such a position as a result of their conservatism.

As a consequence, economic development was a replica of traditional economic models based on cost/benefit analyses, distorted however by the absence of any knowledge of Greek conditions, peculiarities and infrastructure.

2. Universities

Greek universities functioned in a similar way. Staff and knowledge (furnished through the translation of text books) were imported and scientists merely assisted in the implementation of the centrally-made decisions. Any type of intervention by younger scientists was impossible, not only due to the traditional barriers of administrative bureaucracy, but also due to the lack of information and data concerning Greece and its problems. Collection of information dealing with problems arising from the special situation of Greek society and environment was almost nil.

Bearing in mind that even in countries which did possess such information, economic development had serious side effects of a social and environmental nature, it will be easy for anyone to understand the current situation in Greece, with more than 40% of the population concentrated in the Athens Metropolitan area.

3. Urbanization

As a result of agricultural mechanization, a great part of the population was transformed from rural to urban.

The dramatic population decrease observed in rural areas constitutes a great problem. For exemple, 6000 persons lived in the Prespa area 40 years ago, producing everything they needed; today only 1500 are left and they need to import many items, Gavdos, a 30 km^2 island to the south of Crete supported a population of 500 persons 60 years ago. Today, only 60 persons remain, obliged to import almost every essential.

4. Decentralization

Greek administrators tried to stop this 'human desertification' of the rural areas many times in the past. The usual attitude was that new investment was the only solution. Reality showed the application of these methods, developed for other systems in Greece, to be inefficient and erroneous.

D.O. Hall, N. Myers and N.S. Margaris (eds.), Economics of ecosystems management.
 ISBN-13: 978-94-010-8928-9

Everybody seemd to forget the fact that the Greek periphery had been supporting a much larger population in the past, without any need for electricity, cement or steel. All efforts and investment to create jobs were based on urban models (factories, roads, etc.). A new kind of danger appeared; transformation of the rural areas into new developments.

5. The urban environment

At the same time the 'developed' center has been experiencing problems of another type. Athens, the capital, and Thessaloniki, the second town of Greece, are confronted with a deteriorating environment. What kind of economic development we are going to have where children grow up without playing and where we have smog? What kind of an increase in productivity are we going to have in such an intolerable and unacceptable, from any point of view, atmosphere? Should we start to reexamine our values in the light of such failures as soon as possible? The restrictions imposed on the opening of new factories in the Athens area is the result of decisions not being taken seriously in the past.

6. The position of youth

Recently another problem was added to the above, that of unemployment among young people. Without doubt this is another symptom of te inappropriate developmental policies followed in the past. Again, spasmodic actions are considered in an attempt to solve the problem. New jobs are going to be created in the future with procedures based on the same models responsible for the present situation. Is it possible to build a fish factory in Prespa where, due to eutrophication, fish catches are very low? Tens of millions of drachmae were spent on nothing, since after a few years of operation, the factory closed. Is it a sound policy to consider and promote tourism as the basic source of foreign currency with all the dependence and environmental problems it creates? What will be the impact on tourism of the continual increase of jelly fish in our sea or of sea pollution? Should tourist places like Skiathos island come alive only in summer, as if they were hibernating snakes? Did the people who take decisions wonder about the living conditions (economic, social, cultural) of the Skiathos people during the winter? Are we satisfied with an active period from June to September?

7. Crucial points

Taking all the above into consideration we realize that the consequences of the strategies followed up to now are:

- desertification of the rural periphery;
- social and environmental degradation of the urban center;
- inefficiency in producing and disseminating information by the universities;
- lack of innovation in plans for regional development;
- absence of youth from the productive infrastructure;
- increase of unemployment among the young.

8. The rural Greek periphery and young people

It seems, therefore, that we have to deal with a problem with two main facets, the deserted periphery and the inactivation of the youth. The point from which our basic strategy was developed was: 'Can we find a combined solution to these problems? For example, can we send young people directly to the rural areas, ignoring the traditional bureaucratic ways of the administration and the Universities?'.

The above working hypothesis was also based on the hope that such an initiative would act as a catalyst towards a new reality, one with fewer errors and failures. The reasons for this hope are:

- students will come in direct contact with the outlying rural areas and form at least a rough idea of the real situation and problems;
- students will transmit their experience to the university, questioning the validity of traditional ways of teaching and research;

- they will create pressure groups which may motivate university people to work on greek problems and disseminate the information thus obtained;
- working for the State in a direct manner, they will have the ability to transfer their proposals without having to go through bureaucratic channels, where changes are often made;
- the validity are seriousness of their work concerning problems definition and remedial proposals, will give them a strong argument in the defense of their capability to produce sound work;
- their enthusiasm, open-mindness, and – above all – their functioning beyond the status quo, will permit them to be innovative when seeking solutions, something which is very difficult for people already assimilated by the traditional administrative system;
- their environmental sensitivity may lead them to sound and acceptable proposals, both from the economic and ecological points of view (soft energies, endogenous development etc.);
- without the superiority syndrom of the expert they can develop a better understanding of the local people and their problems;
- being in close collaboration with the local population they could easily transfer to them existing information concerning problems and solutions, the flow of information to the periphery being a prerequisite for any type of sound development;
- their presence gives local people satisfaction and hope that the 'center' has not forgotten them. The feeling that the State is interested in them can activate their abilities;
- with their dynamism they can press decision makers to transform their proposals into reality.

9. The theory turns into action

Taking into consideration the above, the Ministry of Youth proceeded to call for tenders, describing the philosophy of the attempt. Proposals came from local institutions, and university people as well as from some independent groups. Twenty-three were selected on the basis of general interest among them, those addressing the mose problem-

Region	Department	Special interest
1. Gavdos island	Chania	– Deserted island; social deterioration; ecotourism; bioagriculture; biosphere reserve
2. Skines	Chania	– Botanical garden
3. Mytilene	Lesvos	– Vegetation map; desertification; occupation; agriculture
4. Mt.Pelion	Magnesia	– Agrotourism; fertilizers; pesticides; perturbations
5. Prespes	Florina	– Population deterioration; National Park; ecotourism
6. Achaia	Achaia	– Soil stabilization
10. Skiathos-Skopelos islands	Magnesia	– Tourism and environment; agrotourism; protection of biotopes
11. Zagori	Ioannina	– Population deterioration; alternative occupations; summer schools
12. Evros	Evros	– Environmental protection and development in a National Park
13. Messolongi	Aetoloakarnania	– Protection of a wetland; development of water cultures
14. Ikaria	Samos	– Population deterioration; alternative ways of development
15. Lefkas	Lefkas	– Fish cultures; educational tourism
16. Rodopos	Chania	– Local initiative of onyx production
17. Patras	Achaia	– Pollution of the Patraic Gulf
18. Achaia	Achaia	– Garbage problem
19. Rhodes	Dodecanes	– Medieval town; conservation; protection
20. Tziverniana	Chania	– Restoration and conservation of the deserted settlement
21. Halki-Tilos islands	Dodecanese	– Restoration of the settlement; alternative development; ecotourism
22. Castellorhizo	Dodecanese	– Population deterioration; ecologically sound economic development

atic regions of the country were preferred.

The next step was the formation of a coordination body. Coordinators asked for scientific directors for each group. More than 70% of these came from the universities and were young scientists with experience in the area under study. Both the coordinators and the scientific directors participated in the task of selecting the participants. Attention was paid to form groups of students and young scientists from many different disciplines, in order to deal with the problems from as many aspects as possible and so give a better output. All this work was elaborated in only 15 days.

Each group consisted of 6–12 members. Discussion started inside each group for undertaking responsibilities, collecting material, etc. before going to the place of study.

A 2-day seminar was organized in Athens in early July. On the first day all members of the groups were gathered together and totalled more than 250. The philosophy of the project was presented and analyzed by the coordinators and general discussions on various aspects of the project followed. Since all the work, from the conception of the central idea of the project to the seminar, was done in just in 3 months, almost everybody involved had some kind of panic concerning future events. On the second day of the seminar three different groups were formed, which worked separately according to the subjects they had been dealing with:

(1) management of the natural environment (10 groups);
(2) environmental crisis and non-conventional ways of development (9 groups);
(3) anthropogenic environment and alternative ways of touristic development (4 groups).

Some serious problems have arisen, however. Though funds were available, it seemed impossible to mobilize them, since a similar project has never existed before and the bureaucracy could not escape the conventional methods used up to now. This point can be considered as crucial, because the central idea – youth to the rural periphery – was outside the traditional way of thinking and acting. Fights against all existing administrational structures became an everyday routine.

At the stage the Ministry of Youth – being a new institution without established or traditional modes of action – proceeded, almost in a piratic way, to solutions totally new to the Greek government. Collaboration with the elected local authorities resulted in a green light for further action.

10. Young people in the rural periphery

After the Athens Seminar each group travelled to their place of study. Constant collaboration with local people resulted in a re-examination of the working hypotheses developed previously. The majority of the groups surpassed our expectations.

In general each group was divided into sectors on the basis of the subject treated and the scientific disciplines relevant to it. Every night discussions among all members of the group resulted in the exchange of data and experience and the plan for the next day's work was formed. The groups started from a description of the existing situation and proceeded to proposals. These proposals were then discussed with the local people in order to reach a general consensus.

11. Conclusion

In early September the writing of the reports started. The work of each group was fully presented in a Seminar held in Halkidiki on the 28th, 29th and 30th of September. With the presentation of reports an almost explosive atmosphere was created. Critisism, self-critisisms, a suspicions of political exploitation of the project, but also dynamism, and stress on the experience gained and feelings developed was the atmosphere at the end. The quality of the reports was above all expectation.

It seems that it is too early to analyze every aspect of the output. But if we take into consideration:

(1) that the young scientists' and students' product was of a high quality and completed in just 3 months from the idea's conception;
(2) the cost of the reports, generally 10 times less that of a 'normal' one which would need at least 14 months to complete;

(3) the innovative and feasible proposals went far beyound our expectations and have the advantage of the local peoples acceptance;
(4) the organization of efforts inside the university for promoting a sound development of the periphery; and
(5) the pressure put on both university teachers and on the Ministry of Youth for the quick realization of the proposals,

we believe that the experiment was successful. There is no going back – nobody can stop this 'snowball'. Tens of new proposals come every day to the Ministry from the universities and the rural periphery. We believe that forces hidden or suppressed up to now are freed and the hope for a better future has more possibility of becoming a reality now than before.

CHAPTER 8

Med-Terra

A. DAVAKI

Today everyone knows that there are critical environmental problems. Environmental protection has become the concern of governments and planners, and nowhere is that concern more justified than in the Mediterranean, where fifty centuries of hard use have reduced land and water resources to a bare minimum.

These problems have been recognized since the 1970s through the formulation, by the United Nations Environment Programme, of a Mediterranean Action Plan. Member countries have set priorities and promised action. This coordinated effort is heartening, but the important aspect of cooperation is missing. Nowhere in the many references to the Plan is there explicit recognition of information transfer; that is, how scientific and technical information generated or required by this effort will be stored, retrieved and disseminated. Presumably, the results of research will be available through normal channels, scattered among the agencies of a dozen countries. For practical purposes, though, those research results would be inaccessible.

Normal channels are no longer sufficient. The costs in time and energy in lack of productivity, in shear frustration are too high. Information can and must become more accessible.

We are proposing to make such information readily accessible by establishing in Athens a multinational network and center called Med-Terra: a center that could become a mechanism for regional cooperation in the exchange of scientific and technical information essential for the tasks of environmental protection.

The center will be primarily concerned, at least initially, with bibliographic information. It will search for, acquire, store and deliver pertinent sources of information contained in publications of any kind from around the world.

By pertinent sources we mean those which might prove useful in dealing with environmental problems. This is, of course, a very broad category, since 'Environment' is not a single-subject discipline.

Environmental studies comprise many fields and subjects – social, biological and physical sciences. To find answers to questions one must search the world literature, which is now accessible through computerized data bases numbering millions of items.

The center network is established to help protect the environment in Greece and the Mediterranean region and facilitate and enhance scholarly research and higher education. To improve the exchange of scientific and technical information among Mediterranean countries, and to encourage, with the wise use of knowledge and information, orderly economic development.

To achieve these goals, we plan to establish a reliable, comprehensive information service that will include:

- collecting and storing on microform documents of the Mediterranean Action Plan and other pertinent publications (the center will not store a

D.O. Hall, N. Myers and N.S. Margaris (eds.), Economics of ecosystems management.

large library collection of publications on paper; most of its holdings – especially the deposited documents from the region – will be available to borrowers/users/applicants on microfiche; we strongly endorse the use of microforms to save postage and storage costs and to improve delivery time);

- searching the center's files and the world literature through computerized data bases, to answer questions;
- delivering documents from the center's collection or obtained from other sources, promptly;
- creation of a computerized data base of bibliographic references pertaining to the Mediterranean environment; and
- maintaining a file of experts and information sources to whom inquiries can be referred.

The service will communicate the findings of environmental management and research to appropriate individuals by publishing bibliographies and state-of-the-art summaries for students, researchers, and managers.

It will improve the exchange of scientific and technical information within the Mediterranean region by helping to coordinate information services among government and international agencies and private institutions, and by contributing its experience in handling all scientific and technical information exchange in the region. Direct users will be individuals actually engaged in research, management, or regulation. Indirect users will be libraries, research institutes, businesses, government agencies, and other organizations.

We expect the center will succeed by offering good basic services: answering questions, acquiring materials, delivering documents, and creating a data base.

This project is built around a particular computer hardware–software combination. The *software* for the date base and network is called MINISIS and was developed by the International Development Research Centre in Ottawa, Canada, from an earlier computer program, ISIS or Integrated Scientific Information System, itself a product of the International Labor Organization and subsequently of the United Nations Scientific and Cultural Organization.

ISIS runs on a large IBM computer while MINISIS runs on the Hewlett-Packard 3000 minicomputer. The MINISIS program as used with the HP-3000 computer is an extra ordinarily versatile and powerful combination. Its main features include:

(1) flexibility: while primarily developed for bibliographic information systems, it is flexible enough for other applications;
(2) data base creation and management facilities and information retrieval in one package;
(3) compatibility with the widely-used ISIS computer program;
(4) easily-learned instructions, even for people with little or no experience with computerized information systems;
(5) a complete set of programs to support library management and information retrieval, within the library or on network;
(6) a powerful search processor that will support use of an on-line thesaurus, multi-lingual recognition of thesaurus terms, and Boolean and arithmetical operations;
(7) the production of bibliographies, library catalogs, and indexes in varied formats;
(8) the direct preparation of computer output microfiche (COM);
(9) a full range of utilities to assist the data base administrator in creating and maintaining data bases;
(10) the exchange of data bases by accepting and producing magnetic tapes conforming to ISO Standard 2709;
(11) the encoding of French and Spanish diacritics
(12) thesaurus recognition in four languages – English, Spanish, French and Arabic; with Greek and Cyrilic under development;
(13) a complete set of application programs;
(14) ease of writing specialized application programs for unique situations using MINISIS instrinsics.

Because MINISIS accepts bibliographic data in the ISIS format, a significant portion of the center's data base can be created by selecting and loading appropriate references from data bases already maintained by FAO, IDRC, UNESCO, ILO, and UNEP. Those thousands of references represent an asset of enormous value, not only for the direct labor costs of entry saved thereby, but for the years of work they represent. Building on that work, the center can put a data base and network into operation almost immediately.

MINISIS is available from IDRC under license; factors taken into consideration when the license is used is whether the organization is non-profit or for profit, and whether the organization is in a developed or developing country.

The software, system support, and the training of two people, is offered to us by IDRC free of charge. A shared resources program will be developed and membership will be offered to Mediterranean and other European countries.

This program will help support the center and will give members the opportunity to take advantage of the center's services on a regular basis, both as depositors of pertinent literature and as users. A particularly attractive feature of the MINISIS system is the capability to handle Arabic. This, we believe, will provide a unique opportunity for regional cooperation in the field.

An invitation for membership will be extended to scientific institutes, universities, research centers, and libraries.

This effort we believe will help promote international cooperation in the field, and will lay the groundwork for the broader exchange of scientific and technical information in the Mediterranean region.

Med-Terra has been approved by the Ministry of Physical Planning, Housing and the Environment and has been included in the Ministry's Five Year Plan.

CHAPTER 9

Public policy for coastal areas

HARRY COCCOSSIS

Introduction

As UNESCO's Second Medium Term Plan (1984–1989) indicates, two persons out of three today live in coastal regions, that is, near the sea, on the shore of continents or on islands (UNESCO, 1983).

The problems and prospects of coastal areas have attracted considerable attention in the last decade or so. Politicians, planners, scientists, public administrators and the public at large gradually realized the need for public policy to alleviate pressures and resolve conflicts in the coastal zone. This paper will briefly review some of the salient aspects of coastal planning and management from a planner's perspective.

The need for coastal policy

Public interest in coastal policy stems from the environmental, economic and social/psychological value of coastal areas. From an environmental perspective, coastal areas can be seen as highly productive and unique ecosystems sustained by the interaction of land and sea. Rocky shores, beaches, estuaries, wetlands and nearshore areas provide habitat, nesting and feeding grounds for a variety of marine and associated organisms (flora and fauna). Also, dunes, beaches and barrier islands serve as natural buffers against environmental hazards, such as tidal flooding, erosion and storm wave damage (Economidou,1982).

From an economic perspective, coastal areas are valuable for their resources and the advantages they offer as the location of economic activities.

The economic value of coastal marine resources rests primarily in fishing, mining and oil drilling. Coastal land resources have also played a critical role in economic development. Due to locational advantages a range of human activities is attracted to the coast such as trade, industry, transport and tourism. The concentration of economic activities in coastal areas is closely interrelated with urban growth and expansion along the coast.

Throughout history, trade has developed earlier between coastal settlements and across the sea than inland. The growth of trade and prosperity has concentrated people and activities in trade centers along the coast. Access to transport and the economies of bulk shipment have attracted additional activities such as industry to ports and coastal urban areas. Two types of industries benefit from coastal locations, those that are sensitive to transport in costs, mainly bulk products (refineries, grain silos, mills etc.) and those that need large quantities of water for processing and cooling (chemicals, iron and steel complexes, power plants etc.). Agglomeration and urbanization economies cause additional concentration of people, industries, services and other economic activities around existing nodes of activities resulting in urban expansion along the coast.

The social/psychological value of coastal areas rests in the basic human need for leisure and recre-

D.O. Hall, N. Myers and N.S. Margaris (eds.), Economics of ecosystems management.

 ISBN-13: 978-94-010-8928-9

ation. Tourism is a typical expression of this need. Coastal areas, due to the environmental diversity and the opportunities for recreation they offer, attract a large proportion of leisure activities worldwide. The recent expansion of water related recreation is a typical case of the increasing demand for coastal recreation.

The pressures for coastal development (tourist, industrial, urban, transport etc.) have often had adverse, and occasionally detrimental, impacts on the environmental, economic and social/psychological value of the coast. Examples of such impacts are seen on every continent:

- pollution, resulting from urban and industrial waste effluents; the increasing use of chemicals in agriculture (pesticides, fertilizers), oil spills and dumping of waste from ships;
- incompatibility of uses of land and water, such as recreation and industrial development, urban development and agriculture, etc.;
- loss of valuable resources, such as fishing stock, underground water of coastal land, from pollution or overexploitation;
- disruption of ecosystems and destruction of habitats from development activities, such as dredging and filling;
- negative or diminishing returns in development itself, as in the case of pollution or overcrowding in recreation areas, or from industrial waste effluents on fishing and aquaculture.

'The essential coast character is its intricate, indisputable interconnection.' (Simon, 1978)

The proliferation of conflicts in the use of coastal resources and coastal space and the problems created both for development and protection of the environment has led several countries and international organizations to opt for public policy in regard to coastal areas. Worldwide, there is a large variety of responses to the necessity for coastal policy, a variety due to differences in the levels of socio-economic development, culture and attitudes, the state of the environment and technology, as well as, the institutional framework of each country. Variation in policy response is also due to differences in problems encountered and issues addressed. Nevertheless beneath the apparent differences in coastal policy lie some basic similarities in the approach and its characteristics.

Coastal planning efforts

Coastal policy is complex, comprehensive and conciliatory. It is complex since it attempts to satisfy multiple objectives and intervene in complex processes; comprehensive since these objects address a wide range of sectors and activities; and conciliatory since its basic premise is the mitigation of conflicts between development and protection. The complexity of conflicts is aggravated by the inherent dynamic interrelationships of the ideological, economic, social, technological and political dimensions of the process of coastal policy-making.

Some countries have opted for an explicit policy towards coastal areas as in the United States, France and Greece, while others have attempted to integrate coastal policy into their physical planning tradition as in Sweden and the Netherlands. Selected examples of both cases are briefly presented below.

Great Britain was the first to launch a special program for coastal areas in 1966, with the aim to protect and preserve the country's outstanding coasts. It was not until 1972 however that coastal protection and coastal management was integrated with the existing planning system. County authorities can adopt guidelines for their area although local authorities retain the power to prepare and implement management plans. The Nature Conservation Council, the National Trust and especially the Countryside Commission play a significant role in the protection of coastal areas at the national level (Waite, 1982).

The United States is credited with the first piece of national legislation dealing explicitly with coastal areas: the Coastal Zone Management Act of 1972. Actually the initiative for coastal policy rests earlier with the State of California and its pioneering effort to preserve the coastal zone, an effort which led to the California Coastal Plan of 1975. The key to California's approach was the transfer of land use control from local government to a special commission at the state level. The Plan itself states that coastal open space, farmland, wetlands, beaches and marine ecology of the state's coasts are to be preserved wherever possible, development concentrated in areas where it exists already and public access to the shore maximized (State of California, 1975).

In a similar manner the Coastal Zone Management Act of 1972 as amended in 1976 puts the management of coastal lands under the auspices of the federal government for planning and management and sets up a framework of procedures in which the federal government, states, local government and private citizens decide together the future of the shore (AIA, 1978). Some thirteen coastal states have voluntarily developed coastal management programs.

France with a traditionally strong role for central government, adopted several measures to address the pressures for development in coastal areas. These include the creation of a land banking authority (Conservatoire du Littoral), the preparation of coastal plans (schémas régionaux du littoral) and of sea use plans (schémas, d'attitude et d'utilissation de la mer), the latter being an innovative feature of French coastal policy (French Ministry of the Environment, 1980).

Greece, a middle income country and a latecomer in terms of coastal policy making, started from the basic intention to coordinate public policy in coastal affairs. In 1981 the National Council for Physical Planning and the Environment – a ministerial cabinet level coordinating committee – set the basic framework of coastal policy, recognizing its sectoral and spatial character (Camhis Coccossis, 1982b).

The Netherlands have a tradition of management of coastal areas. Although coastal planning and management, as comprehensive planning specifically for coastal areas, does not exist, most elements are integrated into the land use/environmental planning system of the country (Wiggerts Koekebakker, 1982).

Sweden with a similar tradition in physical planning has integrated its coastal management policy into the existing system of land and water management. A special commission has been formed to coordinate and draft an overall program for marine resource activities (Carlsberg & Grip, 1982).

Interest in coastal policy also appeared on the part of international organizations, in recognition of the transnational character of coastal problems, such as pollution, and in their similarities to other issues involved, as for example land use conflicts. UNEP's Regional Seas Program is among the more ambitious efforts to bring a variety of national interests together in a coordinated approach to promote protection and development. UNESCO has a special program (X.5) dealing with the integrated management of coastal zones and island regions. Other international organizations promoting coastal management are IUCN (International Union for the Conservation of Nature), NATO (North Atlantic Treaty Organisation) through its Committee on the Challenges of Modern Society and OECD (Organization for Economic Cooperation and Development).

At the European level the Council of Europe has brought coastal problems to the attention of the Planning and Environment Ministers of its 21 member countries. The European Charter for Physical Planning (adopted on February 10, 1983) stresses the necessity for a specific policy for coastal and island regions in order to assure balanced and coordinated development.

Last but certainly not least is the work of the Environment Directorate of the Commission of Economic Communities which supports coastal management primarily through the Conference of Peripheral Maritime Regions of EEC. A special Coastal Charter has already been adopted by the European Parliament in that respect (June 1982).

The process of planning

More than ten years have passed since the emergence of the first policies and programs specifically designed for coastal areas. Although coastal problems have not disappeared, this experience in policy making has often contributed to the alleviation of pressures, but also, and probably of equal importance, to a better understanding of the phenomena and processes involved. From a planning standpoint it would seem fruitful to delve into the process of coastal planning itself in order to identify some critical issues involved in the development of coastal management programs.

The analytical framework employed for the discussion of issues is the three phase process of planning: intelligence, design and choice. These correspond to Herbert Simon's key stages of decision making (Simon, H., 1959).

The basic feature of this process is that diagnosis of the problem is a very critical phase influencing the whole approach to the problem.

Intelligence involves the activities of data collection, analysis and problem identification. Design consists of policy formulation and program design, the phase where alternative courses of action are considered. Choice refers to the selection of the program and includes implementation and evaluation.

Although this model of the planning process appears to imply linearity, this is for sake of simplicity in presentation. In essence planning is cyclical involving strong interaction among its various phases.

Contemporary planning theory accepts the principle of limited rationality in decision-making, in view of the uncertainties involved with the future, the limited capabilities of the human mind to deal with complex problems and less than adequate information available on the issues involved. It is a more pragmatic approach and a shift away from the deterministic long range physical plan of the past establishing planning as a process.

Planning for coastal areas, as identified earlier, deals with complex issues, multiple objectives, several factors and requires a careful treatment of all three phases: intelligence, design and choice.

Intelligence in coastal planning revolves around the issues of problem identification, that is, the specification of the problems for which a special policy is considered. The pecularities of each country in terms of environmental characteristics and development activities set the basic framework for problem identification.

Furthermore problems are also contigent upon the ability of the particular culture and society to perceive them as problems that need immediate attention.

Environmental characteristics relevant for coastal policy include coastal marine and land resources, assets, risks, etc. Development activities refer to the type, scale and rate of growth of activities, the level of technology employed etc.

Correct problem diagnosis depends on two things, availability of data and adequate understanding of the phenomena and processes involved.

Data needs are not independent of policy aspects such as the level of decision-making and basic policy orientation. If fishing is a major activity in the national economy, data needs on fishing stock and fishing grounds are more extensive than other activities or environmental factors.

Data needs are also contingent upon their use in decision-making. So the level of detail required is different if considering siting a specific power plant or designating areas for protection. A few examples illustrate the point.

France has developed a permanent data base (inventaire permanent du littoral) covering French coasts on maps of 1:100 000 scale. Natural areas, resources, property status and uses of land and water are included together with administrative boundaries and regulation areas. Italy has a more elaborate recording system with information on physical characteristics, the use of land, seismicity demographic characteristics, indices of industrial agricultural and tourist development on a scale 1:250 000. In Greece, qualitative information on the state of the equipment and the type of conflicts and opportunities for development have been recorded on maps of 1:200 000 and 1:1 000 000 scales. In the U.S. the Office of Ocean Resources Coordination and Assessment (Office of Coastal Zone Management) has developed a three-level information system. Level 1 is a descriptive and qualitative analysis to identify spatial and temporal relationships among physical environments, living environments, the life histories of various living marine species, economic activities and management jurisdictions. This analysis is used to refine the spatial and temporal extent of problems requiring additional analysis.

Level 2 is a simple quantitive type of analysis that utilizes simple natural systems models, pollutant transport models and simple relationships to make 'order to magnitude' estimates of pollutant concentrations in the marine environment. This type of analysis further defines the demensions of a problem and provides a preliminary quantitative appraisal of alternative solutions.

Level 3 is a detailed regional, site or resource specific quantitative analysis, as for an industrial facility siting or for single species fisheries management plans.

Level 1 and 2 are useful for strategic level decisions while Level 3 is used in tactical decisions.

The existence of a data base by itself is not enough for problem identification. More important is an understanding of the phenomona and processes involved. Although a great deal of scientific knowledge has been accumulated in past decades about coastal problems, there is still a wide field for exploration, primarily as to the interrelationships of processes and the key variables for policy action. The adoption of some successful planning tools towards that end might seem fruitful in the future, as the experience of UNESCO has indicated with gaming simulation, for example.

The decision phase involved key decisions that shape the policy in question. It is the phase where alternative courses of actions are explored and where national policy intervention desired. Coastal policy has been weighted more towards environmental protection than development planning. This is perhaps due to the basic orientation of planning in the seventies when most coastal programs emerged. In the eighties, as economic recession pervades national policies, development aspects seem to gain the forefront of policy priorities.

Two interrelated issues dominate the design phase: level of policy making and type of policy selected. Both deal with the choice of the appropriate policy space.

The level at which policy-making occurs is subject to tradition in public administration and existing legislation, but also to the nature of the problems addressed. Three basic levels can be identified: international, national, and local. International policy in general is more oriented towards policy directives and guidelines as in the case of the EEC, or towards voluntary cooperation as in the case of UNEP. In general it is intended as a supplement or a catalyst, rather than a surrogate national policy, due to inherent limitations in implementation (differences in natural priorities, cultures etc.).

Within a country the critical question is the split between central and local government policy. The existing legal framework and the distribution of jurisdiction and authority determine the policy options available.

Type of policy refers to the range of actions that can be employed for coastal areas. It reflects not only the legal framework but also the goals and priorities in policy-making. Policies can be sectoral or spatial or both. Sectoral policy options for coastal areas serve to introduce coastal issues into each sector's policy. This is an approach mostly relevant at the national level. For example sectoral policies deal with the environment, offshore oil exploitation, aquaculture, public works, construction, power plant operations, and so on.

Spatial policy reflects the need to vary policy measures to match the pecularities of area and site. Its extent can be limited, as for example in a plan for critical habitat areas or cover the whole teritory, like a Regional Plan. Spatial policy is relevant at both national and local levels.

The spatial dimensions of coastal policy are the ultimate test of policy coordination and conflict resolution. A key feature of spatial policy is the area or zone over which it applies. Institutional, physiographic or functional criteria can be used. Using multiple criteria a series of different zones can be identified and a multi-tiered approach to policy formulation employed, allowing for a more flexible system of management and control.

So, for example, over a narrow zone — a few hundred meters wide – restrictive measures can be applied on all development activities as in France and Italy, while for planning purposes a wider zone can be identified with restrictions on selected activities. Administrative (like the Massachussets Coastal Management Program) or topographic features (like the British examples) can be used for boundary delineation. At the same time it might be necessary to extend controls beyond the planning zone in special cases such as pollution from upriver sources.

Choice as a phase of the planning process involves evaluation of alternative courses of action and selection of the most appropriate and desirable course. Critical issues of choice revolve around the basic questions of what controls and incentives are to be adopted, what resources are to be utilized and how the costs/benefits are to be distributed. Choice is strongly linked to design and in practice it is very hard to distinguish between the two since the elaboration of alternatives is not independent of the criteria for evaluation, which in turn depend on policy objectives.

Several analytical tools and techniques have been developed as aids to decision-making in the choice phase. They range from cruder Planning Balance Sheets (Litchfield, 1960) or Goals Achievement Matrices (Hill, 1968) to elaborate ones for policy impact analysis such as large-scale computer models or linear programming. So far there is very little empirical evidence on the application of such tools and techniques for coastal planning.

Recent developments in planning and regional science deserve special attention in view of their relevance for coastal policy. Conflict analysis and conflict management are attempts to find a synthesis of analytical approaches to policy analysis (formulation) and policy impact analysis (Isard & Smith, 1982; Isard & Nagao, 1983).

In this discussion of the planning process it is pedantic to state that public participation plays a key role in both design and choice. In traditional land use planning, public participation has demonstrated impressive results where adopted. It has helped raise public awareness on issues, improve understanding of policy measures and ultimately achieve public support, all preresquisites for effective planning. Effectiveness in policy-making is achieving the objectives set forth. Making plans and programs is hardly enough if they lack implementation. Implementation means primarily political will and commitment to carry out the chosen course of action. Besides political will, often neglected in evaluating planning efforts, policies adopted have to be feasible and acceptable to the public. Feasibility is a function of the available financial and planning instruments for policy implementation but also of the pragmatism of the selected course of action. Pragmatism is a function of the institutional, economic and political framework. The fallacy and pitfalls of overly ambitious, wide-sweeping environmental programs, that ultimately fail, have created more harm than benefit to the environmental protection economic development issue.

Experience with coastal policy suggests that building up a management program step by step through the existing institutional framework is a viable, realistic approach especially for countries with limited experience in planning. The final approach notwithstanding, what really counts at the end is the effective management of change in a time of great uncertainty about the environment, the economy and man himself.

References

AIA (American Institute of Architects). Coastal Zone Management; Balancing Protection and Growth. AIA, Washington DC.

Camhis, M. & Coccossis,H., 1982a. Coastal Planning and Management perspectives. Ekistics 49: 92–97.

Camhis, M. & Coccossis,H., 1982b. The National Coastal Management Program of Greece. Ekistics 49: 131–134.

Carlsberg, E.C. & Grip,K., 1982. Coastal policy in Sweden: Use and Protection of Marine Resources. Ekistics 49: 137–142.

Economidou, E., 1982. The Ecological Value of Coastal Ecosystems. Ekistics 49: 98–101.

French Ministry of the Environment, 1980. La politique du Littoral. Notes Vertes, No.13. Actions Nouvelles, Paris.

Hill, M., 1968. A goals-achievement matrix for evaluating alternative plans. Inst. Planners, pp. 19–29.

Isard, W. & Smith,C., 1982. Conflict Analysis and Practical Conflict Management Procedures. Ballinger, Cambridge, Mass.

Isard, W. & Nagao,Y. 1983. International and Regional Conflict. Ballinger, Cambridge, Mass.

Litchfield, N., 1960. Cost-benefit analysis in city-planning. Am. Inst. Planners, p. 273.

Simon, A., 1978. The Thin Edge. Harper and Row, New York, N.Y.

Simon, H., 1959. Theories of Decision-Making in Economics and Behavioral Science. Am. Econ. Rev.: 253–283.

UNESCO, 1983. Second Medium Term Plan (1984–1989). UNESCO, Paris.

U.S./OCZM (Office of Coastal Zone Management), 1981. U.S. Coastal Regions Strategic Assessment Program. OCZM: National Oceanic and Atmospheric Administration, Washington, DC.

Waite, C., 1982. Coastal Management in England and Wales. Ekistics 49: 124–127.

Wiggerts, H. & Koekebakker, P. 1982. Coastal Planning and Management in the Netherlands. Ekistics 49: 143–149.

CHAPTER 10

Modelling of the socio-economical and ecological consequences of high animal waste application (MOSEC)

B. THOBER, H. LIETH and S. FABREWITZ
with the assistance of
N. MÜLLER, K. NEUMANN and T. WITTE

1. Introduction

1.1 *General statements*

One of the greatest problems for the understanding of societal problems is the lack of scientific communication between the social and natural sciences. The two areas of research have developed so independently from each other that completely new avenues must be developed to bring both fields together.

The necessity for the close cooperation is clearly needed. The major clashes between different groups in our society, the discrepancy between opinions about the quality of life, the unprecedented destruction of nature with technological development, all these are well discussed and documented problems.

In spite of these facts the means of communication between natural science and socio-economic science are still greatly lacking. This is the reason for the development of the MAB program by UNESCO. This program, now underway for eleven years, still has the problem that social and economic scientists continue to work separately from natural scientists. It seems that major barriers have to be removed before any progress can be made.

This is the reason why a group of scientists at the University of Osnabrück is working towards an interdisciplinary program which, if successful, should bring about such a mode for cooperation. The best possible way to develop such a mode is the common work on a model in which all scientific fields involved have the same level of interest. The Osnabrück MAB group found such an object nearby: the intensively farmed area of South Oldenburg, where only-recently-mechanized animal breeding developed with effects of spectacular dimensions. When we designed our project a few years ago, we foresaw problems which are surfacing right now. This paper will describe the problem again and then demonstrate the way we undertake the modelling, which will serve two purposes:

(1) discovering the best ways in which to solve immediate problems, and
(2) to generate coherent models to simulate the co-development of the economy and the ecology of the South Oldenburg region.

By doing so and if completely successful, our model will identify means of cooperation between the scientific fields of the authors involved in this paper.

1.2 *The problems of the South Oldenburg area*

The county of Vechta in South Oldenburg is a world leader in pig breeding and fattening, and same is true for chicken and broiler raising. No other region in Europe carries so many chickens and pigs per hectare of cultivated land as Vechta county.

At the present time the South Oldenburg region is reaching a new type of threshold in its food

D.O. Hall, N. Myers and N.S. Margaris (eds.), Economics of ecosystems management.

ISBN-13: 978-94-010-8928-9

production, one which is becoming apparent in other sectors as well. Neither capital, market, or labor are limiting factors but the stress placed on the environment is becoming a limiting factor to production.

The side effect causing problems in battery farming is the tremendous amount of liquid manure produced, called 'Guelle'. Liquid manure is considered by the farmer as a natural fertilizer and is spread on the fields. The average content of key minerals and nutrients in 'Guelle' is shown in Table 1 (after Vetter & Klasink, 1978). Problems arise in this area from overfertilization during dry spells and from seepage of nitrates into ground water. In many cases the groundwater is now unfit for drinking purposes, which caused the Minister of Agriculture of the State of Lower Saxony to put a limit on the amount of 'Guelle' to be spread on each ha of farmed land.

As perceived by the various cooperating disciplines the following problems appear the most pressing at the moment in the counties of Vechta and Cloppenburg.

Ecologically:

- The nitrate content of 2/3 of the private wells sampled so far supercedes the maximum level allowed of 90 ppm NO_3 (see Table 2 and Fig. 1).
- The 'Guelle density' (during production) in the county of Vechta has significantly surpassed the maximum level of 3 dung units per ha. In some problem townships the level has reached 4 or more dung units/ha (see Tables 3.1, 3.2 and Fig. 2).

Economically:

- Regulations limiting the total amount and reducing the time span for the application of 'Guelle' as fertilizer will adversely affect the economic situation of many farms. Table 4 shows some of the negative effects which it is anticipated may arise from regulations.

Socio-politically:

- Public concern about the nitrate levels in tap water and their effect on human health, especially that of babies, is increasing. This is clearly reflected in several public discussion sessions and in press coverage.
- Local agricultural interest groups have protested violently against the limitation of 'Guelle' usage.
- The regional waterworks are demanding measures to stop groundwater contamination.
- The Minister of Agriculture announces the regulative order which limits the quantity of manure application to 3 dung units per ha.

In order to deal with these problems one faces a number of difficult interconnected questions which

Table 1. Nutrient content in liquid manure (Guelle) with different dry matter content values (mean values taken from 80–90 samples per animal species (after Vetter and Klasink, 1978).

% Dry matter	Organic matter	kg/cbm					
		N	P_2O_5	K_2O	CaO	MgO	Cu
Cattle 'Guelle'							
5	36.5	2.3	1.2	3.0	1.3	0.3	0.003
10	73.0	4.7	2.4	5.9	2.5	0.6	0.005
15	109.5	7.0	3.7	8.9	3.8	0.9	0.008
Pig 'Guelle'							
5	37.0	3.3	2.9	1.8	2.3	0.4	0.014
10	73.9	6.7	5.8	3.7	4.5	0.8	0.028
15	110.9	10.0	8.7	5.5	6.8	1.2	0.042
Hen 'Guelle'							
10	65.0	7.1	6.3	3.2	10.6	0.6	0.006
15	97.5	10.7	9.5	4.8	16.0	0.9	0.009
20	130.0	14.2	12.6	6.5	21.3	1.2	0.012

Table 2. Nitrate in private wells.

		Townships		
	Mean (%)	Damme (%)	Bakum (%)	Neuenkirchen (%)
Total:	598/100	151/100	317/100	130/100
0– 20	76/12.7	35/23.2	33/10.4	8/ 6.2
20– 50	42/ 7.0	8/ 5.3	21/ 6.6	13/10.0
50– 90	105/17.6	26/17.2	54/17.0	25/19.2
90–150	236/39.5	33/21.9	147/46.4	56/43.1
150–200	60/10.0	20/13.2	31/ 9.8	9/ 6.9
200–300	73/12.2	26/17.2	30/ 9.5	17/13.1
300	6/ 1.0	3/ 2.0	1/ 0.3	2/ 1.5

Source: Township – means from the county office Vechta. Date: 17.01.1983.
mg/l = Number of wells/Percentage.

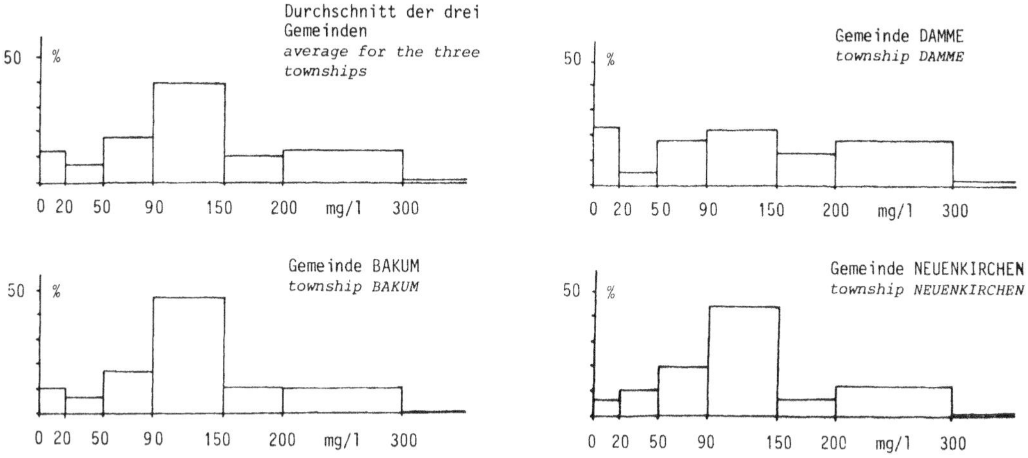

Fig. 1. Nitrate levels in private wells of several townships. Percentage of wells within different nitrate concentration classes. Data source: Township means from the county office Vechta. 17.01.83.

Table 3.1. Number of animals counted for one dung unit according to § 15 Abs. 5 of the waste removal law from Nds.Mbl. 22/1983, p. 438.

Animal or groups of animals	Number/ Dung unit
Cattle over 2 years	1.5
Young cattle 3 months–2 years	3.0
Calves up to 3 months	9.0
Sow with piglets up to 20 kg	3.0
Pigs over 20 kg	7.0
Egg laying hens	100.0
Young chicken	200.0
Broilers	300.0
Ducks	150.0
Turkeys	100.0

The dung unit consists of waste containing up to 80 kg nitrogen or up to 70 kg phosphate.

the individual disciplines are still trying to solve. Table 5 lists the formulated questions together with the disciplines in which they are dealt with.

1.3 *The approach of the Osnabrück working group*

Since the problem possesses so many aspects, we have tried to draw from many disciplines. So far we have inputs from biology, ecology, geography, economy, law, sociology, and psychology (see Witte, 1983). The joint effort to construct a consistent overall model for the agro-ecological system of the county of Vechta demands precise statements of objectives from all the scientists involved so that these can be cast into a model structure. In tran-

Table 3.2. Dung units and farmed area in South Oldenburg (1982).

Counties	Dung units acc. to definition in 3.1	Farmed area ha	Du/ha farmed area
Cloppenburg	202 578	99 504	2.04
Vechta	236 828	58 029	4.08
South Oldenburg	439 406	157 533	2.79

Source: County offices of Cloppenburg and Vechta.

scribing these into a formal system of relations we have found it possible to overcome the barrier of different languages and approachs existing between natural and social sciences.

For the ecological submodel we have compiled and structured the environmental parameters which affect the material flow in the agro-ecosystem. The first phase will yield the status quo for species, populations, and communities of the rele-

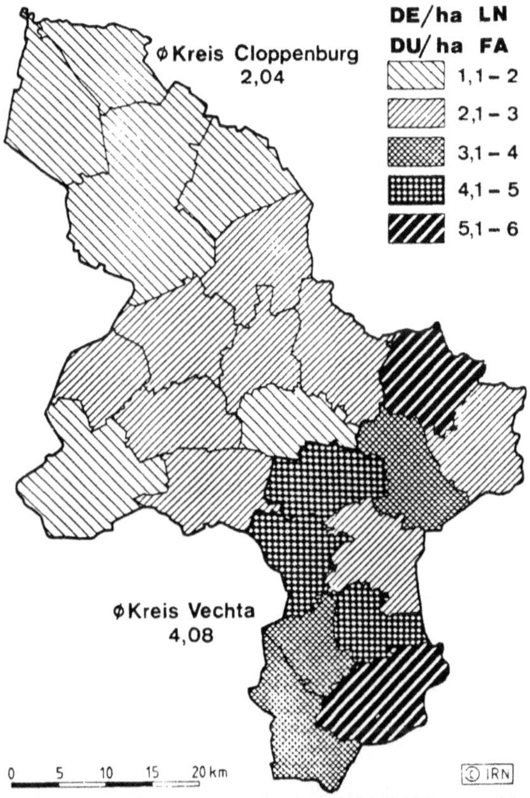

Fig. 2. Dung units per ha farmed area (DU/ha FA) in townships of Vechta county (1982). Calculation base: Runderlaß des Landwirtschaftsministers of 13.04.83 as shown in Table 3. Source: Windhorst, 1983 (in press).

Table 4. Major cost-effecting reactions to restrictive regulations for the utilization of guelle.

1. Expansion of storage tank size to avoid full tanks at restricted times.
2. Rental of more farm land, which increases in cost and becomes scarcer.
3. Stagnation of stock capacities when sales prices decrease.
4. Lowering of stock capacities.
5. Need to export 'Guelle' to neighboring counties.

Note: Position 5 may become a resource when an appropriate technique is available to market surplus 'Guelle' to other regions.

Table 5. Some of the main problems to investigate for our study.

Question	Discipline concerned
1. What changes of the natural environment will occur?	Ecology
2. How can we detect them?	Methodology
3. What impact do these changes have upon man, animal, and plant?	Ecology, human and veterinary medicine, sociology
4. How important are they for man?	Economy
5. Which technological and economic conditions lead to stressful situations?	Economy
6. Are these economic means for an improvement of the situation?	Economy
7. How do the inhabitants of the area react in face of the situation?	Sociology, social-psychology
8. What measures are taken by government and offices?	Political science, organizational sociology
9. What results do they have?	Sociology, economy, ecology

vant areas. It will also yield concentration gradients for the material flows and list their influence upon vegetation and soil biota within the agro-ecosystem as well as in the adjacent landscape. Of primary interest are the solid and liquid waste substances from animals.

For the economic submodel we have developed two strategies:

(1) The prototype of a simulation model for an agricultural enterprise (farm) with animal batteries will allow us to simulate the economic decisions and behavior of a farm. Using this, it will be possible to demonstrate the resulting economic and ecological consequences (Neumann, 1983).

(2) The model of the agrarian structure and total regional economy will be used to test the influence of environmental protection measures upon the overall regional economy.

For the socio-political submodel measures and methods to investigate and formulate the behavior of the people involved and living in the project region are applied. Actions undertaken by non-economic institutions will be analyzed. This includes investigations of social psychology, sociology and political science which we will use to show how agro-economical as well as ecological information is received and utilised. It includes also investigations into the way in which decisions are made and enforced.

Simultaneously with the construction of a series of such models we have developed coupling strategies which allow us to simulate the overall system under a variety of different options and to demonstrate different possibilities for cooperation of the involved scientific disciplines. Figure 3 shows an outline of such a total system. For the three main disciplines (ecology, economy, and socio-political sciences) the major approaches are demonstrated in the following chapters.

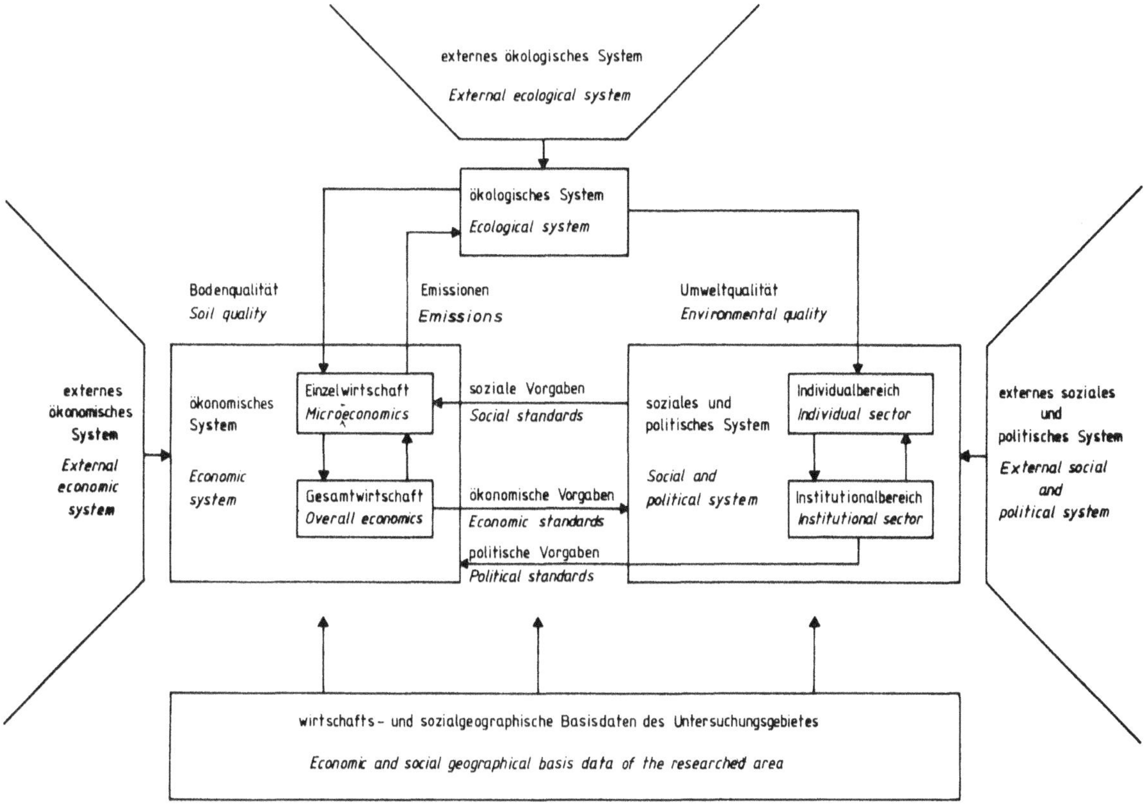

Fig. 3. Structure of the overall agro-ecosystem.

2. Modelling

2.1 *General remarks on modelling*

As we stated in Section 1, our ultimate goal is to model the dynamics of a regional agro-ecosystem with all the actual specifications provided by the South Oldenburg region. A model which refers to this overall system may prove to be a single model consisting of part-models or it may be a system of models, a set of coupled submodels which may be used to simulate the total system by running the whole set.

In either case it will take several stages of constructing and testing smaller models and/or submodels until the total system simulation model is designed. The model must be constructed in such a way that two levels of simulation can be implemented: one on the micro scale where individual farms or pieces of land can be modelled as one unit, one on the macro scale where properties of the entire region in a European or German context can be simulated. Major differences in the state variables must be expected in the socio-political and economic submodels whereas the ecological submodel will remain similar, with only the numerical input values changing. The option to generate the macro-model from the sum of possible micro-models needs further discussion within our group. At the moment the capacity of our computer is unsatisfactory for such an attempt, nor are we ready to determine the number of micro-models needed to describe the region adequately in this manner.

2.2 *Ecological modelling*

Ecological models of an agro-ecosystem need to describe a multitude of natural processes. The most logical approach to a model that will fit our need is to structure all processes and variables into the subsystems soil, groundwater, surface water, flora, and fauna. The latter require further subdivision into 'crop- and livestock-related species' and 'wild and wildlife'.

In order to fit the overall model structure we aim to develop a macro- and a micro-model. The ecological model will describe the 'Guelle' effect on the subsystems mentioned and the feedback to the socio-economic (part)systems on a regional basis. Such models will incorporate variables like species and habitat conservation, soil conservation, recreational value, wildlife management, surface and subsurface hydrology etc. The micro-ecological model will deal with variables like mineral and energy flow and transfer in a Soil-Vegetation-Livestock-Groundwater-System. Such micro-models may easily be coupled with the agro-economic system, especially with its transformation subsystem. This will be explained in Section 2.3.

The following steps will be taken to construct the two model types.

The macro-modelling will start with the assesment of the present state of biological species, populations, and community types in the South Oldenburg region, especially Vechta county.

The stepwise construction of the micro-models will start with the analysis of 'Guelle' dominated nutrient flow and transfer through soil, groundwater, vegetation and livestock with special attention to the nitrate movement.

Our pilot model will initially be concerned with the analysis of the nitrate in well water, a problem which has become very apparent in the region over recent years. The nitrate concentration in one well of a major public water supply system more than tripled between the years 1968 and 1983 (see Table 6). It seems obvious that the increase is due to the high animal population in the county. In order to show the impact that 'Guelle' has on the nitrate concentration of the groundwater we can use the micro-model of our agro-ecosystem. Into the ecological subsystem of this model we can enter numbers for nitrogen in 'Guelle' provided by the economic submodel and try to follow it on its way through the agro-ecosystem. In order to arrive at realistic values, we have taken results from experiments done in the region as well as some values from agricultural summeries. At the present level of funding the group was unable to conduct field measurement. The exercise presented here is necessary, however, to plan and locate future experiments correctly, since the previous ones had shortcomings for their use in our model.

Figure 4 shows the flow chart of nitrogen

Table 6. Average nitrate concentrations from wells in the Holdorf waterworks.

Wasserwerk Holdorf[a] Year	Nitrate concentration		Well No.
	mg/l	Max.	
1968	8,1	32,6	11
1978	12,6	33,7	5
1979	10,9	40,1	11
1980	16,1	63,0	6
1981	19,2	67,0	6
1982	23,6	77,7	6
1983 (incomplete)	29,6	108,8	6

[a] Calculated from measurements of the Oldenburgisch-Ostfriesische Wasserverband.

through the ecological subsystem until it reaches the groundwater where social institutions raise the alarm when it reaches values above 90 ppm. In order to balance the nitrogen flow, we need the inputs of 'Guelle' and precipitation, the output by vegetation and microbial activities, the run-off and the seepage.

The numbers for nitrogen input with 'Guelle' and the nitrogen concentration in seepage water at 2–3 m depth were taken from a paper by Vetter & Steffens (1977). From his Table 7 (p. 627) we have selected experiment 2 in which he investigated the effect of increasing amounts of 'Guelle' on the nitrogen concentration of seepage water. No distinction was made between nitrate and ammonia nitrogen.

Our Table 7 shows the conversion of numbers

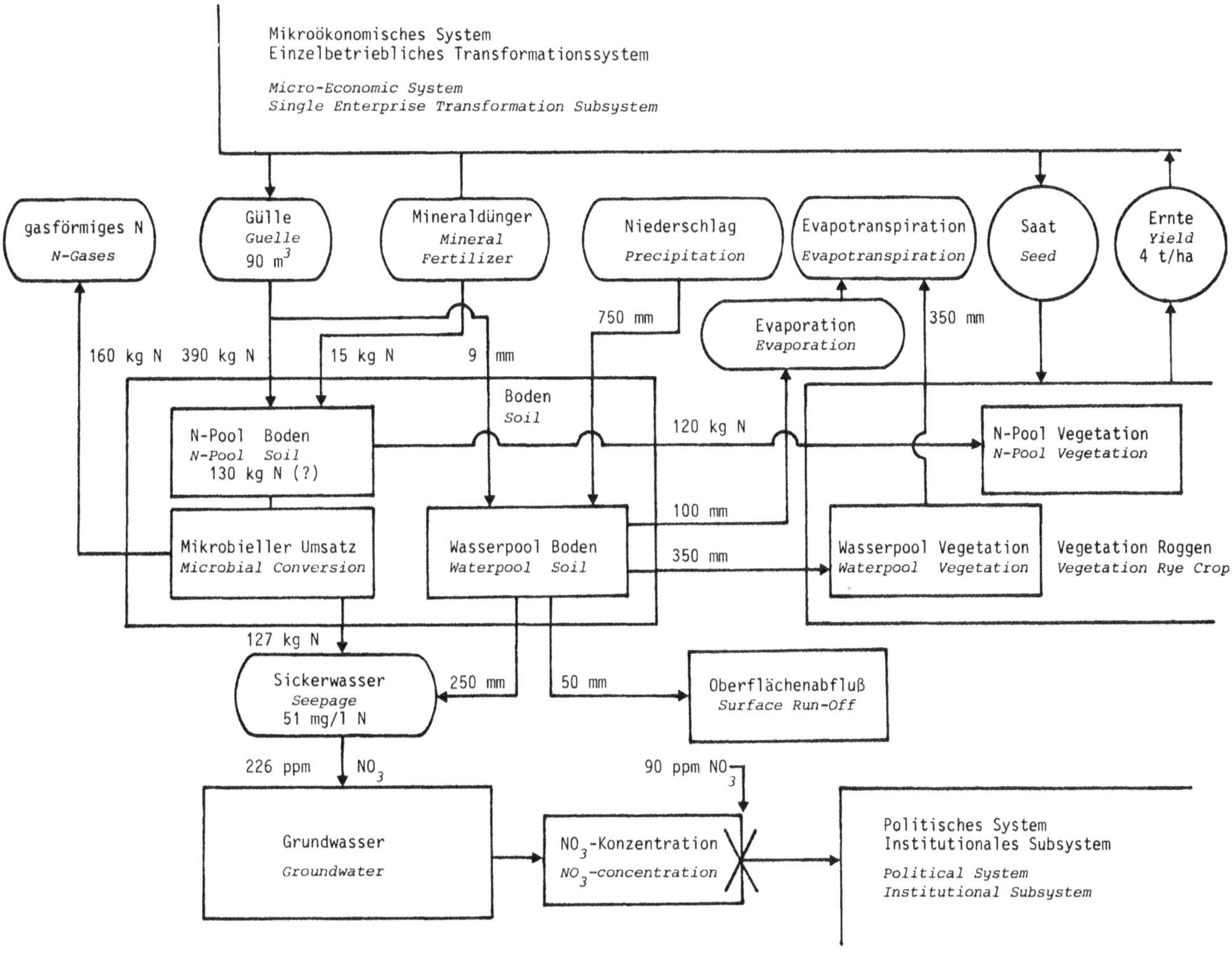

Fig. 4. Generalized model of the nitrogen flow through the ecological subsystem.

Table 7. Nitrogen movements through a rye field on loamy sand. Data: Vetter & Steffens (1977).

Guelle application in m^3 + 15 kg fert.N	Rye yield T/HA	N concentration mg/l	N input kg/ha	N withdrawal through yield kg/ha	N lost by leaching kg/ha
0	3,3	27	15	105	68
30	3,4	31	145	104	78
60	3,8	38	275	114	95
90	4,0	51	405	120	127

from the table of Vetter & Steffens. In order to see whether the numbers would balance, we took yield and dry matter analyses from Finck (1979), Fischbeck, Heyland & Knauer (1975), and Kuntze *et al.* (1981) for nitrogen uptake of rye and corn. The experiment was done with rye, but the most common crop in the high 'Guelle' level farming is corn.

For the conditions chosen we see that for 90 m^3 'Guelle' and some 15 kg N as starter fertilizer 405 kg N/ha yr is applied with the volume equivalent to 9 mm water. The annual precipitation is around 750 mm. The soil initially had 130 kg N/ha. We assume that the soil held about 100mm rain in a 2 m layer. The rye crop with a yield of 4 t/ha may have extracted 120 kg N/ha and used up about 350 mm of rain by transpiration. The equivalent of 250 mm rain and 127 kg N were calculated for a depth below 2 m on the basis that 51 mg N were measured in 1 l of water. 51 mg total N of NO_3 would give 226 ppm NO_3 which would cause immediate alarm at the water control institutions.

There is one uncertainty in that groundwater hardly ever contains ammonia. The groundwater NO_3 levels reported in Tables 2 and 6 and the seepage water from Tabel 7 which entered our model have some 20–30 m depth differences. We do not know yet what happens to water when it passes through some 20 m of subsoil in the South Oldenburg region.

When we examine our balance model, we find that 405 kg N input + 130 kg N soil reservoir stand against 120 kg N output through yield, 127 kg N seepage, about 160 kg N loss through microbial denitrification and an unknown amount leaving with run-off water. Under these circumstances the annual budget is roughey balanced and the N reservoir in the soil rises only slightly.

The questions that arive from this calculation are, how much 'Guelle' can the land take without spoiling the groundwater, which crop keeps the nitrogen from the soil water most profitably, and whether the soil microflora can be stimulated to convert more nitrogen. Solving the problem of how to maintain such a high-nitrogen-level agro-ecosystem without spoiling the well water is a task in itself.

For the purpose of our integrated study we have shown in Figure 4 that the gap between the economic subsystem and the socio-political subsystem can be bridged via the ecological subsystem.

Figure 5 shows the relation between the amount of 'Guelle' applied and the N levels found in seepage water. Under the assumption that 2/3 of the nitrogen at 2 m depth reaches well water as nitrate we could determine the point at which the institutional system would have to intervene and stop the economic system. The fact that the minister of agriculture has reacted ('Guelle' Erlass) shows that this point has already been reached.

2.3 *Economic modelling*

The economic part-model covers two main objectives: the macro-economic submodel and the micro-economic submodel.

The macro-economic part describes the functioning of the regional economy and the microeconomic part the functioning of a single enterprise typical of one of the farms in the area.

The macro-economic model is to be based in part upon the successful modelling of the micro-economics. We restrict the following description to the micro-model.

The prototype model of a single enterprise de-

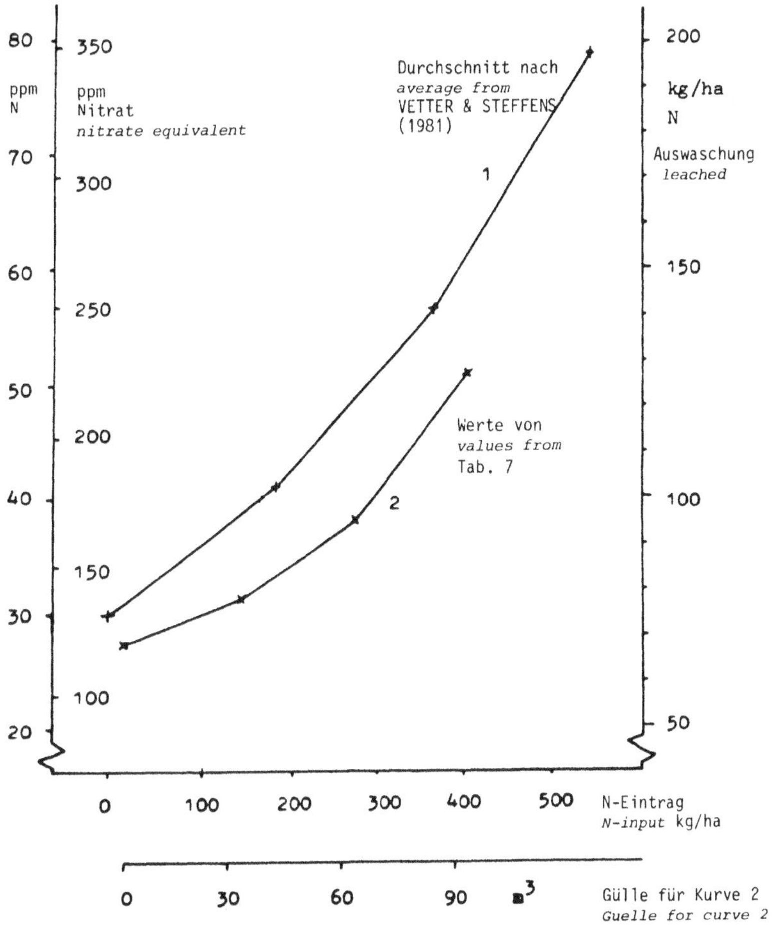

Fig. 5. Impact of increasing quantities of 'Guelle' upon the nitrogen content of seepage water. Data source: Vetter & Steffens, 1977 and 1981.

scribes the functioning of a meat producing farm with battery stables and crop production. To be described are the effects resulting from actions of individual farmers upon the social system as well as the ecosystem. Similarly, influences from these systems upon the single interprise are to be analyzed.

In order to demonstrate our modelling strategy we choose one activity: the application of liquid manure as farm produced fertilizer and its consequences for economy and ecology in the case of overfertilization.

To judge the economy of the 'Guelle' utilization we start with its value in the context of joint production within a farm system. A joint production system is defined as one in which, besides several cash products like livestock and crop, residuals like 'Guelle', smell or noise are produced in significant quantities. The residuals may be classified as waste or as resources for further utilization.

Waste is treated economically so that disposal costs are minimized. Utilizable residuals, however, are employed with optimal economy. 'Guelle' produced on the farm may save commercial fertilizer. The optimal amount of manure for a specific farm can be determined by the 'law of diminishing returns'. If the amount of manure produced on the farm surpasses this optimum level, the additional amount applied on the field can be considered as waste disposal and needs to be economized as such, unless markets for this resource can be found.

Ecologically the 'Guelle' becomes a stress problem if it is employed in quantities which cause

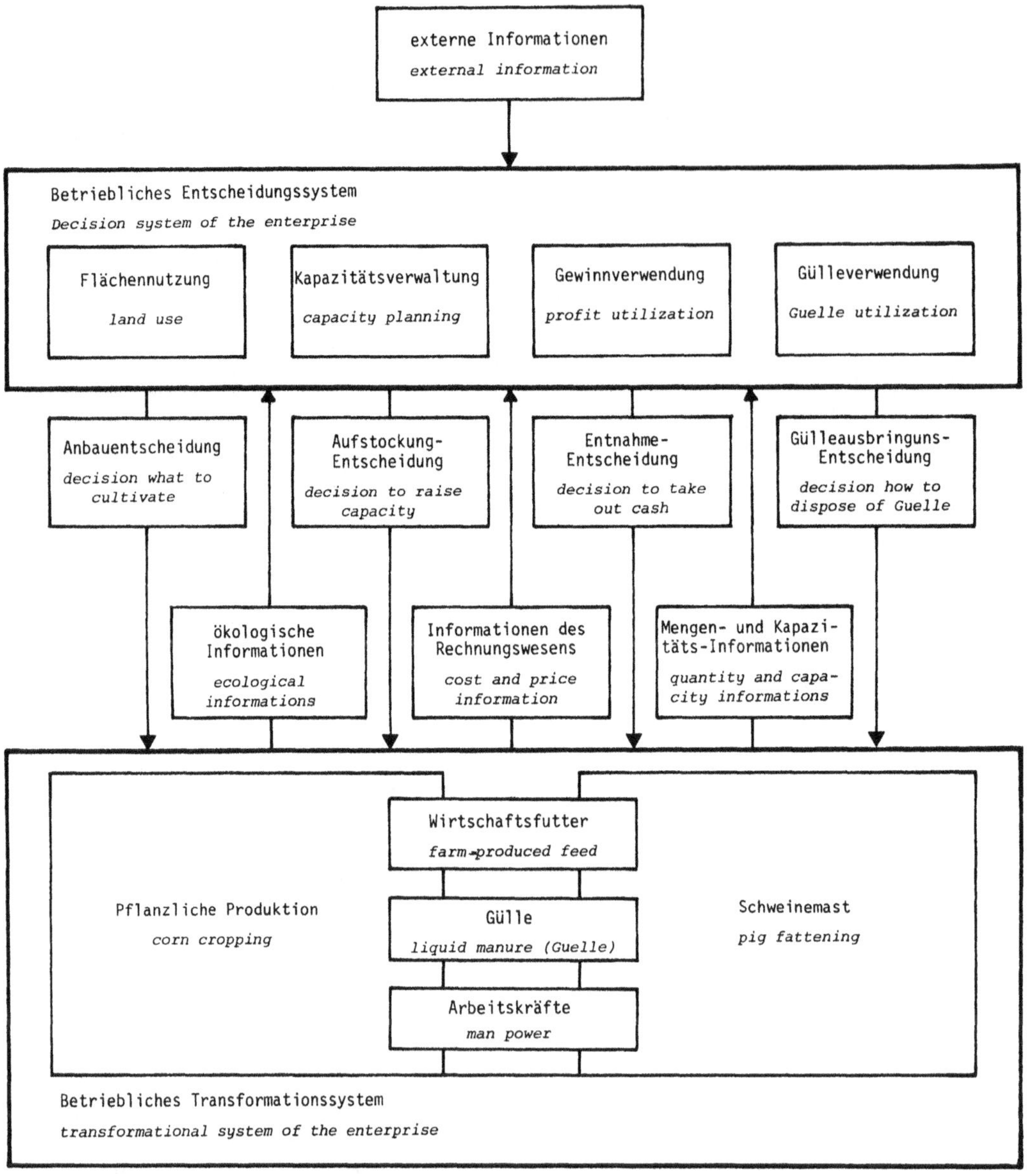

Fig. 6. Single enterprise model of a pig-rearing farm with corn cropping (cited from Neumann, 1983).

overfertilization. The seepages and run-off portions of the mineral nutrients content are especially important because they cause pollution of ground and surface water. This may lead, among other problems, to elevated levels of nitrate in water (see Fig. 1 and Table 2).

Our agro-ecosystem model will be to determine the levels at which production-related byproducts may lead to environmentally stressful situations. The kinds and quantities of byproducts may vary with local conditions and procedures, and these have to be specified in the model. From eco- and social systems stressed in such a way, the enterprise may expect significant feedback. Only four of the possible negative feedbacks will be mentioned: possible yield reduction of crops, reduction of soil fertility, impact upon well water quality and restrictions on the expansion of the enterprise. For more details see Tables 2 and 4.

The economic micro-model will contain two subsystems, the decision system and the transformation system. The latters contain all physical transformation processes of the farm production. It transforms physical inputs into physical outputs. The decision system contains all those processes which steer the transformation system. All informational inputs are digested here. The contents of the two subsystems and their linkages are shown in Figure 6. It is logical that the major connecting points of the economic system to the social system exist with the decision subsystem and the major connections with the ecological system in the transformation subsystem.

Technically speaking, the prototype model of the economic micro-system is a combination of continuous and event-oriented simulation processes. The simulation language SLAM supports such combined models and is therefore used for our model.

2.4 *Socio-political modelling*

The social science modelling covers 3 areas:

(1) The non-economic institutions system which will be depicted with one or more function models simulating the behavior of the institutions in the region.
(2) Use of environment-related information by individuals, which will be covered by two kinds of model types:
    (2.1) The behavior of the 'decision-makers' in farming enterprises are constructed as function blocks which will be connected as submodels into the single enterprise economic model.
    (2.2) The interaction between the total population, economic system, and ecological system, which will be constructed as a macro-model with the capability to depict mass processes.

The following sections give more details about the foundations of the model theory, our special concept, the filter models, and the present testing of the mechanisms of filter models in a pilot model.

The social sciences unfortunately lack a well-founded descriptive basis which can be used for research projects. The approach starts therefore with elementary descriptive questions. With respect to our subproject the question to ask is: How do the non-economic institutions function, especially those dealing with ecological problems?

In order to answer such questions the concept of real structure modelling was developed (Müller, 1982; Müller, *et al.*, 1982. For this we first need to identify the significant decision-making centers and their importance. For these decision-making centers we must develop one or more descriptive models. We may simplify an institutional function model to a black box in which inputs are transformed to outputs.

In order to understand how non-economic institutions function it is first necessary to identify the mechanisms of transformation and then to describe them.

For processes within institutions it is often the case that fact-related inputs, either influences or information, are transformed intrainstitutionally to power-struggles and are solved as such (Müller, 1979). In a descriptive model we therefore need to combine generally technical and 'power-specific' aspects.

In the proposed modelling approach this is done by specifying two filters: (a) a selection filter at the input side; (b) an activity filter at the output side of the model:

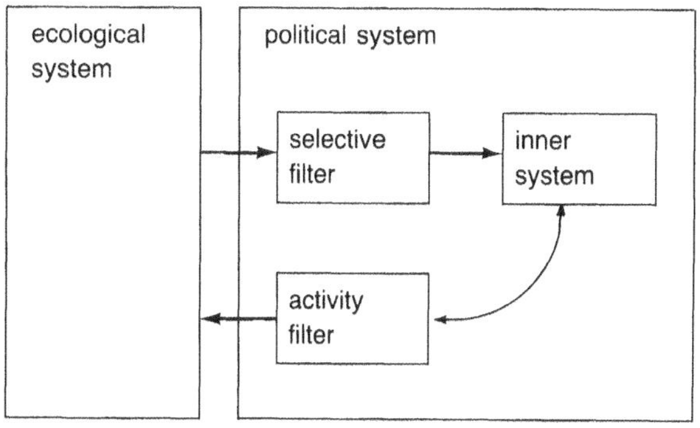

At the present time we are developing a pilot model in which these filter models are specified for the problem area: 'nitrate level in groundwater'. In this pilot model micro and macro area of the agro-ecosystem are looked upon as if a so-called indicator farm were the single source of emissions and simultaneously the object of sanctions from the institutional system. Changes in the fate of this indicator enterprise steer the overall development of the macro agro-ecosystem.

The central state variable in the selection filter model is the acceptance of NO_3 concentrations in drinking water. Numerically equal measurements of nitrate contents in private wells are differently valued according to the state history of the system.

In the area of activity action filters are activated when nitrat levels reach certain levels which are determined by law or by institutions (e.g. 90 ppm NO_3) In this case the central variable named 'waiting until an activity' is influenced by a series of factors one could call 'soft' parameters of the system model (e.g. strenght of ecological fervour, power of the agropolitical infrastructure).

As soon as the state 'waiting' becomes critical, the institutional system initiates one or more actions selected from a whole catalogue of possible options. Among others, possible options include:

- deeper drilling of existing wells
- drilling new wells
- lowering of livestock capacity
- closure of the enterprise

Any one of these options has economic consequences as discussed in section 2.3. Before any of them can be implemented, the final decision of the individual decision-maker is influenced by political and governmental institutions.

3. Plans for future development of the model

Full development of the regional agrosystem model is a remote goal, which will take the combined effort of the newly founded Systems Research Group of the University of Osnabrück to reach. As outlined in preceding chapters, the group intends to build the model in small steps. A few throughts about future development follow.

3.1 *Present state of the project*

At the moment we possess for a number of submodels structural concepts which need refinement and testing. The prototype model of a farm enterprise already contais many aspects of the transformation system. The pilot model for the social system allows the testing of the first version of the specification of the filter models. Flow charts have been elaborated for the ecological submodels. For many parts of the models the date gathering, in both field and literature, is presently the chief concern. This is needed to quantify the model parameters for future tests.

3.2 *The next steps to be taken*

The next steps consist of the completion of the various submodels and subsequently, a thorough testing. These subroutines will then form the basis for their coupling.

The experience gained with systems simulation using the first versions of our coupled models will help us to improve upon the individual models.

As a result we expect an improved overall model from which it will be possible to calculate on a micro-scale the various options for single farm activities, their effect upon the socio subsystem and the ecosubsystem as well as existing feedbacks.

Simultaneously we shall construct the first version of a micro-model which will allow us to observe the effects of regional politics upon the agro-ecosystem.

3.3 After several micro-models are ready and interlinked, one can try to link them with the macro-models. The most feasible way to do this is still an open question. But even the pilot model in which simple models from different disciplines are interlinked, is a great step towards bridging hitherto isolated working disciplines. It will hopefully lead to a better future understanding of social processes initiated by new technology or other innovations. Besides solving local problems the modelling effort may provide a basic concept of how to undertake interdisciplinary modelling.

Acknowledgement

This work was supported by CEC grant No. ENV-596-D.

References

Finck, A., 1979. Dünger und Düngung. Grundlagen und Anleitung zur Düngung der Kulturpflanzen. Verlag Chemie, Weinheim.

Fischbeck, G., Heyland K. & Knauer, N., 1975. Spezieller Pflanzenbau. Ulmer Stuttgart, (UTB, 111).

Kuntze, H., Niemann, J., Roeschmann, G. & Schwerdtfeger, G., 1981. Bodenkunde, 2nd ed. Verlag Eugen Ulmer, Stuttgart (UTB, 1106).

MAB 7, 1982. Deutsches Nationalkomitee (German National Committee), MAB-Mitteilungen: Interactions between ecological, economical and social systems in the regions of intensive agriculture.

Müller, N., 1979. Empirische Herrschaftstheorie. Westdeutscher Verlag, Opladen.

Müller, N., 1982. Realstrukturmodellierung – Probleme entscheidungsorientierter Beschreibung großer sozialer Einzeiten – einige Ergebnisse des Forschungsprojektes HSDMEL. In: H. Krallmann (ed.), Sozio-ökonomische Anwendungen der Kybernetik und Systemtheorie, pp. 141–155. Schmidt, Berlin.

Müller, N., Schön, P. & Thober, B., 1982. Projekt HSDMEL. Abschlußbericht, Universität Osnabrück, University Press, Osnabrück.

Neumann, K., 1983. Betriebliche Simulationsmodelle zur Planung der Umweltschutzes am Beispiel eines landwirtschaftlichen Betriebes. Beiträge des Fachbereichs Wirtschaftswissenschaften der Universität Osnabrück, Nr. 8305, Osnabrück.

Runderlass, 1983. Vollzug des § 15, Abs. 5 des Abfallbeseitigungsgesetzes; hier: Maßnahmen gegen die Überdüngung mit Gülle und Geflügelkot. RdErl. d.ML. of 13.4.1983–38.2–03 14 55. In: Niedersächsisches Ministerialblatt 22: 437–439.

Vetter, H. & Steffens, G., 1977. Untersuchungen über den Einfluß gestaffelter Güllegaben auf Pflanzenertrag, Pflanzenqualität und die Reinheit des Wassers. Bericht über Landwirtschaft 55/4: 620–632.

Vetter, H. & Klasink, A., 1978. Nährstoffgehalte in Gülle und Kot. In: H. Vetter (ed.), Wieviel düngen?, pp. 189–194. DLG-Verlag, Frankfurt/Main.

Vetter, H. & Steffens, G., 1981. Nährstoffverlagerung und Nährstoffeintrag in das oberflächennahe Grundwasser nach Gülledüngung. Kulturtechnik und Flurbereinigung 22: 159–172.

Windhorst, H.-W., 1983. Zur gegenwärtigen Struktur der Veredlungswirtschaft in Südoldenburg. Jahrbuch für das Oldenburgische Münsterland, Vechta (in press).

Witte, T., 1983. Gülle – wertvoller Dünger oder gefährlicher Abfall? Umweltforschung in einer multidisziplinären und standortübergreifenden Arbeitsgruppe. In: Universität Osnabrück intern – Sonderheft: Forschung an der Universität Osnabrück (WS 83/84) (in press).

CHAPTER 11

Natural productivity and regional carrying capacity for human activities on the island of Gotland, Sweden

ANN-MARI JANSSON

1. Introduction

Observations of island fauna and flora have long been important in ecology and have given rise to theoretical speculation. Theories originally developed for islands have proven to have a much wider applicability (MacArthur & Wilson, 1963, 1967). Island biogeography research has also generated insights into the nature of man – environment relationships (Rappaport, 1963). A study of the Fiji Islands elucidated the vulnerability of ecosystems to natural disasters as well as to stresses imposed by man (Brookfield, 1981). Recently the importance of indigenous and imported energy sources to the economic future of an isolated region was studied in the island state of Hawaii (Shupe, 1982). It seems that several advantages can be gained by selecting islands to collect necessary empirical information and to generate new insights into problems related to regional planning.

This presentation deals with the Swedish island of Gotland in the Baltic Sea. Since 1976 this island has been the subject of an interdisciplinary research project using a systems approach to investigate the causal relationships between energy flow in nature and society and between energy and economics (Jansson & Zucchetto, 1978a, b; Zucchetto & Jansson, 1979, 1980, 1981). The aim of our study is to try to find some 'mutually rewarding' combination of man's activities and natural processes which allows the long-term survival of both. The intent of this paper is to place some management problems of today in a historical framework which furnishes explanatory information about the island's carrying capacity for human activity.

Gotland occupies about 3000 km^2 and has a resident population of 55000. It is separated from the Swedish mainland by some 80 kilometers of Baltic waters. The island is built of Silurian limestone and marls with a thin layer of glacial deposits on top. Natural ecosystems include coniferous forests, deciduous woodlands, heathlands, mires, lakes, beaches and coastal waters. The economic activities generating the major exports are agriculture and the food industry, limestone quarries and cement production. Over the last decades tourism has become increasingly important with about 300000 visitors to the island every summer. Administratively Gotland functions as a county but is an integral part of the Swedish economy which gives considerable financial support to the island. Gotland is strongly dependent on a steady inflow of petroleum, electricity, goods and services to maintain its structure and function. Almost half of the population and most of the industrial and government infrastructure are concentrated in Visby, the main urban center.

Historical background

The transformation of land has been going on ever since man came to Gotland. The oldest remains of human colonization date back to about 5000 BC. A

D.O. Hall, N. Myers and N.S. Margaris (eds.), Economics of ecosystems management.

 ISBN-13: 978-94-010-8928-9

slash-and-burn type of shifting agriculture which started in the deciduous forests gave way to stationary and intensified farming during the Bronze and early Iron Ages. Remains of fencing systems and stone house foundations indicate that Gotland at the time of the birth of Christ had about 800 collective farms (Carlsson, 1979) cultivating a total of roughly 70 km^2 of the most fertile soils. Later the territories were split up into smaller units and, until the beginning of the Viking period, the area covered by farm land seems to have doubled in size. Surplus production by agriculture and improved shipbuilding techniques made it possible for the Gotlanders to take part in international trade. This brought "luxury goods" to the island from distant places as can be seen from the many archeological finds on the island.

During the 13th century German tradesmen settled in Visby and made the town a wealthy center of Baltic trade. Gotland had about 13 000 inhabitants at that time with more than half living in Visby. Antagonism between rural and urban populations seems to have weakened the town's resistence to foreign attacks and in 1361 Visby was captured by Danish forces. Gotland became a Danish protectorate for almost three hundred years. During this period the island was almost totally isolated and forced to rely upon its own resources. Visby, cut off from outside support, became extremely poor and unable to maintain and repair its buildings. The economic decline also affected the rural areas and many farms were abandoned.

A new period of growth and foreign exchange began after the introduction of new energy technologies. Water wheels and wind mills became common in Gotland. Production for the export of food products, hides, lime and tar stimulated local employment and income which gave rise to a vigorous growth of the human population. An almost three-fold increase in the population number between 1700 and 1880 AD caused an escalating demand for food, housing, firewood and other essential goods which had to be satisfied by intensified exploitation of the island's natural resources. The total area of arable land was about 195 km^2 in 1700 AD (Moberg, 1938). Large-scale drainage of wetlands and the clearance of deciduous forests have increased the area under cultivation to about 830 km^2. Forested areas were clear cut to supply wood for heating houses, burning lime and the export of timber. By the end of the 19th century the yearly exploitation of forest products was more than twice the regrowth (Jonsson et al., 1981). Then the import of fossil energy suddenly reduced the demand for firewood which rescued Gotland's forests from further destruction. Conifers colonised much of the area lost but some forested heath had disappeared permanently (Fig. 1).

The transformation of land has continued in the present century. About 200 km^2 of natural grazing land and deciduous meadows have been transformed to cultivated pasture and the urbanization process considerably increased the size of urban areas. Limestone quarrying and cement production, with enough capacity to supply the whole of Sweden with cement, have irreversibly affected certain heath areas in the northern part of the island. In retrospect land use has changed relatively little since the period of extensive land reclaimation but the natural landscape has continuously become less diversified. The island's image of an old-fashioned traditional landscape with Visby as an historic monument has changed to that of a modern industrial society increasingly dependent on imported energy.

Environmental problems

What has this transformation of land meant to the overall functioning of Gotland's collective ecosystem? One way of estimating the impact would be to compare the total amount of solar energy fixed in gross photosynthesis before and after the land use changes took place. From reasonable assumptions about past productivity it seems that the present solar energy fixation by the whole terrestrial system – although channelled into food and forestry production – is 3–4% lower than in 1700 AD. Furthermore the present production in agriculture depends on extra energy inputs of about 1.3 PJ of fossil fuels, fertilizers, pesticides and machinery (Zucchetto & Jansson, 1979). Without this extra input the productivity would most probably

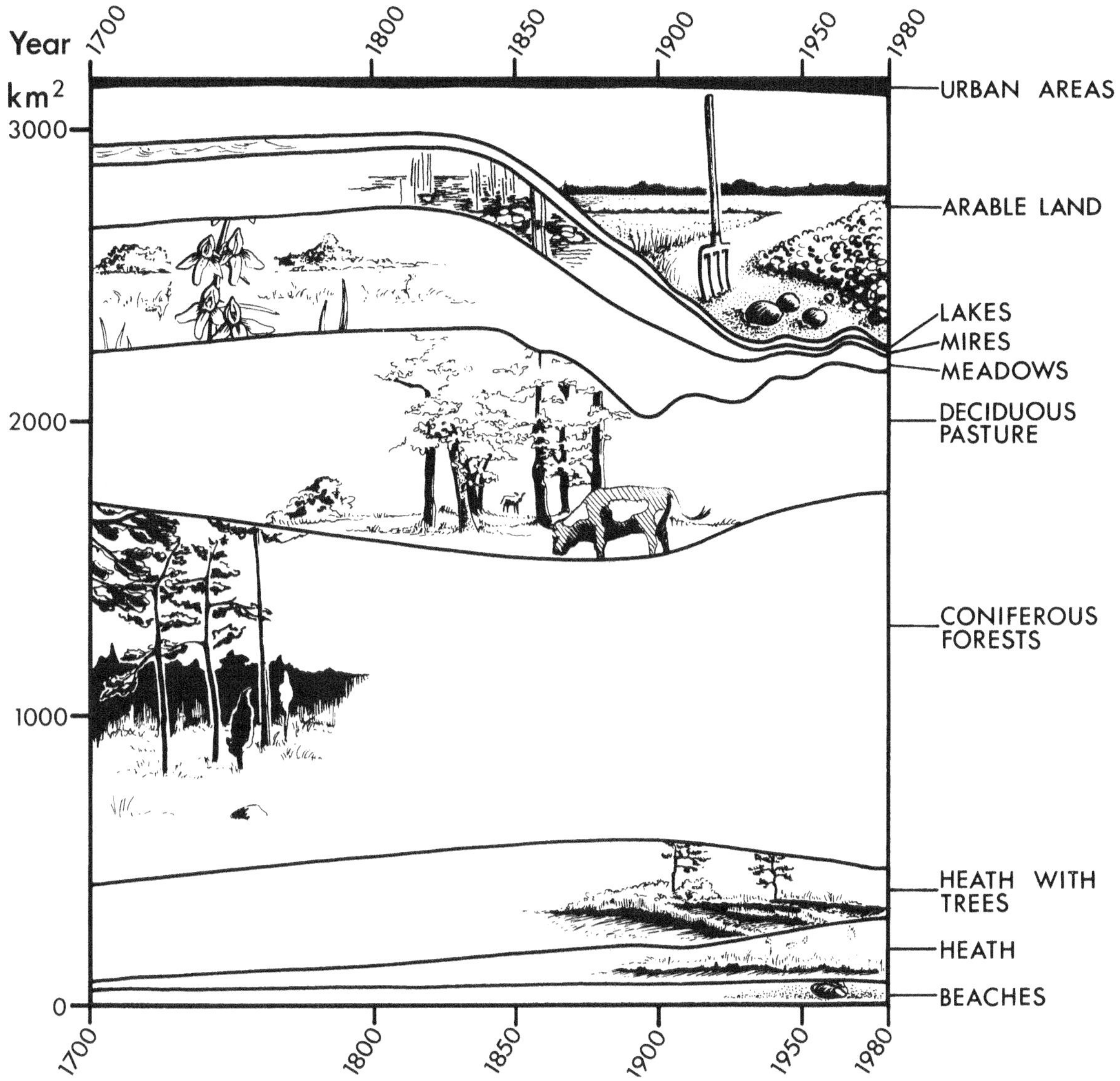

Fig. 1. Estimates of the change in land use on Gotland 1700–1980.

drop to a much lower value. One reason for this is the decreased level of organic matter in the soils. Ancient production was low due to technologically primitive cultivation techniques, but the maintenance of the organic content of the soils was ensured by the application of manure from a large number of cattle (Table 1). The number of cattle per km^2 of arable land decreased from 133 in 1800 to 38 in 1900. By then the use of imported fertilizers had begun and made manure less important. The field area has become so large that the application of manure recycles only about 20% of the energy in the organic matter taken away by the harvest (Zucchetto & Jansson, 1979). The most substantial loss of organic soils, however, has been through the oxidation of peat in drained mires. The peat was between one and two meters thick at the turn of the century and is now about 0.3 meter thick. If this decrease is assumed to be valid for the whole area of drained wetlands then the island has lost in the order of 7700 PJ of energy stored as peat during the last hundred years. This is equivalent to several hundred years of present import levels of fossil energy.

Table 1. Some measures for Gotland's agricultural and urban development.

	1700	1800	1850	1900	1950	1960	1970	1980
Population (thousands)	18	31	45	53	59	54	54	55
Rural/urban population	6.2	7.5	9.7	5.3	1.6	1.0	0.9	0.9
Arable land (km^2)	195	150	244	650	830	810	815	830
Arable land (ha/capita)	0.6	0.5	0.5	1.2	1.4	1.5	1.5	1.5
Fertilizers (kg/ha)	–	–	–	46	370	580	650	600
Yield (grain) (tons/ha)	?	1	1	1.2	1.9	2.4	3.4	2.0
Cattle (thousands)	15	20	19	25	43	50	50	57
Cattle/km^2 arable land	150	133	78	38	52	61	61	67
Firewood (PJ)	1	1.8	2.6	2.6	–	–	–	0.3
Coal and oil (PJ)	–	–	–	0.7	3.3	4.5	13.9	11
Electricity (PJ)	–	–	–	–	0.25	0.4	1.1	1.9

Another form of environmental damage caused by agriculture is the increasing amount of nitrates in ground water. The average nitrate content is now 8 mg/liter (Spiller et al., 1981). The close relationship between land use and water quality has gradually become recognized. The drainage of wetlands followed by the depletion of peat deposits and an extensive application of artificial fertilizers are the major reasons for the high nitrate concentrations. The decrease in peat layers has also affected the water storage capacity of the soils to such a degree that a shortage of water is now experienced in the summer months. Investments have to be made in building freshwater reservoirs and irrigation systems to replace nature's storage capacity and ability to buffer any lack of rain during the growing season. Attempts to evaluate in economic terms the yearly costs for society to compensate for the benefits produced by the mire ecosystem indicate that a drainage enterprise affecting 800 ha of mire would lead to a yearly cost of about 5 milj. Sw.Cr. for water management and diminished attractiveness of the area (Folke, pers. comm.).

The shortage of water in dry years has become limiting for economic production on Gotland. A shadow price of water was determined via an input–output model and optimization analysis on the assumption that the use of freshwater is as large as the yearly recharging of groundwater (Andréasson et al., 1983). The shadow price at present consumption rates was 22 Sw.Cr. per m^3 compared to the actual price of 6 Sw.Cr. per m^3 that industries pay for water. Competition for water between different sectors in society has just began. Two activities which are not thought of as having any considerable effect upon the freshwater reserves are limestone quarrying and cement production. These large-scale activities in Gotland have a negative impact upon the potential groundwater build up by increasing run-off. The consumption of water in cement production is 2.7 liters per Sw.Cr. of economic output if both direct and indirect consumption are taken into account (Andréasson et al., 1983). Next to food production and tourism this is the highest consumption among the economic sectors in Gotland.

Energy future

The significance of water and other environmental resources as limiting factors for production in Gotland have to be considered in all future planning. One important planning question deals with the problem: how can Gotland be supplied with energy at the end of this century when it is likely that oil will no longer be easily obtainable? The possible composition of an energy supply system for 1995 can be judged from existing knowledge about the dimensions of the natural energy flows and possible technical developments to make part of these fluxes available for society. The following energy sources have been considered in an optimization analysis of an alternative energy supply system for

Gotland (Ahlbom, 1982): electricity supplied by an existing cable from the mainland; electricity produced in local coal and oil power plants as well as by wind and wave generators; imported fossil fuels (fuel oil, diesel, gasoline, coal); biomass from wood waste, straw, peat and energy plantations produced locally and used for heating purposes; solar collectors and heat pumps likewise supplying low temperature heat locally. The possibility of producing liquid fuels for transport was not considered. The principal environmental constraints that were taken into account in this analysis were: the availability of wind, water, biomass and suitable land and an upper tolerable level for air and water pollution. The total demand for fuels and electricity was assumed to be the same in 1995 as in 1975 while the import of fossil fuels was restricted to 75% of the value of 1975. This corresponds to the plans for reducing regional oil consumption which have recently been drawn up by Swedish authorities.

If one wants to maximize the supply from indigenous energy resources optimization analysis gives the following solution for Gotland's future energy supply (Fig. 2). With the set of constraints and capacity limits used it is possible to cover only about 60% of the island's fuel energy demands by means of local resources. Intensively-grown energy plantations – fertilized and irrigated fast-growing willow trees – deliver 2.7 PJ while 1.5 PJ come from extensive (i.e. less resource-consuming) plantations. Straw, wood waste and peat contribute 1.8 PJ together, while heat pumps and other techniques are only used to a small degree. Imports consist of 1.6 PJ of oil besides the fixed amount of 1.7 PJ of gasoline and diesel. Sixty wind generators of 2 MW each manage to cover the island's total demand for electricity.

If one instead wants to minimize the costs for producing energy a completely different solution is obtained (Fig. 2). The results show that the local sources are not used beyond the 25% of fuel which is produced to satisfy regional energy plans. Of this amount straw provides the most substantial contribution. In this 'minimal cost' scenario imported coal has become the most important energy source.

The total energy cost for the first alternative is 40–50% higher than the alternative which minimizes energy costs. But if the effects upon the total

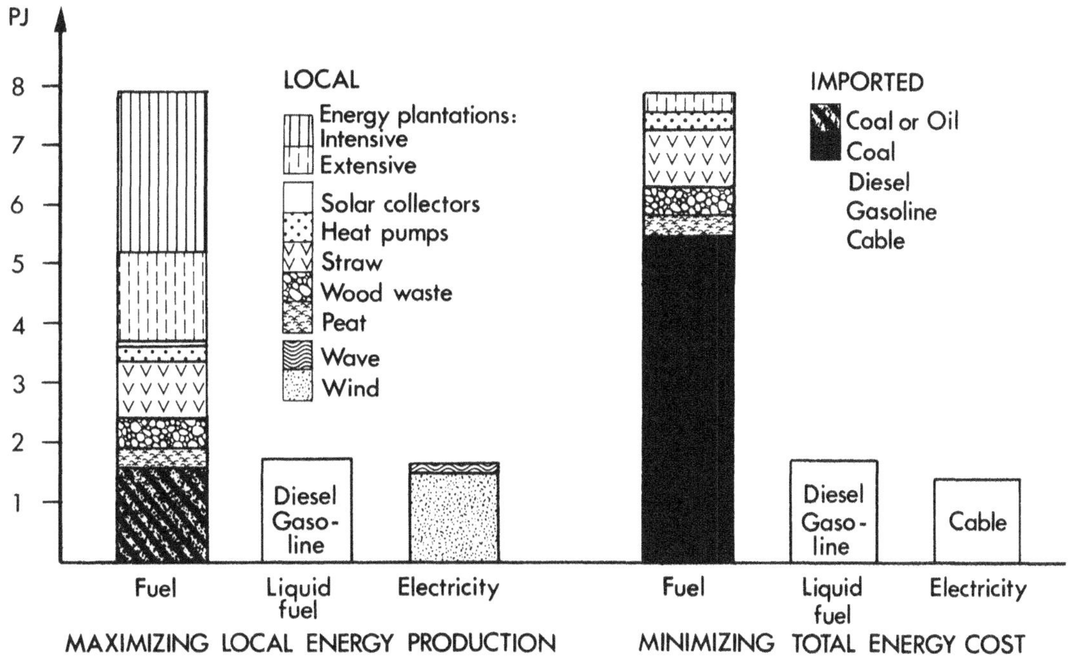

Fig. 2. Two scenarios for Gotland's energy supply system in 1995.

economy are considered, it turns out that the system which is nearly self-sufficient with energy would also increase the gross regional product for Gotland. This is because the production of local energy also generates more income and jobs (Andréasson, pers. comm.). One way of obtaining fuel self-sufficiency would be to let the energy plantations compete for land with food production by agriculture. Since the present food production is large enough to support almost seven times the resident population this does not seem to be an unrealistic development. To supply Gotland's fuel demand a total area of 600 km^2 is needed half of which comes from energy plantations on cropland. The economic consequences of this change would, however, be quite extensive since agriculture and the food industry contribute 28% to the gross regional product (Jansson & Zucchetto, 1978).

Discussion

From the historical description and the analysis of the environmental and energy situation it is clear that Gotland has changed from a subsistence economy based on indigenous natural resources to an open economy affected by events on the Swedish and foreign markets. Urbanization has led to a concentration of people in Visby. The ratio of the rural to the urban population has dropped from almost 10:1 during the middle of the last century to 0.9:1 (Table 1). The rural to urban ratio seems to have been below one once before in Gotland's history, namely during the fourteenth century. At that time the population in Visby received most of its income from foreign trade. Later came a long period of decline when the island was cut off from outside support and forced to rely upon its own resources. A critical period in Gotland's history occurred in the last century when the regrowth of forests was severely curtailed by demands for fuel, until imported coal became available. This coal import restricted forest cuttings and made possible a reestablishment of the forest productivity. The import of coal and oil also made possible an expansion of the industrial and urban sectors which in turn increased the demand for fuel and electricity. The per capita consumption of energy has increased about five times in the last hundred years.

When comparing present energy consumption and the exploitation of natural resources with past levels it is hard to believe that self-sufficiency can be possible in the future. The pre-industrial society which existed until the middle of the last century was supported by the renewable energy flows of sun, wind and rain and by the fertility of the soils. In 1850 AD the size of the human population had reached 45 000 and was obviously close to the carrying capacity limit of the region. Further population growth supported by the large stores of biomass energy is forests and peat layers would probably have stopped without the advent of new sources of fossil fuel energy. Today the de facto population size is more than 60 000 when a weighted number of visitors to the island is included. The increased per capita consumption and intensified use of environmental resources have affected the function of the island's collective ecosystems. Much of the biomass production has been channelled into human food production and depends on extra input of fossil energy. It remains to be seen if the overall decrease of 3–4% in solar energy fixation and biomass production can be compensated for by modern technology. There is great hope that more energy efficient technologies in industry, transportation and buildings together with energy savings will reduce energy end use in Sweden by 35% (Johansson et al., 1983).

The local politicians have made no serious attempts so far at easing the island's dependence on imported energy. By means of a new 130 MW cable Gotland is now connected to the electric grid on the Swedish mainland. This reduces interest in pursuing a development towards any alternative energy supply system. It is possible, however, that the access to electricity from the mainland might help Gotland to develop renewable energy alternatives. An economic crisis in the western industrial world, leading to a decreasing demand for building materials would directly influence the cement industry on Gotland. If cement production goes down part of the imported electric power would, for a period at least, become available for other purposes. This might foster an advantageous climate for the development of an energy technology which makes possible a transition to an energy supply system based on local energy sources including wind generators and biomass plantations, of the type that has been proposed in this study.

Acknowledgement

The Gotland study was made possible by grants from the Swedish Natural Science Research Council, the Energy Research and Development Commission and the Swedish Council for Planning and Coordination of Research.

References

Ahlbom, H., 1982. Gotländsk energi (Energy on Gotland). Askö Lab., Univ. of Stockholm, Mimeographed (in Swedish).

Andréasson, I-M., Hilding, T. & Jansson, A-M., 1983. Vatten som begränsande faktor för ekologisk och ekonomisk produktion på Gotland (Water as a limiting factor for ecological and economic production in Gotland). Nordic Seminar on Waterstatistics, Nordisk ministerråd, Miljörapport 2: 240–249.

Brookfield, H., 1981. Man, environment and development in the outer islands of Fiji. Ambio 2-3: 59–67.

Carlson, D., 1979. Kulturlandskapets Utveckling på Gotland. En Studie av Jordbruks- och Bebyggelseförändringar under Järnåldern (The Development of the Cultural Landscape in Gotland. A Study of Changes in Agriculture and Settlements during the Iron Age). Press Förlag, Visby (in Swedish).

Jansson, A-M. & Zucchetto, J., 1978a. Energy, economic and ecological relationships for Gotland, Sweden. A regional systems study. Ecol. Bull. 28: 1–154.

Jansson, A-M. & Zucchetto, J., 1978b. Man, nature and energy flow on the island of Gotland. Ambio 4: 140–149.

Johansson, T.B., Steen, P., Bogren, E. & Fredriksson, R., 1983. Sweden beyond oil: the efficient use of energy. Science 219: 355–361.

Jonsson, R., Lindberg, U., Lundgren, T., Mattsson, B., Olsson R. & Zerne, B., 1981. Kalksten, en gotländsk naturresurs (Limestone, a natural resource of Gotland). Dept. of Human Ecol., Univ. of Gothenburg, Mimeographed (in Swedish).

MacArthur, R.H. & Wilson, E.O., 1963. An equilibrium theory of insular zoogeography. Evolution 17: 373–387.

MacArthur, R.H. & Wilson, E.O., 1967. The Theory of Island Biogeography. Princeton Univ. Press, Princeton.

Moberg, I., 1938. Gotland um das Jahr 1700. Eine kulturgeographische Kartenanalyse (Gotland AD 1700. A cultural-geographical mapanalysis). Geogr. Ann 1-2: 1–112 (in German).

Olofsson, A.G., 1945. Gotlands Län Hushållningssällskap 1791–1941, Minnesskrift (Country Agricultural Society of Gotland 1791–1941, Memorial Publication). Almqvist & Wiksell AB, Uppsala (in Swedish).

Rappaport, R., 1963. Aspects of Man's influence upon island ecosystems. Alteration and control. In: F.R. Fosberg (ed.), Man's Place in Island ecosystems. Honolulu Bishop Museum Press, Honolulu.

Shupe, J.W., 1982. Energy selfsufficiency for Hawaii. Science 216: 1193–1199.

Spiller, G., Jansson, A-M. & Zucchetto, J., 1981. Modelling the effects of regional energy development on groundwater nitrate pollution on Gotland, Sweden. In: W.J. Mitsch, R.W. Bosserman & J.M. Klopatek (eds.), Energy and Ecological Modelling. Elsevier Scientific Publ. Co., Amsterdam.

Zucchetto, J. & Jansson, A-M., 1979. Total energy analysis of Gotland's agriculture: A nothern temperate case study. Agro-Ecosyst. 5: 329–344.

Zucchetto, J. & Jansson, A-M., 1980. Optimization of economic and ecological resources for regional design. Res. Manage. Optim. 1 (2): 111–143.

Zucchetto, J. & Jansson, A-M., 1981. System analysis of present and future enery/economic developments on the island of Gotland, Sweden. In: W.J. Mitsch, R.W. Bosserman & J.M. Klopatek (eds.), Energy and Ecological Modelling. Elsevier Scientific Publ. Co., Amsterdam.

CHAPTER 12

Ambelakia, a rational ecosystem management in the 18th century

R. MANIDAKI

Ambelakia, a small mountainous settlement in continental Greece – Thessaly – can be considered as a unique production and marketing phenomenon of the 18th century. The effective management of the environmental resources led to a rational management and impressive enlargement of the economic system over a period of almost sixty years (1750–1812).

The decisive factor was the development of a technique for dyeing red cotton yarns using the roots of the wild madder (alizari, *Rubia tinctorum* L.) growing in the area. As the dyeing business in Ambelakia expanded, the quantities of madder growing locally were not insufficient, and supplementary quantities were imported from Asia Minor. The entire population of Ambelakia joined in 1780 to form a socio-economic cooperation that was named Syntrofia (meaning company, association) which operated successfully until the beginning of the 19th century (1812).

1750–1760	Development of colouring methods
1770–1780	Small private units competing with each other and tending to integrate
1780–1795	Formation of Syntrofia
1912	The end of the unification after the development of new techniques and the disclosure of the secret method

Syntrofia was a peculiar unification of producers, workers, masters, merchants, and entrepreneurs in a quasi company that had no precise legal form but a comprehensive loyalty and support from its members. The main occupation of the people living in Ambelakia was the spinning of yarns from cotton bought from elswhere (Seres, Larissa Smyrna in Asia Minor) and the dyeing of cotton yarns. The product was exported all over Europe. The spinning and dyeing business was carried out in households, some of which only specialized in the dyeing business and turned graduelly into dyeing houses, developing a sophisticated method that was kept secret with great concern. The quality of the colours was famous and was appreciated throughout Europe, especially in France, Germany and Austria.

The main advantage of the cotton yarns from Ambelakia was their indelible and shiny colours which were maintained by using madder according to a very complicated method (see above). Twenty-four dyeing houses operated in Ambelakia, with a production of 2500 balls every year (200 oks each) exported to Pesti, Vienna, Dresden, Anspach, Bayreuth, Koningsberg, and Petroupolis. About the quality of cotton yarns dyed red we read in the book of Robert Paris, Histoire du commerce de Marseille, V, de 1670 à 1789, Le Levant '. . . there are numerous qualities of yarns. The colours constitute the most important difference. The yarns coloured red have higher prices (pp. 513–514). The red yarns are two times more expensive and their

[1] W.M. Leake, Travels in Northern Greece, London, 1835, pp. 386–387. Travels, pp. 288–289.

D.O. Hall, N. Myers and N.S. Margaris (eds.), Economics of ecosystems management.
 ISBN-13: 978-94-010-8928-9

colours have been for a long time an inaccessible secret to the French manufactur . . .'[2]

During that period of time it was easier to learn the processes of spinning and weaving than to know the secrets of India or Levant colouring. During the first part of the 17th century they were attempting to get the secret of red colouring according to the method of Andrinople, a town in Thrace.[3]

This written evidence about the exclusive secret and successful technical method of red colouring, the exploitation of this monopolistic element and the extended business activities abroad constitute an example of ecological and economic equilibrium, an equilibrium between natural resources and the people's livelihood. The rational management of natural resources provided not only a living for the people but a sophisticated socio-economic system of cooperation and a range of external entrepreneurial activities of a system that can be analyzed in terms of modern international economics. The improvement of the dyeing method was the main variable which led to rapid system enlargement in man–environment interactions. Natural forces and human factors produced and enlarged a system of ecological, economic and social factors.

Syntrofia, the quasi company, with all its external activities and its internal capacity was able to satisfy individual needs and to solve commom problems.

I will not cover in detail the production organization, internal economic social structure and function of the system or the socio-economic relations of the people belonging to the association. However, I shall try to explain the scale relationship that is, how the economic system ranged in scale from the household unit through the community and local area to the international levels. I shall try to explain the behaviour of the system as a unit in the context of the European economy of that time. I shall try to focus on the external dynamic process, on the external entrepreneurial activities and analyze them in terms of modern international economics.

Their production organization and the whole system of collaboration is unique and attracted the attention not only of European contemporaries but a number of present-day historians and economists, too. There is a lively debate as to whether the character of production and distribution was capitalist or was one of the first forms of cooperative planning. The legal status of the undertaking is considered either as a cooperative, or as a union of cooperatives, or a company that has no precise legal form, or as a joint stock company. Whatever its legal form, Ambelakia developed activities which have characteristics in common with a 20th century type of business activity, the internationalization of firms.

Syntrofia of Ambelakia carried out business in a very effective way. The cooperative or firm developed business activities in more than one country: international, transnational and supranational activities. It operated affiliated branches in Vienna, Trieste, Dresden, Hamburg, Odessa and Lyon, charged with marketing Ambelakia's export production of dyed cotton yarns. In order to relate these external business activities to recent features of international business, we need to specify briefly which enterprises can be considered as multinationals from the economics point of view.[4] What are the significant criteria for selection? The decisive factor is geographical, and applies to the extent of the area in which the firm develops its range of activities that are controlled and directed from its headquarters. When this area is extended over

[2] Felix Beaujour, Tableau du commerce de la Crece, Paris, 1800, pp. 273–274.

[3] Paul Masson, Histoire du commerce Francais dans le Levant au XVIIIe siecle, Paris, Librairie Hachette, 1911, pp. 438–440.

[4] H.V. Perlmutter, L'enterprise internationale trois conceptions, Revue economique et sociale, Vol. 23, No. 2., Mai 1965, pp. 151–165. H.V. Perlmutter, The Tortous Evolution of the Multinational Corporation, Columbia Journal of World Business, Vol. 4., 1969, p. 11. John H. Dunning, The Multinational Enterprise: The Backround in the Multinational Enterprise, edited by John H. Dunning, George Allen and Unwin, Ltd., London, 1971, pp. 16–17. Aharoni Yair, On the Definition of Multinational Corporation, in the Multinational Enterprise in Transition, Selected Readings and Essays, Edited by A. Kappor and Phillip, F. Darwin Press, Princeton, New Jersey, 1972, pp. 3–20.

more than one country, and at the same time the firm has an interest in integrated marketing with prospects in more than one country, then this can be considered a multinational enterprise. Among the variety of multinational companies we recognize those in which the main business activity consists in the international trade of locally produced products.

When the *Syntrofia* of Ambelakia '. . . decides to renew the collaboration and fraternity in order to carry out the industry of red dyeing . . . the trade of red yarns . . ., and at the same time . . . elects the directors of supervisors in Germany . . .', as we read in their charter of the year 1795,[5] we recognize international trade activities in the same way that the multinational corporation sells internationally the products produced locally.

Subsidiaries were established in Thessaloniki, Yiannina, Trieste, Dresden, Aachen, Hamburg, Leipzig, Odessa and Lyon.[6] The office of the supervisory director, the head of business abroad, was in Vienna. This director of the Vienna branch at the same time supervised all other branches operating abroad, reviewing their performance. This is clearly mentioned in Article 9 of the 1795 charter. The branch directors were generally citizens of Ambelakia. In the 1780 charter, Article 8, are mentioned the names of Arsenis Hatziparaskevas, Nicolaos Pediou, Georgios Schwarty and Nicolaos Michael as board members supervising Ambelakia, who at the same time were directors of the branches abroad. The positions of foreign branch directors were filled by periodic rotation among the Ambelakia board members themselves. Later, according to the charter of 1795, directors of branches abroad were appointed by the central administration in Ambelakia. Ambelakia's operations had the following characteristics:

(a) business activities in more than one country;

(b) appointment of directors of the subsidiaries by the mother company;

(c) interest in global-scale marketing.

This is clear because all these activities resulted in '. . . increased profits by avoiding intermediaries and undertaking the trading of local production . . .'. Its characteristic feature, described in manuscripts of that time, consisted of branches aimed not only at selling products from Ambelakia but also at supplying Ambelakia with various products from abroad such as iron razors, housewares, and jewelry. This is reminiscent of modern Japanese trading companies such as Mitsui, Mitsubitsi, and Dumitomo which belong to the well known system of *Sogo Sho Sha* that has as its main objective the combining of manufacture and international trade (import and export). These Japanese companies have a mainly commercial character, although they combine manufacturing activities developed in Japan with international trade. They import raw materials and export final goods with a smoothly operating extensive network of branches abroad.

In the case of Ambelakia, raw materials were imported by the branches and final goods, produced locally, were exported in the same way. During the decade 1760–70, similar international trading patterns developed in other areas of the world. The East India Company and the Massachusetts Bay Colony developed commercial trading houses operating in more than one country. But the growth of multinational activity is sometimes considered a phenomenon of the second half of the 20th century, resulting mainly from improvements in the means of communication. We can easily imagine the importance and difficulties of transnational activities at a time when communication was not developed. In spite of distance and obstacles, collaboration and communication were developed efficiently at the scale required. But the feature of Ambelakia that makes me think of its similarity to modern multinational companies concerns the theory which explains the internationalization of firms. According to this theory, in broad terms, a strong incentive for the internationalization of firms is the possession and application of new high technology. The availability of new high technology makes the extension of business ac-

[5] The charter was saved by F. Boulanger in French in his book, Ambelakia ou les Associations et les Municipalités Hellènique, Paris, 1875, p. 53.

[6] W.M. Leake, Travels in Northern Greece, London 1835, p. 387. Ed. Dodwell, A. Classical and Topographical Tour Through Greece, London, 1819, Vol. 2, p. 108.

tivities abroad attractive. The period of time for business success depends on the 'product life cycle' or 'method life cycle', according to Professor Raymond Vernon's theory which explains international trade and international investment in terms of technological growth.[7]

The inhabitants of Ambelakia, like the inhabitants of many other small towns such as Larissa, Tyrnavo, Rapsani, Farsala, Baba, Ayias or mountain villages such as Portaria and Makrynitsa, were all well known for cotton spinning and dyeing with a very exclusive and successful method using wild madder. The wild madder growing in the area was supposed to be of better quality than the cultivated variety. But the inhabitants of Ambelakia were especially well skilled, and developed their own methods of dyeing the cotton yarns red. Their yarns were bright, shiny and unrivalled in all European markets. The red yarns from Ambelakia were not of such fine quality as the English products, but they were preferred in Europe ... 'because of their solid colours'.

The workers of Ambelakia developed a very exclusive method of dyeing. The stages in the dyeing process were divided among them and the workers continuously improved their specialized techniques, as everyone specialized in a specific stage of the process. The dyeing houses of Ambelakia can be considered as innovative firms. In the 18th century, about twenty-four dyeing houses were operating for the production of red cotton yarn. The annual output was about 3000 bales (125 ka each bale). The application of their relatively advanced technology in the dyeing process made possible the successful development of international trade activities from Ambelakia. The development of high technology products was the *sine qua non* for the internationalization of their business. The undertaking of Ambelakia involved a social functioning that exploited the monopolistic technological advanage they had developed. They were very conscious of the importance of technological innovations for international trading, and of the effectiveness of a 'social security' system. We can see how important the technological factor was for all their economic activities if we read carefully their charter of 1795: '... they decided to renew the *Syntrofia* for the business of red dyeing ...'. They refer to the red dyeing business only, not to any spinning or textile business, because their interest is concentrated in the dyeing and in the technology of dyeing; not in the textile industry per se. The dyeing process was a technological secret well kept by the workers of Ambelakia. Typically, the unification of the entrepreneurs, workers, etc. occurred when they became conscious of the importance of their technical knowledge and skills. The Ambelakia Syntrofia disappeared when the method life cycle ended with the appearance of new methods in England. The quality of the red yarns from Ambelakia deteriorated in the 19th century with application of new methods of red dyeing in England. The English yarns were lower in price because of lower transport costs and economies of scale. The life of the unique multinational undertaking of Ambelakia came to an end when a more advanced technology overtook the market.

In summary, the Ambelakia undertaking can be related to the modern multinational enterprise in two ways:

(1) internationalization in both cases is determined by a new technological process, by high technology;
(2) business activities are in both cases extended to more than one country on a permanent basis and the international market is treated as a single integrated entity.

The entrepreneurs of Ambelakia started from household production, based on local natural resources operated on a national scale and finally had transnational perspectives when they understood the importance of market integration across

[7] R. Vernon, International Investment and International Trade in the Product Cycle. The Quarterly Journal of Economics, LXXX, May 1966. R. Vernon, Sovereignty at Bay, Basic books, New York, 1971, Ch. 3, The Location of Economic Activity in J.H. Dunning edit., Economic Analysis and the Multinational Enterprise, George Allen and Unwin, London, 1974. Louis T. Wells, Test of a Product Cycle Model on International Trade: U.S. Exports of Cousumer Durables, Quarterly Journal of Economics, February 1969, the Product Life Cycle and International Trade, Harvard Business School, Division of Research, Boston, 1972.

Europe as necessary for their larger success and survival.

The system's equilibrium was finally disturbed by external factors. The advanced new technology in England, and the acquisition of their technological secrets by French technicians in Montpelier were forces that destroyed the effectiveness of the national ecosystem management and led to the expiration of the Ambelakia Syntrofia.

CHAPTER 13

The mercury problem and fishing in Finland

MARTIN LODENIUS

Abstract

The extent of mercury pollution in Finnish watercourses was first realized in 1967 when significantly elevated mercury concentrations were found in fishes of prey from several lakes and rivers. The mercury originated mainly from the pulp and chlor-alkali industries, and fishermen sustained heavy losses when both fish consumption and the price of fish declined.

The use of mercury in the pulp industry ended in 1968 and the discharge of mercury from the chlor-alkali plants has decreased significantly, but the decrease in the mercury content of the fish has been slow. The mercury problem was recently accentuated when elevated mercury levels were also found in many lakes without any known industrial mercury discharge. The highest concentrations occurred in man-made lakes, where the water level fluctuations are great. In most of the man-made lakes in northern and western Finland the fishes of prey (pike, perch and burbot) are unsuitable for human consumption because of their high mercury content. The construction of drainage ditches in forests and bogs is also assumed to be responsible for elevated mercury concentrations in Finnish lakes. The fishermens' losses caused by decreased sales, reduced prices and the costs for unused equipment amount to several million Finnish Marks.

Introduction

The industrial use of mercury has caused severe contamination in fish and man in several parts of the world. It was first recognized in the 1950s and 1960s in Japan, where more than 100 people from the small fishing villages of Minamata and Niigata died and several hundred were handicapped for life.

Mercury poisoning is caused by methyl mercury either from industrial sources after an accumulation in aquatic food chains or from mercury-dressed seed. In the whole world more than 800 persons have died and more than 8000 have been afflicted because of methyl mercury poisoning (D'Itri & D'Itri, 1978).

The mercury problem has been characterized as a human tragedy because attention has been focused on the victims of this poisonous metal. However, little is known about the economic consequencies caused by this kind of pollution. In the late 1960s Canadian fish heavily polluted with mercury caused extensive economic losses for both sportfishing and the native Indians, who worked as guides for the fishermen. The consumption of local fish was banned, the incomes from fishing licenses and supplies dropped, and unemployment increased. From the Kenora area in Ontario alone a loss of $6–10 millions has been calculated (D'Itri & D'Itri, 1978).

D.O. Hall, N. Myers and N.S. Margaris (eds.), Economics of ecosystems management.
 ISBN-13: 978-94-010-8928-9

Fishing in Finland

Only 0.2% of the Finnish population are classified as professional fishermen. Most of them live along the coast. The most important fish species is the Baltic herring (87% of the catch). In addition there are approximately 1 million recreational fishermen in Finland and the value of their catch is as high as that of professional fishing. A great proportion of the amateur catch consists of perch, pike and burbot, i.e. species containing the highest amounts of mercury. The total value of the Finnish fisheries in 1981 was 356 million FIM (US$ 64 million or 0.3% of the Gross Domestic Product; Table 1).

Mercury pollution from the pulp and chlorine industries

Phenyl mercury was widely used as a slimicide in the Finnish pulp industry until other chemicals were taken into use from 1968. During the 1960s more than 500 t of mercury was used in Finland for this purpose. The chlor-alkali industry uses mercury as a catalyst in the production of caustic soda and chlorine. The annual release of mercury into the water from the three Finnish chlor-alkali plants has been reduced from 1–4 t to less than 100 kg as a result of more efficient waste water treatment. Mercury is also used for seed dressing (7.5 t), in the dentistry (2 t), in technical equipment etc. However, mercury emissions into the environment are nowadays essentially smaller than in the 1960s.

In 1967 very high mercury contents (several mg/kg) were found in fish from many polluted areas (Häsänen & Sjöblom, 1968; Nuorteva, 1971). The mercury accumulation in the moderately polluted Lake Päijänne, the second largest lake of Finland, has been described by Särkkä et al. (1978) (Fig. 1). When it became known that there were polluted fish, this gave rise to a strong public debate, which caused a rapid decrease in the fish consumption and economic losses for thousands of fishermen. Locally a sharp decline in the fish price could be noticed (Fig. 2).

Approximately 10% of the inland waters were more or less mercury polluted, with the highest concentrations occurring in fishes of prey caught in certain polluted areas. The decrease in fish consumption affected not only the polluted fish but also fish species with low mercury contents, e.g. the Baltic herring, and fish caught in unpolluted waters.

In 1971 the use of mercury polluted fish was at last restricted by official directives. Fish containing more than 1 mg/kg of mercury was considered unsuitable for human consumption and fish containing 0.5–1 mg/kg were not to be eaten more often than once a week. In 1974, seven years after the extent of the mercury pollution was first realized, the parliament finally passed a law concerning compensation to fishermen for the economic losses caused by mercury and other pollutants.

New sources of mercury – new restrictions

The water level is regulated in one third of the Finnish lakes. In 1980 high mercury levels were found in fish from man-made lakes in northern and western Finland. Hair samples from people eating

Table 1. The fisheries in Finland in 1981 according to the Fisheries Division of the Finnish Game and Fisheries Research Institute.

		Professional	Amateur	Total
Fishermen		7 000	1 000 000	
All species	Amount (10^6 kg)	94.4	33.2	127.6
	Value (10^6 FIM)	174	182	356
Pike, perch & burbot	Amount (10^6 kg)	2.2	17.6	19.8
	Value (10^6 FIM)	12	96	108

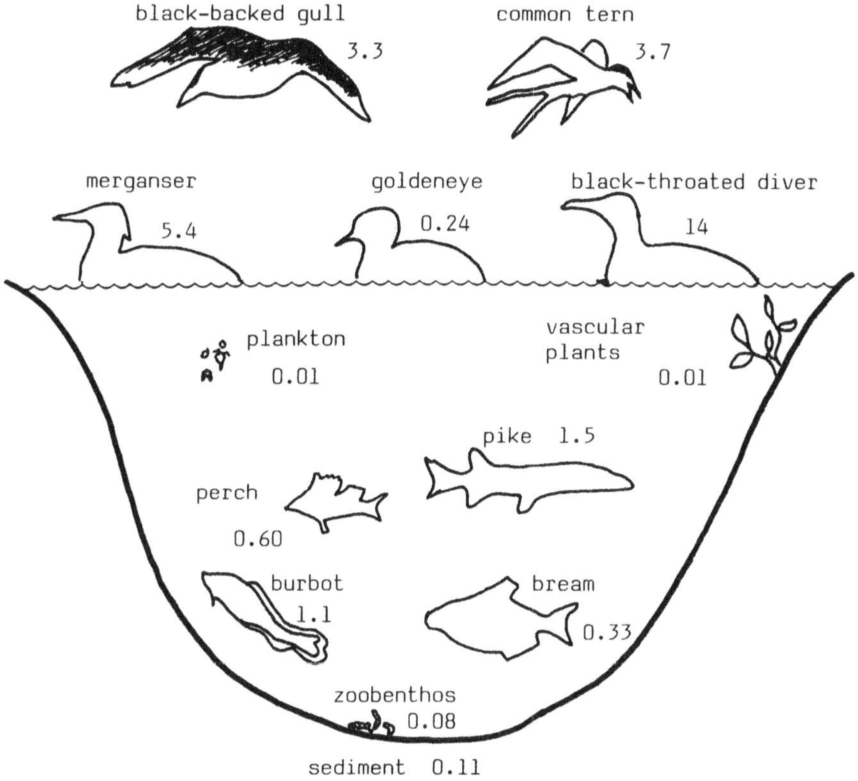

Fig. 1. The accumulation of mercury (average contents in mg/kg) in Lake Päijänne, Central Finland (according to Särkkä et al., 1978).

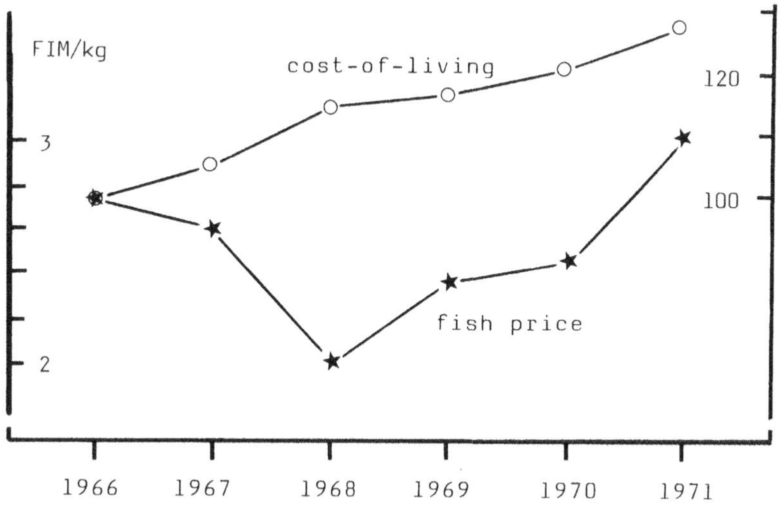

Fig. 2. The price of pike in Kotka, S. Finland, 1966–1971 in relation to the cost-of-living-index (1966 = 100).

fish from these impoundments were found to have significantly higher mercury concentrations than normally (Lodenius & Seppänen, 1982; Lodenius et al., 1983). A new public debate arose and fish consumption declined drastically in these areas. In Finland 15% of the total area has been drained in order to promote forestry. However, drainage of peatlands may also elevate mercury levels in lakes (Simola & Lodenius, 1982). The extensive regulation of waterlevels and the drainage of peatlands may explain the surprisingly high mercury values from many seemingly unpolluted lakes.

The mercury levels in watercourses polluted by the pulp industry decreased quite slowly due to the great amounts of mercury deposited in sediments and circulating in the aquatic food chains. In the badly polluted watercourse of Hämeenkyrö no clear decrease was observed in the mercury content of smaller pike while the annual decrease for bigger specimens was only 0.13 mg/kg (Nuorteva et al., 1979).

New restrictions were necessary and in 1981 a new official list of mercury polluted waters was published. This was adjusted in 1982. In this list the area of heavily polluted waters has decreased significantly while the areas with fish containing 0.5–1 mg/kg of mercury has almost doubled. The number of mercury polluted watercourses had increased in ten years from 22 to 66, the total area remaining at approximately 4000 km^2 (Table 2). The value of the potential catch of fish from the mercury polluted watercourses may be estimated to be 10.4 million FIM (US$ 1.9 million).

The public debate and the new restrictions affected fish consumption in the polluted areas, leading to decreased mercury levels in the hair of fish consumers (Lodenius & Seppänen, 1982). The mercury problem was managed more effectively this time.

Economic consequences

According to the governmental committee set up to evaluate the losses to professional fishermen caused by mercury and other pollutants, total losses amounted 1.6–1.7 million FIM in 1971. The losses consisted of:

- unused fishing equipment;
- reduced sales;
- lower fish prices;
- rents for fishing areas and interest on loans.

The losses to amateur fishing were not estimated. Considering that the value of the amateur catch of predatory fish is eight times as high as that of the professional catch (Table 1), the loss may be estimated to be more than 10 million FIM.

As a result of the 'compensation law' of 1974, 100 000–280 000 FIM has been paid annually to professional fishermen as compensation for the pollution. At the beginning, the major part of the compensation was paid for losses caused by mercury but in the last few years compensations for other losses have dominated. The amount of compensation decreased steadily from 1977 to 1981 (Fig. 3). In any case only a small part of the economic losses

Table 2. The mercury polluted watercourses in Finland 1971 and 1981–82 (according to the directives of the National Board of Health the sale of fish is banned when the mercury content exceeds 1 mg/kg and the consumption of fish should be restricted when it is 0.5–1 mg/kg).

	Inland waters	Sea areas	Finnish total
1971			
Sale banned	1 410	760	2 170 km^2
Restriction of use	1 710	200	1 910 km^2
Total	3 120	960	4 080 km^2
1981–82			
Sale banned	260	60	320 km^2
Restriction of use	3 070	560	3 630 km^2
Total	3 330	620	3 950 km^2

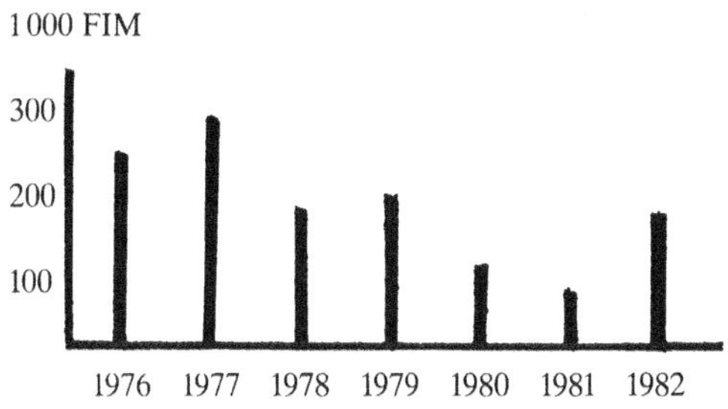

Fig. 3. The compensations paid to professional fishermen 1976–82 for economic losses caused by pollution (Ministry of Agriculture and Forestry).

caused by mercury pollution has been compensated.

Conclusion

An estimation of the economic consequences of the mercury problem is very difficult. This is true especially when considering recreational fishing, which is usually not considered at all in economic calculations. A successful management of the mercury problem requires consideration of medical, environmental, social and economic factors. The protection of public health should of course be of primary importance in cases like this.

When the first signs of the mysterious 'Minamata disease' appeared in the 1950s medical experts and the authorities were nonplussed. After many years of laborious scientific research work the disease was shown to be methyl mercury poisoning. However, when the mercury problems occurred in the late 1960s and early 1970s in Canada, Iraq and Scandinavia, the authorities were almost as inefficient as in Japan. The research capacity was completely insufficient, information to the fishermen and fish consumers was scarce and poor, and restrictions came late.

Today thousands of new, synthetic chemical compounds circulate in our environment. Among these many certainly cause severe environmental problems. Our knowledge about mercury and other environmental pollutants is considerably better now than 20 years ago. Let us hope that the history of mercury pollution will teach us how to manage environmental problems more effectively in the future.

References

D'Itri, P.A. & D'Itri, F.M., 1978. Mercury contamination: a human tragedy. Environ. Manage. 2: 3–16.

Häsänen, E. & Sjöblom, V., 1968. Mercury content of fish in Finland in 1967 (English summary). Suomen kalatalous 36: 1–24.

Lodenius, M. & Seppänen, A., 1982. Hair mercury contents and fish eating habits of people living near a Finnish man-made lake. Chemosphere 11: 755–759.

Lodenius, M., Seppänen, A. & Herranen, M., 1983. Accumulation of mercury in fish and man from reservoirs in northern Finland. Water, Air, Soil Pollut. 19: 237–246.

Nuorteva, P., 1971. Methylquecksilber in den Nahrungsketten der Natur. Naturw. Rdsch. 24: 233–243.

Nuorteva, P., Lodenius, M. & Nuorteva, S.-L., 1979. Decrease in the mercury levels of Esox lucius (L.) and Abramis farenus (L.) (Teleostei) in the Hämeenkyrö watercourse after the phenylmercury ban in Finland. Aquilo Ser. Zool. 19: 97–100.

Särkkä, J., Hattula, M.-L., Paasivirta, J. & Janatuinen, J., 1978. Mercury and chlorinated hydrocarbons in the food chain of Lake Päijänne, Finland. Holarct. Ecol. 1: 326–332.

Simola, H. & Lodenius, M., 1982. Recent increase in mercury sedimentation in a forest lake attributable to peatland drainage. Bull. Environ. Contam. Toxicol. 29: 298–305.

CHAPTER 14

The role of cyanophytes in the nitrogen economy of rice fields: a review

ALEXIS TSANGRIDIS

Introduction

The origin of rice cultivation goes back to the remote past. It is reported (Hogan, 1970) that, according to the Archaeological Survey of India, four terraces built for rice cultivation on the banks of the Ravi River in South-West Kashmir date to the Pleistocene or Ice Age. The country of origin of rice is unknown, but South-East Asia seems to be the center of rice cultivation. It is interesting to note that in Chinese and other languages of South-East Asia, agriculture and rice or food and rice are synonyms (Copeland, 1924).

It is very well known that rice is the main, traditional and most popular food for the people of the Far Eastern countries; nowadays, ninety per cent of the world's rice is grown in those areas. The percentage distribution, by region, of area and production of paddy, is shown in table 1. Under these circumstances, the interest of many scientists in the regions of South-East Asia concerning rice yield, is significant and quite comprehensible.

Among the factors which determine paddy yield, certain cyanophytes possessing the ability to fix nitrogen, are of considerable importance, for they can be used as biofertilizers. Their importance becomes more crucial in most of the Far Eastern countries where commercial fertilizers are relatively expensive. In Europe, the potential of cyanophytes as green manure has been generally neglected. Some Russian researchers are the exception to this rule, focusing their studies on the relation between cyanophytes and soil fertility (Goryunova & Orleanskii, 1967, 1968).

Nitrogen-fixing ability of cyanophytes and their beneficial role

Frank first reported in 1889 that some cyanophytes are able to fix nitrogen (Arora, 1969; Metting, 1981). Fifty years later, De (1939) pointed out that this ability can be important for agriculture and ecology. The research carried out up to now with regard to the nitrogen-fixing ability (NFA) of the blue-green algae (BGA), has been developed in two directions.

(A) The first one, examines the NFA of BGA in

Table 1. Distribution, by region, of area and production of paddy (percentage).

Region	Area		Production	
	1961–65 average	1971	1961–65 average	1971
Europe	0.4	0.6	0.7	1.1
N. America	0.6	0.6	1.2	1.2
S. America	4.2	4.7	3.5	3.5
Near East	0.5	0.5	0.6	0.6
Far East	91.7	90.8	91.5	90.0
Africa	2.6	2.8	2.3	2.5
Oceania	0.04	0.05	0.2	0.1

Source: Grist (1975).

D.O. Hall, N. Myers and N.S. Margaris (eds.), Economics of ecosystems management.
 ISBN-13: 978-94-010-8928-9

the laboratory and in situ, i.e. the species ability and capacity as well as the factors which affect the rate of nitrogen fixation. One of the methods used to measure the NFA is to study the $^{15}N_2$ uptake. This method did not become very popular, for it is time-consuming, expensive and requires a mass spectrometer. Stewart et al. (1967), based on the fact that the nitrogen-fixing complex (nitrogenase) reduces acetylene to ethylene (cf. Burris, 1975), suggested that the rate of acetylene reduction may be used as an index of the rate of nitrogen fixation. They, moreover, established an in situ method which can be applied in aquatic environments, in soils and with nodulated plants. This in situ method has been widely used and has contributed significantly to our knowledge concerning the NFA. Besides, it gives data closer to the natural environment than the isotope $^{15}N_2$ method (Yoshida & Ancajas, 1970). More than eighty taxa of BGA are now known to possess the ability to fix nitrogen (cf. Tsangridis, 1982). The factors regulating nitrogen fixation have been thoroughly studied. It has been found that nitrogen fixation is inhibited by ammonia when present in high concentration, low pH and temperature values, desiccation, some pesticidal compounds, some inorganic elements such as cadmium, lead and zinc in high concentrations and long dark periods (El-Nawawy et al., 1968; Henriksson, 1971; Fogg et al., 1973; Dasilva et al., 1975; Stewart et al., 1975; Englund, 1977; Henriksson & DaSilva, 1978; Starks et al., 1981, Antarikamonda & Lorenzen, 1982; Al-Mousawi & Whitton, 1983). The capacity of the nitrogen fixation performed in situ is dependent on the external conditions and Henriksson et al. (1972) correctly point out that the determinations of the maximal capacity should be measured under controlled conditions in the laboratory.

(B) The second direction of research related to nitrogen fixation is concentrated on the beneficial role of certain cyanophytes to paddy yield. The applied agricultural aspect of these micro-organisms mentioned by De (1939), was followed by Singh (1961). He gives most impressive data, reporting that inoculation of paddy soils in districts of India with *Aulosira fertilissima* increases the paddy yield to 114%. A considerable number of references is now available dealing with the beneficial role of BGA in paddy yield. Most of the research has been carried out in India (Relwani & Subrahmanyan, 1963; Subrahmanyan et al., 1964, 1965; Sankaram et al., 1966; Mudholkar & Sahay, 1968; Goyal & Venkataraman, 1971; Selvaraj & Robinson, 1971; Agarwal, 1979; Kulasooriya et al., 1980) but there are also references from Japan (Watanabe et al., 1951; Watanabe, 1962, 1966), Australia (Bunt, 1961), Saudi-Arabia (Aboul-Fadl et al., 1967), the Soviet Union (Shtina, 1965; Goryunova & Orleanskii, 1967, 1968) and China (Ley et al., 1959; Huang You-xing et al., 1981). In most of these studies, the importance of BGA in paddy yield is compared with that of commercial fertilizers using pot and field experiments. Agarwal (1979), summarizing the results of the experiments made in the Indian Agricultural Research Institute, reveals that:

1. In the total absence of chemical fertilizers, the application of BGA can fix about 25–30 kg of nitrogen per hectare per cropping season, producing a 10–15% increase in grain yield.
2. Where chemical fertilizers are used, the dose can be reduced by about one third and supplemented by BGA to get the same grain yield.
3. Even when the full recommended levels of chemical fertilizers are used, complementary BGA is beneficial, resulting in higher grain yield.

The systems applied when using BGA as a biological input in agriculture are reported in detail by Venkataraman (1972). He suggests five systems, based on:

(a) stirring tank cultures;
(b) closed circulation;
(c) open bubbling;
(d) dual purpose open circulating systems;
(e) culturing on moist volcanic gravels.

The open bubbling system was adopted by Watanabe et al. (1959) who used natural resources in their efforts to minimize the cost of production. They produced mass-cultures of *Tolypothrix tenuis,* a well-studied nitrogen-fixing species (cf. Watanabe, 1951; Kiyohara et al., 1960; Watanabe, 1962), using hot spring water and natural gas(methane) as heat and carbon dioxide sources respectively. The growth rate of the alga was estimated to be 6.4 g (dry weight) per square meter per day, i.e. a yield of 7.0 tons of the alga per annum; this is equal to the quantity of the alga necessary for inoculating about 750 acres of paddy field.

The beneficial effect of algal inoculation cannot be considered as universal since certain soils and climatic conditions determine to a great extent the results obtained. Soil moisture is a major ruling factor in this algal management and this is why in temperate soils, the use of BGA as an alternative nitrogen source is uneconomic compared to inorganic fertilizers; however, in well-irrigated fields, the desiccation problem could be overcome (Rodgers et al., 1979). The research recommended by the Committee of the International Biological Programme on the ecological importance of BGA, showed that also in temperate regions, nitrogen-fixing algae are common in certain soils, and that, as in the tropics, they are an important link in the nitrogen cycle; since the BGA are photo-autotrophic, they are also of global importance on account of their ability to supply organic matter to the environment (Henriksson, 1978). It has been found that over 20 kg N/ha/yr may be fixed by BGA in certain temperate regions (Henriksson, 1971).

Besides nitrogen fixation, BGA may often benefit rice plants directly of indirectly, for these micro-organisms liberate oxygen which is utilized by the submerged roots of the plants (Selvaraj & Robinson, 1971; Sankaran, 1975) and release growth-promoting, auxin-like substances (Gupta & Lata, 1964; Venkataraman, 1972). This explains the beneficial role of certain non-nitrogen-fixing BGA in the yield of rice (Table 2).

The use of BGA as biofertilizers in rice fields may have one more benefit since it would restrict the amount of nitrogen fertilizers used, thus protecting the environment from certain hazards involved, i.e. eutrophication or health hazards when nitrate-contaminated sources are used for drinking water. Although perhaps rather drastic, it has been predicted that in 25 to 50 years the ultimate crisis in agriculture will occur in the United States; either the fertility of the soil will drop precipitously, because inorganic fertilizers will be withheld, throwing the nation into a food crisis, or the amounts of inorganic nitrates and phosphates applied to the land will be so great as to cause an intractable water-pollution problem (Ehrlich et al., 1973). Commonet (1977) points out that the purpose of evaluating the benefits and hazards of the use of nitrogen fertilizer is to support a rational choice among alternative courses of action that can improve the benefit/risk ratio. The application of BGA, at least in rice fields, could be one alternative action.

Table 2. Effect of *Oscillatoria* on the yield of rice plants.

	Grain yield (g/pot)
Control	5.96
Alga (1 kg/ha)	9.09
Ammonium sulphate (89.68 kg N/ha)	12.59
Alga + Ammonium sulphate	19.37

Source: Venkataraman (1972).

To sum up, BGA may not only benefit rice cultivation due to their nitrogen-fixing ability, their growth-promoting substances and liberation of oxygen but also contribute to the environmental protection policy.

Remarks

The significance of nitrogen in agriculture can be easily discerned from the numerous chemical fertilizers produced throughout the world, representing an investment of 10 billion US dollars (Venkataraman, 1977). There is no doubt that any saving in the consumption of chemical nitrogen without affecting the productivity of paddy, is of paramount importance for the economy of every country. The IARI (Indian Agricultural Research Institute) estimates that while 25 kg of nitrogen would cost about £6, algal material required to provide the same amount of nitrogen would cost about £2 even if it were produced commercially; if the farmer produces his own material, the cost would fall to 50 p or even less (Agarwal, 1979).

However the inoculation method is not as simple

as it seems on paper. Before inoculation, a detailed study of the algae inhabiting rice fields is required as well as a thorough study of the ecological parameters ruling the ecosystem investigated. If algae could be cultured cheaply, they could prove to be a remarkably cost-effective fertilizer complement; storage and transport of the algae would present no difficulties as algae mixed with sand and then dried can remain viable for many years (Rodgers et al., 1979).

The utilization of thermal springs as natural energy resources may be a promising way to achieve cheap cultures of algae.

In Greece, thermal springs are abundant. Nigrita thermal springs, located near Thessaloniki (N. Greece), could be a starting point to establish a pilot plant for the mass culture of nitrogen-fixing cyanophytes, since these springs have the useful advantage of a high content of carbon dioxide. The knowledge of the algal flora from rice fields near Thessaloniki (Anagnostidis et al., 1981; Tsangridis, 1982) as well as the extensive documentation existing from biological investigations of the thermal springs of Nigrita (Anagnostidis, 1959, 1961, 1967, 1968; Anagnostidis & Golubic, 1966; Anagnostidis & Schwabe, 1966; Anagnostidis & Rathsack-Künzenbach, 1967; Economou-Amilli, 1976) would be of great importance to this establishment.

With the promise that algal supplements could result in a saving of about 30% in commercial nitrogen fertilizer (Venkataraman, 1977), research projects towards the understanding of nitrogen economy in rice fields with particular emphasis on the role of BGA, should be more actively considered.

Greece is among the rice-yielding countries of the world, however the analysis of rice field ecosystem with regard to nitrogen economy, has been, so far, completely neglected.

References

Aboul-Fadl, M., Taha, E.M., Hamissa, M.R., El-Naway, A.S. & Shoukry, A., 1967. The effect of the nitrogen fixing blue-green alga Tolypothrix tenuis on the yield of paddy. J. Mikrobiol., VAR 2: 241–249.

Agarwal, A., 1979. Blue-green algae to fertilize Indian rice paddies. Nature 279: 189.

Al-Mousawi, A.H.A. & Whitton, B.A., 1983. Influence of environmental factors on algae in rice-field soil from the Iraqi marshes. Arab Gulf J. Scient. Res. 1: 237–253.

Anagnostidis, K., 1959. Auffindung des Mastigocladus-Thermen-Typus in Griechenland. Biol. Glasnik. 12: 87–88.

Anagnostidis, K., 1961. Untersuchungen über die Cyanophyceen einiger Thermen in Griechenland. Inst. Syst. Bot. & Pflanzengeogr. Univ. Thessaloniki 7: 1–322, Taf. 1–38 (Griech. mit dtsch. Zsf. und Diskussion).

Anagnostidis, K., 1967. Thermale und marine Spirulina-Vegetation in Griechenland: Ein ökologischer Vergleich. Verh. Intern. Verein. theor. angew. Limnol. 16: 1565–1567. Warszawa.

Anagnostidis, K., 1968. Untersuchungen über die Salz- und Süsswasser – Thiobiocönosen (Sulphuretum) Griechenlands. Wissensch. Jahrbuch Physiko-Math. Fak. Univ. Thessaloniki, 10: 406–860, Fig. 1–109+80 Tab. & 1 Karte (Griech.)

Anagnostidis, K. & Golubić, S., 1966. Über die Ökologie einiger Spirulina-Arten. Nova Hedwigia 11: 309–335.

Anagnostidis, K. & Schwabe, G.H., 1966. Über artenreiche Bestände von Cyanophyten und Bakteriophyten in einem Farbstreifensandwatt sowie über das Auftreten von Gomontiella-artig deformierter Oscillatoriatrichoma. Nova Hedwigia 11: 417–441.

Anagnostidis, K. & Rathsack-Künzenbach, R., 1967. Isocystis pallida-Blaualge oder hefeartiger Pilz? In: I.A.C.: 4 Symposium über Fragen der Cyanophytensystematik in Kastanienbaum, 1966. Schweiz. Z. Hydrol. 29: 191–198.

Anagnostidis, K., Economou-Amilli, A. & Tsangridis, A., 1981. Taxonomic and Floristic Studies on Algae from Rice-Fields of Kalochorion-Thessaloniki, Greece, Nova Hedwigia 34: 1–66.

Antarikanonda, P. & Lorenzen, H., 1982. N_2-fixing blue-green algae (Cyanobacteria) of high efficiency from paddy soils of Bangkok, Thailand: Characterization of species and N_2-fixing capacity in the laboratory. Arch. Hydrobiol. Suppl. 63, 1 (Algological Studies 30): 53–70.

Arora, S.K., 1969. Blue-green algae: a source of nitrogen and organic matter in paddy fields. Riso 18: 63–66.

Bunt, J.S., 1961. Nitrogen-fixing blue-green algae in Australian rice soils. Nature 192: 479–480.

Burris, R.H., 1975. The acetylene-reduction technique. In: W.D.P. Stewart, (ed.), Nitrogen fixation by free-living micro-organisms, I.B.P. 6: 249–257.

Commoner, B., 1977. Cost-risk-benefit analysis of nitrogen fertilization: A case history. Ambio 2: 157–173.

Copeland, E.B., 1924. Rice. London, Macmillan (cited in Grist, 1975).

DaSilva, E.J., Henriksson, L.E. & Henriksson, E., 1975. Effect of pesticides on blue-green algae and nitrogen-fixation. Arch. Environ. Contam. Toxicol. 3: 193–204.

De, P.K., 1939. The role of blue-green algae in nitrogen fixation in rice fields. Proc. Roy. Soc. London 127B: 121.

Economou-Amilli, A., 1976. Ecological remarks on diatoms

from some greek thermal springs. Proc. Thermal Waters, Int. Congr. Thermal waters, Geothermal energy and Vulcanism of the Mediterranean area 2: 113–140.
Ehrlich, P.R., Ehrlich, A.H. & Holdren, J.P., 1973. Human Ecology. Problems and Solutions. W.H. Freeman and Company, San Francisco. 304 pp.
El-Nawawy, A.S., Ibrahim, A.N., Aboul-Fadl, M., 1968. Nitrogen fixation by Calothrix sp. as influenced by certain sodium salts and nitrogenous compounds. Acta Agr. Acad. Sci. Hung. 17: 323–327.
Englund, B., 1977. Studies on N_2 fixation by free-living and symbiotic blue-green algae using the acetylene reduction technique. Acta Univ. Upsaliensis, Abstr. Upsala Diss. Fac. Sci. 45 pp.
Fogg, G.E., Stewart, W.D.P., Fay, P. & Walsby, A.E., 1973. The blue-green algae. Academic Press, London, 459 pp.
Goryunova, S.V. & Orleanskii, V.K., 1967. Modern methods for using nitrogen-fixing blue-green algae for increasing the fertility of rice-fields. In: Akad. Nauk SSSR. Inst. Mikrobiol. Biologicheskii azot i ego rol'v zemledelii: 309–315 (in Russian).
Goryunova, S.V. & Orleanskii, V.K., 1968. Biology of some dominant forms of blue-green algae in rice-fields of Kuban. Mosk. Univ. Biol. Poch. Fakul. Mikroorgan. Sel Khoz. 2: 194–202 (in Russian).
Goyal, S. & Venkataraman, G., 1971. Effect of algalization on high yielding rice varieties I. Response of rice varieties. Phykos 9: 137–138.
Grist, P.H., 1975. Rice. Tropical Agriculture series: 1–601. 5th edition Longman, London.
Gupta, A.B. & Lata, K., 1964. Effects of algal growth hormones on the germination of paddy seeds. Hydrobiologia 23: 430.
Henriksson, E., 1971. Algal nitrogen fixation in temperate regions. Plant Soil, Spec. Vol. 1971: 415–419.
Henriksson, E., 1978. Physiological aspects on blue-green algae in association with other plants in soil. Prof. R.N. Singh's Commemoration Volume (S.P. Singh, ed.). Benares. Advances in Cyanophyta Research: 7–13.
Henriksson, L.E., Enckell, P.H. & Henriksson, E., 1972. Determination of the nitrogen-fixing capacity of algae in soil. Oikos 23: 420–423. Copenhagen.
Henriksson, L.E. & DaSilva, E.J., 1978. Effects of some inorganic elements on nitrogen-fixation in blue-green algae and some ecological aspects of pollution. Z Allgem. Mikr. 18: 487–494.
Hogan, J.T., 1970. Footnote facts on rice. Rice J. 73(6): 5 (cited in Grist, 1975).
Huang You-xing, Fang Guang-ru, Yan Yu-zhou, Wang Ting & Su Guo-feng, 1981. Studies on the preparation of dried Inoculum of blue-green algae and its application in paddy field. Proc. Symp. on paddy soil: 356–362. Science Press, Beijing and Springer-Verlag, Berlin, Heidelberg.
Kiyohara, T., Fujita, Y., Hattori, A. & Watanabe, A., 1960. Heterotrophic culture of a blue-green alga, Tolypothrix tenuis I. J. Gen. Appl. Microbiol 6: 176–182.
Kulasooriya, S.A., Roger, P.A. & Watanabe, I., 1980. Relationship between blue-green alga growth and the standing crop in wetland rice fields. Int. Rice Res. Newsl. 5: 18–19.
Ley Shang-Hao, Veh Tsing-Chuan, Liu Fujui, Wang Lihmei & T'sui Shi-King, 1959. The effect of nitrogen fixing blue-green algae on the yields of rice plant. Acta Hydrobiol. Sinica 4 (Cited in Shtina, 1965).
Metting, B., 1981. The systematics and ecology of soil algae. Bot. Rev. 47: 195–312.
Mudholkar, N.J. & Sahay, M.N., 1968. Effect of blue-green algae on rice yield under different soil types. Ind. J. Agron. 13: 115–118.
Relwani, L.L. & Subrahmanyan, R., 1963. Role of blue-green algae, chemical nutrients and partial soil sterilization on paddy yield. Current Sci. 32: 441–443.
Rodgers, G.A., Bergman, B., Henriksson, E. & Udris, M., 1979. Utilization of blue-green algae as biofertilisers. Plant Soil 52: 99–107.
Sankaram, A., Mudholkar, N.J. & Sahay, M.N., 1966. Algae as aids in rice production. Indian Farming 16: 37–38.
Sankaran, S., 1975. Blue-green algae role in rice culture. Farmer and Parliament 10: 15–16.
Selvaraj, P. & Robinson, J.G., 1971. Blue-green algae in rice fields. Farm and Factory 5: 17–18.
Shtina, E.A., 1965. Fixation of free nitrogen in blue-green algae. In: Ecology and Physiology of blue-green algae. Akad. Nauk, SSSR: 160–177.
Singh, R.N., 1961. Role of blue-green algae in nitrogen economy of Indian Agriculture. Ind. Coun. Agric. Res. New Delhi 1–175.
Starks, T.L., Shubert, L.E. & Trainor, F.R., 1981. Ecology of soil algae: a review. Phycologia 20: 65–80.
Stewart, W.D.P., Fitzerald, G.P., & Burris, R.H., 1967. In situ studies on N_2 fixation using the acetylene reduction technique. Proc. Natl. Acad. Sci. 58: 2071–2078.
Stewart, W.D.P., Haystead, A. & Dharmawardene, M.W.N., 1975. Nitrogen assimilation and metabolism in blue-green algae. In: W.D.P. Stewart (ed.), Nitrogen fixation by free-fixing micro-organisms I.B.P. 6. 471 pp.
Subrahmanyan, R., Relwani, L.L. & Manna, G.B., 1964. Observations on the role of blue-green algae on rice yield compared with that of conventional fertilizers. Current Sci 33: 485–486.
Subrahmanyan, R., Relwani, L.L. & Manna, G.B., 1965. Fertility build-up of rice field soil by blue-green algae. Proc. Ind. Acad. Sci. Sect. B, 62: 252–272.
Tsangridis, A., 1982. On cyanophyta and chlorophyta from rice-fields of Kalochorion-Thessaloniki. Taxonomic, floristic, phytogeographical, ecological research. Doctoral thesis, Univ. Athens, Greece. 221 pp. (in Greek with English summary and discussion).
Venkataraman, G.S., 1972. Algal biofertilizers and rice cultivation. Ind. Agric. Res. Inst., New Delhi, Today & Tomorrow's printers & publishers. 75 pp.
Venkataraman, G.S., 1977. Blue-green algae (A biofertilizer

for rice). New Delhi, Ind. Agric. Res. Inst. Division of Microbiology: 1–8.

Venkataraman, G.S. & Neelakantan, S., 1967. Effects of the cellular constituents of the nitrogen-fixing blue-green alga Cylindrospermum muscicola on the root growth of rice plants. J. Gem. Appl. Microbiol. 13: 53–62.

Watanabe, A., 1951. Effect on nitrogen-fixing blue-green algae on the growth of rice plants. Nature 168: 148.

Watanabe, A., 1962. Effect of nitrogen-fixing blue-green alga: Tolypothrix tenuis on the nitrogenous fertility of paddy soil and on the crop yield of rice plant. J. Gen. Appl. Microbiol. 8: 85–91.

Watanabe, A., 1966. The blue-green algae as the nitrogen fixators. Proc. 9th Inter. Congr. Microb.: 77–85.

Watanabe, A., Nishigaki, S. & Konishi, C., 1951. Effect of nitrogen-fixing blue-green algae on the growth of rice plants. Nature 168: 748–749.

Watanabe, A., Hattori, A., Fujita, Y. & Kiyohara, T., 1959. Large scale culture of a blue-green alga, Tolypothrix tenuis, utilizing hot spring and natural gas as heat and carbon dioxide sources. J. Gen. Appl. Microbiol. 5: 51–57.

Yoshida, T. & Ancajas, R.R., 1970. Application of the acetylene reduction method in nitrogen fixation studies. Soil Sci. Plant Nutr. 16: 234–237.

CHAPTER 15

Endangered species, genetic capital and cost-reducing R&D

GARDNER M. BROWN, Jr. and JOSEPH SWIERZBINSKI

1. Introduction

Excessive transactions costs, poor property rights and other institutional and technical impediments make it difficult, on the one hand, to capture the value provided to those who view endangered species and, on the other hand, to prevent species from being destroyed by harvesters, including poachers. In contrast, the value of the alternative domestic use of the habitat often can be readily captured by those who harvest the forests, plant the land with crops or graze domestic animals. Krutilla, Myers and others have capably explained how the substitution of activities with capturable payoffs for those with real but elusive benefits, leads to a too rapid rate of species extinction from a social point of view. An alternative and perhaps more fundamental explanation for the too frequent demise of species is discussed in this paper. We take up the theme, earlier argued by Ciriacy-Wantrup (1952) that species are an inherent source of scientific information as a stock of potential discoveries.

Flora and fauna directly or indirectly are valuable because they provide the essential ingredients for new and improved drugs.[1] They are the source of new and improved varieties of seedlings and seeds.[2] And from them are obtained materials to produce better manufactured goods.[3, 4] For thousands of years mankind has actively selected superior seeds although modern techniques have improved this process.[5] For example, wild varieties of wheat, corn, rice and other major agricultural

[1] Approximately one half of the world's medicines and nearly one half of the prescriptions written in the U.S. contain compounds from a natural origin (Schery, 1972). Illustratively, alantoin from blowfly larvae encourages the healing of deep wounds; canthardin from the European blister beetle is used for certain malfunctions of the uro-genital system. Digitalis produced from foxglove plants is a significant drug for lowering blood pressure. From the Rosy Periwinkle comes vincristine and vinblastine used so effectively to treat leukemia in children, Hodgkins's disease (80% remission) and acute lymphocytic leukemia (99% remission). See Myers (1978) and Oldfield (forthcoming).

[2] About 40–60% of the increased yield in corn in recent decades is attributed to improved varieties. Recently an endangered, very primitive relative of corn was discovered in southwestern Mexico. Since it is a perennial, successful crossing could save the costs of planting. A wild peruvian tomato provided *Fusarium* Wilt resistance to more than 100 tomato varieties in use today. See Oldfield (forthcoming).

[3] Illustratively, plants are a major source of dyes, tannins, fats, waxes and lubricants in industrial processes; alginate compounds obtained from seaweeds are used in shampoos, fire-extinguishing foam and other products. Shellac comes from lac secreted by the insect *Laccifer lacca*. See Oldfield, Myers (1978) and Prescott-Allen (1980).

[4] Additionally, animals whose systems are close substitutes to ours are used experimentally. Armadillos are attractive scientifically because they are the only animal other than man susceptible to leprosy. Chimpanzies are the only animals for which the safety of anti-hepatitis vaccines can be tested (Myers, 1978). Research on the hibernation of black bears has led to an improved diet for patients with kidney failure (Prescott-Allen, 1980).

[5] Lima bean seeds found in Peruvian tombs 2600 years old were ten to one-hundred times larger than wild beans growing in the same area (Allard, 1960). Successive strata of remains in Bat Cave, New Mexico indicated increases in the average size of ears of corn and kernels dating to as early as 2500 BC (Briggs & Knowles, 1967).

D.O. Hall, N. Myers and N.S. Margaris (eds.), Economics of ecosystems management.

 ISBN-13: 978-94-010-8928-9

crops are crossed with existing varieties to produce changed optimal planting and harvest times, higher yields, greater resistance to pestilence and drought and other beneficial characteristics (phenotypes). Indirectly, as the structures of natural plants, animals, insects and other species are discovered, they provide templates to be synthetically mimicked by chemicals, especially drugs, and other products.[6]

Why would entrepreneurs throughout the world permit flora and fauna to become extinct if they are potentially valuable? In many cases, the germ plasm can be and is collected and stored. When storage technology is too uncertain, habitats containing the valuable species can be purchased. One obvious reason why private entrepreneurs exert little effort to preserve some germ plasm is imperfect patent laws at both the national and international level. As a result, too little information will be produced. Nor is this situation easily correctable. Much of the potential knowledge embodied in endangered species is similar in character to that produced by basic research. The technical difficulties involved in establishing property rights to most basic research are well known. In addition, as we discuss in more detail below, even perfect patent laws would not solve the problem.

Two features of genetic capital deserve emphasis. The preservation of additional species or varieties enriches the distribution of varieties available for research. Typically, the discovery of useful information requires a study of many varieties, since it is unlikely that a particular variety can be identified a priori as valuable. Hence the knowledge value of preserving additional species might best be considered to reside in the enriched distribution as a whole rather than in the particular species or varieties themselves.

Second, note that the distribution of varieties provided by an inventory of flora or fauna represents a public good of the undepletable kind. Its exploitation by one researcher does not deny its exploitation by others (assuming that one researcher cannot capture all the members of a particular species). By contrast, a common property resource such as an unmanaged fishery, forest or common represents a depletable externality. Harvest by one precludes the harvest of the fish, tree or grass by others.

Baumol and Oates, among others, have remarked on the importance of the distinction between the two types of externality. A social optimum (with zero transaction cost) is insured if those exploiting a resource with depletable externalities are charged a price equal to the marginal social cost. But the optimal price to the user (each experimenter) of any undepletable common property resource including a stock of knowledge should be zero since there are no marginal social costs arising from individual use.[7] How then could the holder of a private stock of genetic capital market his asset to other researchers? (Baumol & Oates, 1975).

In the analysis below, it is assumed that species preservation enhances the chance of discovering more productive products. Equivalently, we assume that there are industries where these discoveries shift downward the cost of providing the industries' products or services. We model the behavior of such an industry with regard to the stockpiling of genetic capital. In particular, we assume that each firm in the industry has access to the entire stock of varieties and the least cost mode of production.

In recent years the effects of market structure on expenditures for cost-reducing R&D have been an important area of research. It is noteworthy that the formal models in the literature have focused virtually entirely on the case of perfect patent laws and the race to innovate first.[8] For example, essentially all the work treated in a recent survey by Kamien and Schwartz studies market structure in a world of private information (Kamien & Schwartz,

[6] Fireflies utilize chemoluminescence to produce light. A light developed by American Cyanamid Company using this substance does not give off sparks or heat and is therefore valuable in potentially dangerous circumstances.

[7] The distinction between the production of information and use of information is drawn out by Hirshleifer & Riley (1979). Having discovered the new idea, the optimal price is the marginal unit cost of producing the new product in which is embedded the new idea. Since the new idea is also a public good of the undepeletable variety, the marginal cost of its use is zero and should not enter the social account for determining the new product price (see equation (5)). This is just another application of Arrow's (1962) reasoning.

1982). Those models describe different aspects of the race to turn information into useful products and then stave off imitators for as long as possible.

By contrast, the strong assumption we are making – no private property rights in discovery – represents the opposite polar case. Given the characteristics and importance of basic research, this opposite polar case seems worth examining. In essence, the following model and discussion represents the formal acknowledgment of the ideas presented by Nelson and especially by Arrow, set in the context of an industry model of applied R&D (Nelson, 1959, 1980; Arrow, 1962).

In the next three sections we develop a formal model to compare the stock of genetic capital held by a monopolist or, more generally, an N firm industry, with the socially optimal stock. We prove that for every N, the industry holds too little genetic capital relative to the optimum. There are three separate economic effects which cause this result.

First, industry output is lower because consumers' surplus is not completely taken into account by each and all members in the industry. That is, the firms' demand function is not society's (except in the limit). Second, the public goods aspect of genetic capital is reflected in the fact that each of the N firms only takes into account $1/N$ of the cost reduction purchased by increasing the stock of genetic capital. Third, whenever the number of firms is not very large, a firm's ability to profit from cost reduction is eroded by similar responses from like firms.

In their review of the research and development literature, Kamien and Schwartz observe that an intermediate degree of market concentration, N small but not 1, provides the greatest incentive for R&D (Kamien & Schwartz, 1982). Our model focuses on a quite different set of economic forces from those of the models they discuss. Nevertheless, the magnitudes of the three different effects change in different ways and directions when N changes. Hence $N = 1$ is not always the market structure which conserves the most genetic capital. Nevertheless, our model does demonstrate strong forces discouraging investment in R&D as N increases, a point developed more fully in Section 5. Hence for many cases, although not all, we expect that from the point of view of conserving genetic capital, monopoly is the favored market structure when a social optimum is not possible.

2. A model of cost-reducing R&D

We begin our formal model by considering a generalized good or service. Recognizing the importance of research in the medical industry, the good is called 'cures'. The amount or number of cures provided by firm i is X_i and the total amount provided by the 'industry' is X. If $g(X)$ is the price of a single cure, then social welfare from a given level of X is described by the usual benefit index:

$$G(X) = \int_0^{C_0} g(X')\,\mathrm{d}X'.$$

Consider an example which we will return to from time to time. Cures are produced by a combination of concentration (Y) and dosages (D) in the convenient form:

$$X = f(Y)\,D.$$

This signifies that the average and marginal productivity of dosages are the same but vary with

[8] Larry Ruff's paper (1969) is one of the few in which research has public good characteristics. Research results from competitors are obtained 'through' a transmission coefficient which is parametric but can vary from perfect secrecy to completely public.

Reinganum has developed a model of the race to innovate in which the winner may be poorly rewarded due to incomplete patent protection. Her model includes as a limiting case a model constructed by Kamien and Schwartz of the race to innovate in a competitive economy with no patent protection. However both Reinganum's and Kamien and Schwartz's description of technology is very different from ours. In particular, they concentrate on the race to develop a single innovation with known consequences rather than the accumulation of a stock of knowledge as we do. Thus, for example, the only uncertainty in their models is the timing of the innovation. While our models all agree in predicting little or no R&D expenditures in the limiting case of infinitely many firms and purely public knowledge, the implications of the models for a finite number of firms, especially a small number, can be very different (Reinganum, 1982; Kamien & Schwartz, 1972).

concentration.[9] If all the dosages have the same unit cost, C_0, then the total cost of cures is

$$\frac{C_0 X}{f(Y)} = C_0 D.$$

The total cost of cures, $C(X,Y)$, is written more specifically as

$$C(X, Y) = X C_0 \, \mathrm{e}^{-\alpha Y}, \tag{1}$$

after choosing a convenient functional form for $f(Y)$.

We envision research being undertaken to produce higher concentrations of the drug, and thus cheaper cures. Alternatively, if the good X is agricultural, the goal can be a higher yield per acre which combines with acres (D) to produce a harvest (X). Research is conceived to be a sampling process. Technologies are indexed by the available concentrations they produce and the distribution of technologies depends on the stock of genetic capital, M.

To illustrate the general formulation, the available concentration (Y) is assumed to be distributed according to the exponential distribution. Let M be the stock of capital, θ be a scale parameter, $f(Y,M)$ be the probability density function of Y, and $F(Y,M)$ be the comulative distribution function of Y. The equations for the exponential case are:

$$f(Y, M) = \frac{\theta}{M} \mathrm{e}^{-\frac{\theta}{M} Y} \tag{2a}$$

$$F(Y, M) = 1 - \mathrm{e}^{-\frac{\theta}{M} Y} \tag{2b}$$

$$E(Y|M) = \frac{M}{\theta} \tag{2c}$$

$$\mathrm{Var}(Y|M) = \left(\frac{M}{\theta}\right)^2. \tag{2d}$$

More generally, let the set of available technologies for producing X be indexed by a parameter Z. Z is a scalar random variable whose distribution depends on M. The total cost function is restricted to have the following form:

$$C(X, Z) = C(X) \cdot C_1(Z). \tag{3a}$$

Without loss of generality, the random variable can be transformed and (3a) can be written in the form:

$$C(X, Z) = C(X) \cdot Z \tag{3b}$$

with $C'(X)>0$, $C''(X)\geq 0$. This multiplicative form of cost uncertainty is theoretically useful, econometrically convenient and widely used (Newberry & Stiglitz, 1981). In our example, (3b) is obtained from (1) by making the identifications $C(X) = X$ and $C_0 \, \mathrm{e}^{-\alpha Y} = Z$. The cumulative distribution function for Z is easily described in this case.[10]

We assume that Z takes on values in the range 0 to C_0; therefore, costs are always positive and bounded. We will assume for technical convenience that C_0 is finite, though of course it may be very large.

We make three important simplifying assumptions concerning the R&D process and its effects on cost reduction. First, regarding the mechanics of sampling, assume that the actual costs of production are determined by a single, costless sample from the distribution of Z. This is not a major limitation. Sampling plans of a fixed size S can be immediately subsumed into this formalism through the identification

$$Z = \mathrm{Min}[Z_1, Z_2, \ldots, Z_S]. \tag{4}$$

The cost of sampling also is not important for the case of fixed numbers of observations, S. If S as well as M is regarded as a decision variable, then

[9] Since we are describing a market, these assumptions make sense if all those cured require about the same dosage level and respond in an equitable way to variations in concentration.

[10] When $Z = C_0 \, \mathrm{e}^{-\alpha_Y}$ and Y is distributed according to equations (2a, b, c and d), the random variable Z is distributed according to the equations derived from the change of variables integration formula (Mood et al., 1974)

$$f(Z, M) = \frac{\theta}{\alpha M}\left(\frac{1}{C_0}\right)^{\frac{\theta}{\alpha M}} Z^{\left(\frac{\theta}{\alpha M} - 1\right)}, \quad 0 \leq Z \leq C_0$$
$$= 0 \text{ otherwise} \tag{i}$$

$$F(Z, M) = 0 \quad Z<0$$
$$= \left(\frac{Z}{C_0}\right)^{\frac{\theta}{\alpha M}} \quad 0 \leq Z \leq C_0 \tag{ii}$$
$$= 1 \quad Z > C_0$$

$$E(Z|M) = \overline{Z}(M) = \frac{C_0 \theta}{\theta + \alpha M} \tag{iii}$$

the costs of sampling become relevant quantitatively but not qualitatively.[11]

More fundamental are the other two assumptions: the probability density of Z, $f(Z,M)$, is determined by the total stock of genetic capital regardless of how the ownership of this stock is split among firms; and the least cost technology $C(X)Z$ is available to all firms. The above two assumptions make the stock of genetic capital and the information it generates public goods. This formally reflects the point of the earlier discussion: the conservation of endangered species and genetic capital is a contribution mainly to the earlier stages of the research process. The public goods character of basic research long has been recognized, at least implicitly, in discussions of government funding for research. To be sure, genetic capital in particular and cost-reducing research in general is a peculiar kind of public good, because a level of capital M provides a distribution of possible technologies. Alternatively, the two assumptions made above can be interpreted as describing a world where trade secrets are hard to keep and innovations diffuse quickly. As remarked earlier, such a world is at least an important polar case and one which may well be an appropriate description of some industries.

The use of a distribution of available technologies as a model of the R&D process has been previously discussed in the literature. Evensen and Kislev suggested that development could be modelled as repeated sampling from the same distribution while basic research could be modelled as a change in the parameters of the distribution. Some empirical evidence from agricultural research suggested the relevance of this approach. The model below incorporates Evensen and Kislev's description of the technology of R&D into an industry behavioral model of investment in R&D (Evensen & Kislev, 1976).

More recently, Nelson suggested the usefulness and relevance of the distribution of potential technologies approach; discussed the public goods problem; and identified some of the economic effects which occur in our model. Indeed, one of his many suggestions for future work is the incorporation of these effects in an oligopoly model. In this regard our model represents a study of one corner of the sweeping agenda Nelson lays out for the profession (Nelson, 1980, 1981).

Finally, Telser has developed an industry model of the R&D process which incorporates a version of Evensen and Kislev's distribution assumption. However Telser's approach differs in a number of important ways from Evensen and Kislev's model and from ours. Perhaps most importantly, he does not analyze the accumulation of a stock of knowledge as we do. Rather, Telser's distributions describe the outcome of a particular research project. Each separate project has its own distribution of potential results which in particular does not depend on the results of previous projects but only on current R&D expenditures. Learning occurs only through the improvement of the best available current technology. Telser studies the case of private knowledge, perfect patents and the race to innovate (Telser, 1982).

Although our model is similar to a static model, an important sequential element remains. We assume that by the time the production decision regarding X is made, the available technology Z is determined. Each technology determines an optimal level of production X (Z). However, at the time M is chosen, the available technology Z is unknown. Hence the choice of an optimal M involves as well, the choice of an optimal production *strategy*, $X(Z)$. In contrast, a typical model of cost-reducing capital K allows for a simultaneous choice of X and K. Such a purely static model of R&D removes one uncertainty which is a crucial part of

[11] In particular, the existence of two inputs, S and M, provides an additional source of efficiency distortion when comparing the monopolist's outcome with the social optimum. It is well known that the monopolist's output distortion generally will also change the input combination from the optimal mix. However, unless the R&D process is described quite precisely, little is to be gained by introducing this additional complication.

The tradeoff between S and M indeed would be interesting in a dynamic world characterized by sequential output and investment decisions, particularly in the N firm case. Then sampling would be an element of strategic rivalry and it would be necessary to treat explicitly the economies or diseconomies of scale resulting from dividing sampling activity among N firms. While important, neither of these elements seems crucial in the optimal or suboptimal allocation of M. Thus in this first effort a single costless sample is assumed.

the R&D process. Moreover, the static formulation can lead to substantial error in estimating the benefits of a given stock of capital as will become apparent.

3. Socially optimal cost-reducing R&D

We turn now to the maximization problem from which the socially optimal stock of capital, M_{s0}, is derived. Since at the time M_{s0} is chosen, Z is unknown, the optimal choice of M_{s0} depends on expectations about the socially optimal production strategy $X_{s0}(Z)$. Focusing on X_{s0} for the moment, for each given Z, denote the net benefits associated with a production plan $X_{s0}(Z)$ by $B(X_{s0}(Z),Z)$. Then,

$$B(X_{so}(Z), Z) = G(X_{so}(Z), Z)) - C(X_{so}(Z))Z. \tag{5}$$

Thus for each Z, $X_{s0}(Z)$ maximizes expected benefits if it satisfies the equation:

$$g(X_{so}(Z)) = C'(X_{so})Z))Z, \tag{6}$$

that is, the familiar condition that price = marginal cost. As long as $C'' \geq 0$, g/C' can be inverted to find $X_{s0}(Z)$.

Now let $B(M)$ be defined by the equation

$$B(M) = \int_0^{C_0} f(Z,)\, B(X_{so}(Z), Z)\, dZ. \tag{7}$$

$B(M)$ is the expected net benefits associated with a given level of M.

The socially optimum level of genetic capital, M_{s0}, is given by the solution to the problem

$$\underset{M}{Max}\; B(M) = \varphi(M). \tag{8}$$

$\varphi(M)$ is the cost of purchasing the level of capital, M, and with it the distribution $f(Z,M)$. Assume that

$$\varphi(M),\ \varphi'(M),\ \varphi''(M) \geq 0 \text{ for all } M.$$

To derive the socially optimal stock of genetic capital, M_{s0}, rewrite the first part of the objective function, $B(M)$, in a more transparent form. Integrating (7) by parts yields:

$$B(M) = B(X_{so}(C_0),(C_0) + \int_0^{C_0} F(Z\ M) \left(-\frac{dB}{dZ}\right) dZ \tag{9}$$

Equation (9) has an intuitively pleasing interpretation. $B(X_{s0}(C_0),\ C_0)$ represents the benefit floor, the net benefits when the technology Z is the worst possible. The term, $\frac{dB}{dZ}$, represents the increment in benefits when the observed technol ogy goes from Z to $Z - dZ$ (the minus sign comes from the fact that decreasing Z's represent lower costs). $F(Z,M)$, the cumulative distribution function, represents the fraction of cases that the increase in benefits $\left(-\frac{dB}{dZ}\right)$ is actually realized. Hence (9) says that the expected benefits from a stock of capital M is the benefits in the worst case plus all the increments in net benefits, each increment weighted by the fraction of the time it can be expected to occur.

Notice from (5) that $B(\cdot)$ is a function of both X and Z. Therefore,

$$\frac{dB}{dZ} = \frac{\partial B}{\partial X} \cdot \frac{\partial X}{\partial Z} + \frac{\partial B}{\partial Z} = \frac{\partial B}{\partial Z} = C(X_0(Z)). \tag{10}$$

This follows from an application of the reasoning behind the envelope theorem once it is recalled that $X_{s0}(Z)$ is chosen to maximize $B(X,Z)$ for each Z and at that X_{s0}, $\frac{\partial B}{\partial X} = 0$ (see (6)).

Using (10), (9) can be written in the form

$$B(M) = B(X_{so}(C_0), C_0) + \int_0^{C_0} F(Z, M)\, C(X_{so}(Z))\, dZ. \tag{11}$$

Differentiating $B(M) - \varphi(M)$ and setting the derivative to zero provides the necessary condition for the socially optimal stock of genetic capital,

$$B'(M) = \int_0^{C_0} \frac{\partial F}{\partial M}(Z, M)\, C(X_{so}(Z))dZ = \varphi'(M). \tag{12}$$

Equation (12) expresses the necessary condition in a convenient way. $\frac{\partial F}{\partial M}$ and $\varphi'(M)$ summarize the relevant information about the R&D process and $C(X)$ and $X_{S0}(Z)$ summarize the relevant information about the demand and supply of X.

In the special case when $C(X) = X$, the total production cost function has the special form $C(X,Z) = XZ$. Z can then be interpreted as the constant, unit cost of production. Since this case is analyzed in some detail later, we note first that the necessary condition for an optimum M_{S0} is

$$B'(M) = \int_0^{C_0} \frac{\partial F}{\partial M}(Z, M) X_{S0}(Z)\,dZ = \varphi'(M) \tag{13}$$

and second that (6) has the solution $X_{S0}(Z) = g^{-1}(Z)$.

Before turning to the market solution for an arbitrary number of firms, we note that Appendix A compares the necessary condition for the general case (12) with the necessary condition arising from a purely static optimal capital model.

The sequential aspect introduces a strategic element, lost in the static formulation. This firm–industry interaction term alters the conclusions reached in Section 5 concerning the market structures which produce the most R&D.

4. Cost-reducing R&D in a market context

The second model, one of R&D investment in an N-firm industry, is intended to approach a model of a perfectly competitive industry for N large. For $N = 1$ it reduces to a monopoly model. In general, the industry is modelled as an oligopoly formed by N firms. The firms have identical cost functions and each forms its expectations regarding rivals' actions in the same way. They behave in a non-cooperative fashion. The amount of the product produced by firm i, X_i, and the amount of genetic capital purchased by firm i, M_i, are considered to be the basic decision variables of each firm. Hence we will be considering equilibria in quantities and not prices.

Given the sequential nature of the model, each firm i must choose its stock of genetic capital, M_i, based on expectations. These expectations include expectations about $\tilde{M}$, the stock of capital bought by the other N-1 firms, expectations about $\tilde{X}(Z)$, the total production of the other firms for each possible realized technology, and also expectations about $X_i(Z,\tilde{X}(Z))$, the firm's own production if the realized technology is Z and the other firms are expected to produce $\tilde{X}(Z)$.

Define $\prod(X_i(Z),\tilde{X}(Z),Z)$ to be the profits obtained by firm i if the realized technology is Z, the production of the other N-1 firms is $\tilde{X}(Z)$ and the production of firm i is $X_i(Z)$. $\prod(X_i,\tilde{X},Z)$ is the industry analogue of $B(X_{S0})Z),Z)$. Then

$$\Pi(X_i(Z), \tilde{X}(Z), Z) = X_i(Z)g(X_i(Z) + \tilde{X}(Z)) - C(X_i(Z))Z. \tag{14}$$

Each firm is assumed to form its expectations concerning $X_i(Z)$ and $\tilde{X}(Z)$ by assuming that each realized technology Z results in an N-firm symmetric Nash equilibrium in quantities (the Cournot solution). Since the least cost Z is known to all and the firms are symmetrically placed regarding costs, this outcome is possible and is perhaps the most plausible outcome a priori.

To find $X_i(Z)$ we assume each firm takes the production of the other firms, $\tilde{X}(Z)$ and the technology Z as given and maximizes profits. Hence $X_i(Z)$ satisfies the equation

$$\frac{\partial \Pi}{\partial X_i}(X_i, \tilde{X}, Z) = gX_i + \tilde{X}) + X_i g'(X_i + \tilde{X}) - C'(X_i)Z = 0. \tag{15}$$

The symmetry assumption specifies $X_i(Z)$ and $\tilde{X}(Z)$ in terms of industry output $X_N(Z)$:

$$X_i(Z) = \frac{1}{N} X_N(Z) \text{ and } \tilde{X}(Z) = \frac{N-1}{N} X_N(Z). \tag{16}$$

Equation (15) then can be expressed in terms of industry output as:

$$g(X_N(Z)) + \frac{X_N(Z)}{N} g'(X_N(Z)) - C'\left(\frac{X_N(Z)}{N}\right) Z = 0. \tag{17}$$

Let $\prod(M,N)$ be the expected profits to firm i given stock M, where there are N firms.

$$\Pi(M, N) = \int_0^{C_0} f(Z, M)\, \Pi(X_i(Z), \tilde{X}(Z), Z)\,dZ \tag{18}$$

Finally, it is assumed that the firm takes $\tilde{M}$, the capital stock of the other N-1 firms, as a given constant. In this case, the firm's choice of M_i solves the problem:

$$\underset{M_i}{Max}\ \Pi(M_i + \tilde{M}, N) - \int_M^{M+M_i} \varphi'(\tau)\, d\tau \qquad (19)$$

The term,

$$\int_M^{M+M_i} \varphi'(\tau)\, d\tau$$

is the cost to firm i of purchasing M_i given that the other firms purchase $\tilde{M}$. In other words, M is supplied according to the marginal cost curve $\theta'(\tau)$. This form of the cost function for M_i is chosen because it closely corresponds to the classical model of a competitive economy, in which each firm takes the price of each input, $\theta'(M)$, as given.

The equilibrium condition for M_N, the amount of genetic capital held by the N-firm industry, is defined by the necessary conditions for a solution to (19) coupled with the symmetry condition that $\tilde{M} = (N\text{-}1)M_i$. At each stage the firm's decision is assumed not to affect the other firms' behavior *at that stage*. However through its effect on Z, firm i's choice of capital certainly affects the behavior of the other firms at the later production stages and firm i takes this explicitly into account when choosing M_i, via the assumptions about $X_i(Z)$ and $\tilde{X}(Z)$.

With the objective function defined by (18), the equilibrium conditions for M are precisely those for a symmetric Nash equilibrium in the quantity variable M_i. Taken together, the equilibrium conditions for X_i and M_i represent a natural sequential extension of the Cournot solution from 1 period and 1 variable to 2 periods and 2 variables.[12]

[12] In fact, the equilibrium conditions generated by our model represent an example and application of a class of sequential generalizations of the Nash equilibrium called sequential equilibria by Kreps and Wilson and perfect equilibria by Selten (Kreps & Wilson, 1982).

Alternately, the equilibrium conditions for M and X_i (Z) taken together, can be regarded as specifying a non-cooperative Nash equilibrium in strategy space. However we follow Shubik's recommendation and favor the extensive form over the normal form of the game as a vehicle for analysis (Shubik, 1982).

The analysis of the equation for the amount of genetic capital held by the N-firm industry, M_N, proceeds in a similar although more complicated fashion, compared to that for the social optimum. Equation (18) is integrated by parts to obtain the equation for $\prod(M,N)$.

$$\Pi(M, N) = \Pi(X_i(C_0), \tilde{X}(C_0), C_0)$$
$$\int_0^{C_0} F(Z, M)\left(-\frac{d\Pi}{dZ}\right) dZ. \qquad (20)$$

$\prod(M,N)$ represents the expected profits to firm i from using strategy $X_i(Z)$ when the other N-1 firms jointly produce $\tilde{X}(Z)$ and when the stock of genetic capital is M. The expected profits to firm i are the sum of the profits in the worst case when $Z = C_0$, together with all the increments in profits $\left(-\frac{d\Pi}{dZ}\right)$ resulting from cost reductions from $ZC(X)$ to $(Z-dZ)C(X)$, each increment weighted by $F(Z,M)$, the fraction of outcomes in which the increment is realized.

The economic forces causing departures from optimality, will be captured by the differences between $\frac{d\Pi}{dZ}$ and $\frac{dB}{dZ}$ (f. (9) and (20)). Careful consideration of $\frac{d\Pi}{dZ}$ is thus warranted. From the definiton of $\prod(X_i,\tilde{X},Z)$, in (14)

$$\frac{d\Pi}{dZ} = \frac{\partial\Pi}{\partial X_i}\frac{\partial X_i}{\partial Z} + \frac{\partial\Pi}{\partial\tilde{X}}\frac{\partial\tilde{X}}{\partial Z} + \frac{\partial\Pi}{\partial Z}. \qquad (21)$$

From the necessary condition (15), $\frac{\partial\Pi}{\partial X_i} = 0$. Next $\frac{\partial\Pi}{\partial Z}$ can be written as $\frac{\partial\Pi}{\partial Z} = -C(X_i(Z))$.

This is exactly similar to the expression for $\frac{\partial B}{\partial Z}$. Since the N-firm and optimal outcome will be compared in terms of the socially optimal output $X_{S0}(Z)$ and the total industry output, $X_N(Z) = NX_i(Z)$, the last equation, written in terms of $X_N(Z)$, is

$$\frac{\partial\Pi}{\partial Z} = -C\left(\frac{X_N(Z)}{N}\right). \qquad (22)$$

We note parenthetically, that the diferences between $C(X_N(Z)/N)$ in (22) and $C(X_{S0}(Z))$ capture two of the three effects leading to too little conservation of genetic capital by the N-firm economy. The term $\frac{\partial \Pi}{\partial \tilde{X}}\frac{d\tilde{X}}{dZ}$ in (21) captures the third effect. These effects are discussed in greater detail later. The stock of capital in the N-firm industry is found by using (21) and (22) to rewrite (20) in the form:

$$\Pi(M,N) = \Pi(X_i(C_0), \tilde{X}(C_0), C_0) + \int_0^{C_0} F(Z,M)\left[C\left(\frac{X_N(Z)}{N}\right) - \frac{\partial \Pi}{\partial \tilde{X}}\frac{d\tilde{X}}{dZ}\right]dZ \quad (23)$$

The expression in brackets is a function of Z only and not M.

The necessary condition for the stock of genetic capital, M_N, held by the N-firm industry, is obtained by differentiating the objective function in (19) with respect to M_i to obtain the necessary condition for M_i in terms of $\tilde{M}$. Then the symmetry assumption $\tilde{M} = (N\text{-}1)M_i + NM_i$ of $\tilde{M} = \frac{(N-1)}{N} M_N$ is utilized to obtain

$$\frac{d\Pi}{dM}(M_N,N) = \int_0^{C_0} \frac{\partial F}{\partial M}(Z,M_N) \cdot \left[C\left(\frac{X_N(Z)}{N}\right) - \frac{\partial \Pi}{\partial \tilde{X}}\frac{\partial \tilde{X}}{\partial Z}\right]dZ = \varphi'(M_N). \quad (24)$$

Equation (24) is the N-firm industry analogue of (12). Integration by parts has allowed us to write (24) in a form so that M_N and M_{S0} readily can be compared by comparing $C(X_{S0}(Z))$ and $\left[C\left(\frac{X_N(Z)}{N}\right) - \frac{\partial \Pi}{\partial \tilde{Z}}\frac{\partial \tilde{X}}{\partial Z}\right]$. As a preliminary matter, assume $\frac{\partial F}{\partial M}>0$ which conveniently holds for our example.[13] The assumption $\frac{\partial F}{\partial M}>0$ may seem strong. In fact, it is routinely made in the literature, often without explicit recognition.[14]

Returning to the comparison, the economic interpretation of $-\frac{\partial \Pi}{\partial \tilde{X}}\frac{\tilde{X}}{Z}$ is straightforward. In (23), the firm contemplates each potential increase in profits from a reduction in cost from $ZC(X)$ to $(Z-dZ)C(X)$. These increases depend on the forecasted production both of the firm itself, $X_i(Z)$, and of the other firms $\tilde{X}(Z)$. The term, $-\frac{d\tilde{X}}{dZ}$, represents the forecasted increase in the production of other firms due to the cost reduction arising from a decrease in Z. $\frac{\partial \Pi}{\partial \tilde{X}}$ represents the change in profits of firm i due to the change in the production of the other N-1 firms. Hence $-\frac{\partial \Pi}{\partial \tilde{X}}\frac{d\tilde{X}}{dZ}$ represents that part of the change in firm i's profits caused by increased production of the other N-1 firms due to a cost reduction. As long as $g(X)$ is downward sloping, an expansion in the other firm's producion will shrink firm i's profit. Hence $\frac{\partial \Pi}{\partial \tilde{X}}<0$. But a reduction in production costs will normally cause an expansion in the other firms' output. Indeed it can be shown that if the total industry revenue function $Xg(X)$ is concave then this must be so. In subsequent discussion, it is assumed that $\frac{d\tilde{X}}{dZ}(Z)$ Then, $-\frac{\partial \Pi}{\partial \tilde{X}}\frac{d\tilde{X}}{dZ}$ is also negative, in accord with intuition.

From the equation for $\prod(X_i,\tilde{X},Z)$, (14), it is

[13] Footnote 10 provides an explicit formula for $F(Z,M)$ for our example cost function. We note that $\frac{\partial F}{\partial M}$ is always positive in the range 0 to C_0. Since $C(X_{S0}(Z))$ is always positive, this implies at once that $B'(M)$, the marginal benefits of an increase in genetic capital, M, is positive for all M. If $\left[C\left(\frac{X_N(Z)}{N}\right) - \frac{\partial \Pi}{\partial \tilde{X}}\frac{\partial \tilde{X}}{\partial Z}\right]$ is always positive, then the same conclusion holds for $\frac{d\Pi}{dM}$, the marginal profits to firm i from an increase in M. Thus the assumption that $\frac{\partial F}{\partial M}>0$ for all relevant Z is a convenient one.

[14] In much of the economic literature dealing with uncertain production, increased mean production is modelled in terms of a left or right shift of the distribution of the uncertain variables (Sandmo, 1971; Hey, 1979). So for example, in our case, an increase in genetic capital M would be modelled as a left shift of the distribution $F(Z,M)$, a left shift being required since increasing M on average decreases costs. We could have done the analysis in terms of left shifts but this is inconvenient and unnecessarily restrictive. For example, it would then be difficult to model M as a parameter in one of the usual families of distributions (the logistic and normal distributions are exceptions). It would be impossible to peg the minimum cost Z, say at zero, and still use the left shift format. Thus we make the less restrictive assumption $\frac{\partial F}{\partial M}>0$.

straightforward to calculate $-\frac{\partial \Pi}{\partial \tilde{X}}\frac{d\tilde{X}}{dZ}$. Differentiating with respect to $\tilde{X}$, rewriting X_i and $\tilde{X}$ in terms of X_N, and differentiating $X_N(Z)$ yields the equation[15]

$$-\frac{\partial \Pi}{\partial \tilde{X}}\tilde{X}' = -\frac{X_N(Z)}{N^2}(N-1)C' \cdot \left[1 - \frac{1}{N\varepsilon} + \frac{g}{g'}\frac{1}{N\varepsilon^2}\frac{d\varepsilon}{dX} - \frac{C''Z}{g'N}\right]^{-1} < 0. \qquad (25)$$

When there are constant unit costs, $C(X) = X$, $C'' = 0$ and the last term in the bracket is zero. Substituting (25) into (24) yields the necessary condition for M_N in the constant unit cost case. Equation (26) is the industry analogue to (13),

$$\Pi'(M,N) = \int_0^{C_0} \frac{\partial F}{\partial M}\frac{X_N(Z)}{N}\left[1 - \frac{N-1}{N}\left[1 - \frac{1}{N\varepsilon} + \frac{g}{g'}\frac{1}{N\varepsilon^2}\frac{d\varepsilon}{dX}\right]^{-1}\right]dZ = \varphi'(M). \qquad (26)$$

A comparison of (13) and (26) neatly summarizes the three effects at work in reducing M_N below M_{S0}. The marginal cost of purchasing distribution, $\varphi'(M)$, is the same in both cases. However, three factors tend to reduce $\Pi'(M,N)$ below $B'(M)$ for each M and every N (see Fig. 1). Hence it is guaranteed that $\Pi'(M,N)$ intersects $\theta'(M)$ before $B'(M)$ thus insuring $M_N < M_{S0}$ for $M > 0$. (Since $\varphi'(M)$ is an increasing function of M).

[15] From the equation for $\Pi(X_i, \tilde{X}, Z)$, (14), it is straightforward to calculate $-\frac{\partial \Pi}{\partial \tilde{X}}\frac{d\tilde{X}}{dZ}$. Differentiating with respect to $\tilde{X}$ and then rewriting X_i and $\tilde{X}$ in terms of X_N yields:

$$\frac{\partial \Pi}{\partial \tilde{X}} = \frac{X_N(Z)}{N}g'(X_N(Z)) < 0. \qquad (i)$$

Since $\tilde{X}(Z) = \frac{N-1}{N}X_N(Z), \tilde{X}'(Z) = \frac{N-1}{N}X_N'(Z)$, and $X_N(Z)$ is given by the solution to (17). Hence X_N can be computed implicitly by differentiating (17). Defining $\varepsilon(X)$ as the price elasticity of X, we obtain the result,

$$X_N = \left[g\left[1 - \frac{1}{N\varepsilon}\right] + \frac{g}{N\varepsilon^2}\frac{d\varepsilon}{dX} - \frac{C''Z}{N}\right]^{-1} \cdot C'\left(\frac{X_N}{N}\right), \qquad (ii)$$

$$\tilde{X}' = \frac{N-1}{N}X_N'. \qquad (iii)$$

Combining equations (i), (ii) and (iii) yields equation (25).

The first effect is the replacement of $X_{S0}(Z)$ by $X_N(Z)$. $X_N(Z)$ is less than $X_{S0}(Z)$ because the N-firms attach no value to consumers' surplus.[16]

The second effect is the replacement of $X_{S0}(Z)$ not by $X_N(Z)$ but by $X_N(Z)/N$. The division by N is a reflection by the public goods aspect of M; only $\frac{1}{N}$ of the profits due to the cost reduction purchased by increasing M are captured by any given firm.

Finally, there is the factor

$$\left[1 - \frac{N-1}{N}\left[1 - \frac{1}{N\varepsilon} + \frac{g}{g'}\frac{1}{N\varepsilon^2}\frac{d\varepsilon}{dX}\right]^{-1}\right].$$

This factor represents the reduction in any one firm's ability to profit from a cost reduction because of the response of the other firms.

By inspection we see that all three factors reduce $-\frac{d\Pi}{dZ}$ below $-\frac{dB}{dZ}$ for each Z. Since $\frac{\partial F}{\partial M} > 0$, this means that $\Pi'(M,N)$ must be reduced below $B'(M)$ for all M and N. Since φ' is the same in both cases, M_N must be below M_{S0}. Thus in the case of constant unit costs of production, it is clear that the industry stock of genetic capital is less than the socially optimal stock. This must be true in the case of monopoly ($N = 1$) and for all N.

The result that $M_N < M_{S0}$ for all N is easily extended to the more general case of increasing marginal production costs. This requires only a bit of thought about how to model the marginal social costs of production. When marginal costs are increasing, the optimal way to produce X is to split it evenly among the N firms. Hence the social product cost function is $NZC\left(\frac{X}{N}\right)$ and the marginal cost function is $ZC'\left(\frac{X}{N}\right)$, which replaces $ZC'(X)$ in (6) defining $X_{S0}(Z)$. Once this wrinkle is realised, a comparison of (6) and (17) shows that for all N and Z, $X_{S0}(Z) > X_N(Z)$. Hence $X_{S0}(Z) > X_N(Z)/N$ and

[16] That $X_N < X_{S0}$ can be formally proved by regarding $X_N(Z)$ as a continuous function of both N and Z defined by (17). Letting N go to infinity, $X_N(Z)$ approaches X_{S0} for N large and $C' = 1$. Equation (17) can be implicitly differentiated with respect to N to show that $X_N(Z)$ is an increasing function of N for all N and Z. Hence for all N, $X_N(Z)$ is less than $X_{S0}(Z)$.

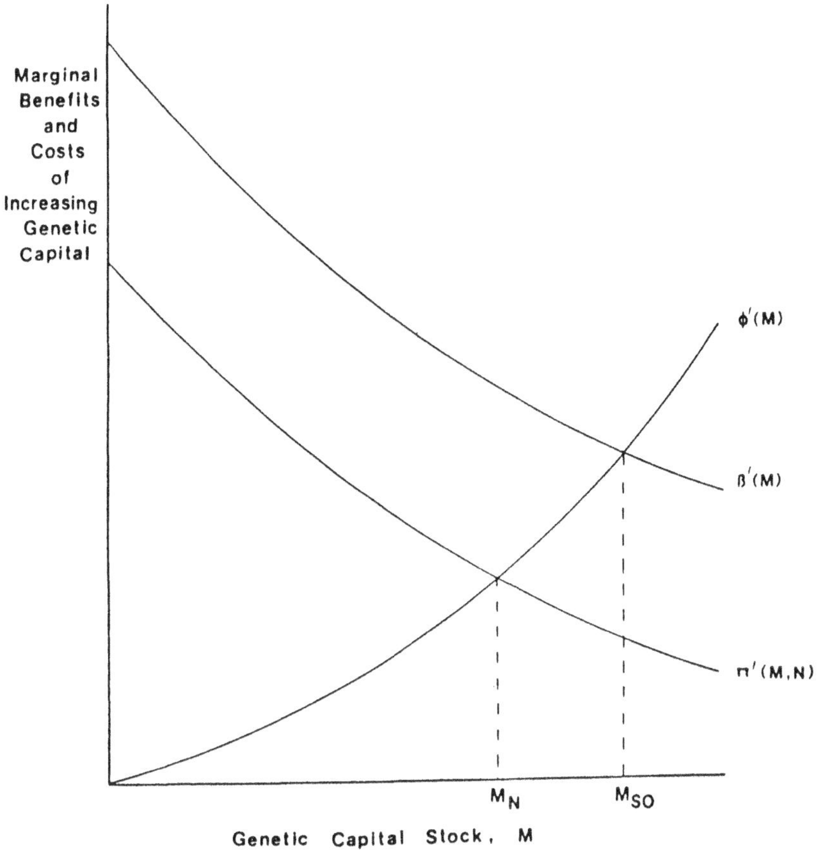

Fig. 1. The socially optimum stock of genetic capital, M_{SO}, and the N-firm industry stock of capital, M_N.

because $C(X)$ is an increasing function, $C(X_{S0}(Z))$ is greater than $C(X_N(Z))$. A comparison of (24) and (12) immediately indicates that $B'(M) > \prod'(M,N)$ for all M and N. Hence, since $\varphi'(M)$ is the same in both cases, $M_{S0} > M_N$ for all N if an equilibrium M_N and M_{S0} exists.

The socially optimum amount of genetic capital is relatively greater whether $-\frac{\partial \Pi}{\partial \tilde{X}} \tilde{X}'$ is included or excluded. Including the negative term only reduces $\prod'(M,N)$ further, thus not changing the result.[17]

In summary, for all the cost functions, demand equations and behavioral assumptions discussed in this paper, an N-firm industry saves less genetic capital or in general invests less in a public stock of knowledge than the social optimum.

[17] Sticklers for mathematical accuracy will note that we have been assuming that $B'(M)$ and $\prod'(M,N)$ plotted vs. M intersect $\varphi'(M)$ from the left and above. This is the sufficient condition for M_N and M_{S0} determined by (27) and (19) to be maxima, as they must be, and not minima.

5. Industry structure and the stock of genetic capital

We have discerned three effects which reduce M_N below the social optimum. The importance of each varies with N in a different way. First, as N increases, the marginal revenue curve becomes flatter and $X_N(Z) \rightarrow X_{S0}(Z)$, thereby moderating the reduction in M_N which otherwise arises when competitive firms give zero weight to consumers' surplus. On this ground, M_N is increasing in M. However, the second and third effects reduce M_N as N increases. Each firm is able to capture less and less of its investment in M and each firm is hurt more and more by the greater and greater size of the other firms' total response as these firms become more numerous.

The third effect is particularly interesting since it is a result of the inherently sequential nature of the decision problem faced by each firm.

The change in the effects with respect to increases in N differ in sign. Their magnitudes depend *inter alia* on $g(Z)$, $\varphi(M)$, $C(X)$ and $F(Z,M)$. Hence unambiguous results do not exist. In the following section we consider the case of constant elasticity of demand and constant unit costs of production. In this special case, the Schumpeterian hypothesis that monopoly leads to the greatest R&D expenditure – the greatest stock of M – holds for elasticities greater than 1. For elasticities between $^1/_2$ and 1 a duopoly retains at least as much genetic capital as a monopoly. In general, as the demand for X becomes more inelastic a greater number of firms is required for there to be incentives to invest in R&D.

Finally we note the effect of the sequential, strategic element captured by the term $-\dfrac{\mathrm{d}\Pi}{\mathrm{d}\tilde{X}}\dfrac{\partial X}{\partial Z}$. A more static, traditional public goods model would consider only the effects on investment due, on the one hand, to the inattention to consumers' surplus and, on the other hand, to the spillover effect of one firm's investment on the other firms' profits.[18] If the strategic element is neglected, then even for the case of elastic demand, it is not always true that a monopolist holds the most M.

Consider the constant elasticity demand curve $g(X) = X^{-1/\varepsilon}$ where ε is the price elasticity. Initially let ε be greater than 1. Let the costs of production be $C(X,Z) = ZX$. Hence the realized technology Z

[18] See Nelson (1980) for an example of this type of analysis.

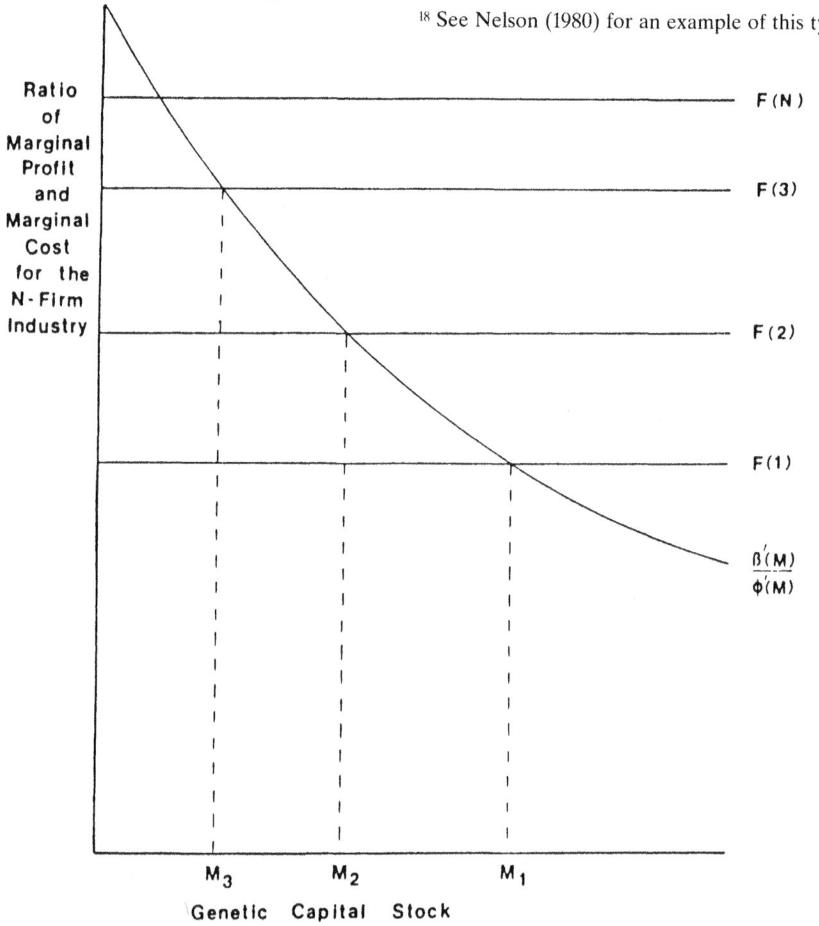

Fig. 2. The N-firm industry stock of genetic capital, M_N, for the case of constant elasticity of demand, $F(N) = N\left[1 - \dfrac{1}{N\varepsilon}\right]^{-\varepsilon}\left[1 - \dfrac{N-1}{N}\left[\dfrac{N\varepsilon}{N\varepsilon - 1}\right]\right]^{-1}$

specifies the constant unit cost of production. These specifications lead to the equations:

$$X_{S0}(Z) = Z^{-\varepsilon} \tag{27}$$

$$X_N(Z) = Z^{-\varepsilon}\left[1 - \frac{1}{N\varepsilon}\right]^{\varepsilon} = X_{S0}(Z)\left[1 - \frac{1}{N\varepsilon}\right]^{\varepsilon}. \tag{28}$$

In addition (25) is rewritten as

$$-\frac{\partial \Pi}{\partial \tilde{X}}\tilde{X}' = -\frac{X_N}{N^2}(N-1)\cdot\left[1 - \frac{1}{N\varepsilon}\right]^{-1}. \tag{29}$$

As we are mainly interested in the effect of N on M_N, the necessary condition for M_N is featured. From (28) and (26) and,

$$B'(M) = \int_0^{C_0} \frac{\partial F}{\partial M} X_{S0}(Z)\,\mathrm{d}Z = \varphi'(M), \Pi'(M, N)$$

is rewritten in terms of $B'(M)$:

$$\Pi'(M, N) = B'(M)\left[1 - \frac{1}{N\varepsilon}\right]^{\varepsilon}\frac{1}{N}\left[1 - \frac{N-1}{N}\left(\frac{N\varepsilon}{N\varepsilon - 1}\right)\right], \tag{30}$$

and the necessary condition for M_N is most usefully written as

$$\frac{B'(M_N)}{\varphi'(M_N)} = N\left[1 - \frac{1}{N\varepsilon}\right]^{-\varepsilon}\left[1 - \frac{N-1}{N}\left(\frac{N\varepsilon}{N\varepsilon - 1}\right)\right]^{-1}. \tag{31}$$

Note that B' and φ' are functions of M only not N. Hence the necessary condition for M_N can be written in terms of a function of M on the LHS of (31) equal to a function of N only on the RHS. Note also that the corresponding equation for the social optimum is marginal benefit = marginal cost ($B'(M_{S0})/\varphi'(M_{S0}) = 1$). It is then easy to show that the stock of capital held by the N-firm industry declines as the number of firms increases, provided investment is positive.[19] See Figure 2.

Equation (31) holds for $N\varepsilon > 1$. When $\varepsilon < 1$ the marginal revenue curve for a monopolist is negative. Hence a monopolist restricts production to very low levels, in the limit zero. Let σ be a lower bound on the monopolist's production independent of Z. If $\bar{Z}(M)$ is the mean of the distribution of Z, given M, then a bound on the benefits the monopolist receives from increasing M is given by $\sigma\bar{Z}'(M)$. As long as $\bar{Z}'(M)$ is bounded (which will be the case except for pathologies), $\sigma\bar{Z}'(M)$ approaches 0 as σ approaches 0, hence the marginal benefits of investing in R&D are very small for the monopolist. The point is intuitively clear. The monopolist produces very little and hence gains very little from reductions in the unit cost of production.

For each firm in a duopoly, however, the anticipated marginal revenue curve, given Z, is not negative for elasticities between $1/2$ and 1. Hence these firms expect to produce in accord with equation (28). Since this production is substantial, the firms may reap non-zero benefits from some investment in M. The amount of investment is given by equation (31). For the range of elasticities $1/2$ to 1 a duopoly invests more in R&D than a monopolist.

[19] For the case of elastic demand, $\varepsilon > 1$, (31) implies that M_N is a decreasing function of N if B'/φ' is a decreasing function of M and

$$F(N) = N\left[1 - \frac{1}{N\varepsilon}\right]^{-\varepsilon}\left[1 - \frac{N-1}{N}\left(\frac{N\varepsilon}{N\varepsilon - 1}\right)\right]^{-1}$$

is an increasing function of N. The second condition is demonstrated immediately by regarding $F(N)$ as a continuous function of N, differentiating and discovering that $F'(N)$ is positive.

The sufficient condition that the firm's choice of M represents a profit maximum implies that B'/φ' is decreasing in a neighborhood of the equilibrium. At the equilibrium $B' = F(N)\varphi'$. The sufficient condition is: $B'' - \varphi''F(N) < 0$. But

$$\frac{\mathrm{d}}{\mathrm{d}M}\left(\frac{B'}{\varphi'}\right) = \frac{B''}{\varphi'} - \frac{B'}{\varphi'^2}\varphi'' = \frac{1}{\varphi'}\left[B'' - \frac{B'}{\varphi'}\varphi''\right] = \frac{1}{\varphi'}[B'' - F(N)\varphi''] < 0. \tag{i}$$

For many reasonable demand and production cost combinations, the marginal benefit function $B'(M)$ is decreasing throughout the range of M. If this is true or if φ'' is sufficiently large, then B'/φ' also will be a decreasing function for all M. Let us now assume that B'/φ' is decreasing in the relevant range of M. Since (31) asserts the equality of B'/φ' and $F(N)$, as N and therefore $F(N)$ increases, M must decrease to maintain that equality. This proves the result.

The argument can be extended to smaller elasticities and larger N. For $1/3 < \varepsilon < 1/2$, the expected marginal revenue curves are negative for the monopolist and duopoly but not for $N = 3$ or greater. Hence in this range a 3-firm industry is likely to produce the most R&D. In the range of elasticities $\frac{1}{N}$ to $\frac{1}{N+1}$ an $N+1$ firm industry will produce the most R&D. We note that as N becomes sufficiently large and ε sufficiently small it may be that there is no industry equilibrium which invests in R&D.

Of course no one expects to see a demand curve which is inelastic in all ranges of production. But the above discussion suggests that relatively inelastic demand is one economic feature which might prevent a monopoly from being the market structure which holds the greatest stock of 'public' knowledge.

Return now to the case of elastic demand and consider a static model which necessarily neglects the strategic effect captured by $-\frac{\partial \Pi}{\partial \tilde{X}} \frac{\partial \tilde{X}}{\partial Z}$. With this term omitted, the necessary condition for industry capital reduces from the expression in equation (31) to the expression,

$$\frac{B'(M_N)}{\varphi'(M_N)} = N\left[1 - \frac{1}{N\varepsilon}\right]^{-\varepsilon}. \tag{32}$$

For elasticities sufficiently close to 1, the right-hand side of (32) is greater for $N = 1$ than for $N = 2$. With B'/φ' a decreasing function, this implies that a symmetric duopoly holds more genetic capital than a monopolist. In this case, neglecting the strategic effect, as occurs in the simple static model, reverses the conclusion about the industry structure which produces the most R&D.

The combination of the constant elasticity demand curve and the distribution of $F(Z,M)$ from our example results in fairly pathological behavior. $B'(M)$ is an increasing function which becomes infinite for finite M. The problem arises because for $X_{S0}(Z) = Z^{-\varepsilon}$, $X_{S0}(Z)$ becomes infinite as Z approaches zero and $F(Z,M)$ does not decline quickly enough near zero for all M. The pathology could be avoided by including a non-zero unit cost floor, by allowing C'' to be greater than zero or by putting a ceiling on demand at very low prices.

Although we describe the above case as a pathology, it represents one example of a special class, one where the probability and value of a spectacular discovery, a pearl of great price, is sufficiently high that we should save all the genetic capital or species we can regardless of cost. Although this situation may seem unlikely, it is perhaps a fair representation of the position adopted by some conservationists. The case of linear demand, constant unit costs and the distributional assumptions of our example is presented in Appendix B.

6. Conclusion

Flora and fauna are usefully regarded as genetic capital which yield knowledge as a result of research and development. If firms cannot deny other firms access to this potential undepletable pool of knowledge, a clear discrepancy between the social and private value of genetic capital is created. Even if there are perfect patent laws for the valuable ideas once discovered, the divergence exists. No single firm will pay appropriately to preserve the right to sample from a richer distribution of species, if competitors also prospectively can make use of the marginal genetic diversity preserved.

A partial antidote to this argument is to remove competition by creating a monopoly. Much of the R&D literature corroborates the Schumpeterian hypothesis that such a market structure is most conducive to R&D; i.e., a monopoly results in preserving the greatest genetic capital in the present context, albeit less than the social desideratum. Indeed we confirm this result for the special cases of linear and constant elasticity of demand (greater than one). However, it is not possible to prove that a monopoly is superior, in general, when there is public knowledge because three forces link market structure (N) with the stock of genetic capital (M) and one is 'perverse' as we discussed in Section 5. Additionally, it is pretty clear that there is little incentive for a monopolist to find lower cost technologies when the demand function is inelastic. Profits are obtained by restricting output. On the other hand, non-cooperative multiple firm industries would produce greater output and have a greater incentive to seek cost-saving discoverties.

We have presented a simple analysis involving genetic capital in order to lay bare some fundamental principles which might have been obscured in a heavily detailed model. In reality, there are many different genetic pools. Firms store different subsamples of selected gene pools because they are expected to combine best with the firm's special knowledge and resources. A monopoly conceivably might invest in species more than is socially optimal in order to complete in a wider arena against dramatically different substitutes – better fertilizers for better seeds; radiology or preventive medicine instead of better drugs. Genetic capital is preserved by individuals and public agencies whose foibles and peccadilloes are not costless. It remains to be seen whether greater realism and empirical research lend further weight to our conclusions that too little genetic capital is preserved by all market structures; and that a monopoly tends to perform the best relative to the social optimum but with important exceptions.

References

Allard, R.W., 1960. Principles of Plant Breeding. John Wiley and Sons, Inc., New York.

Arrow, K.J., 1962. Economic welfare and the allocation of resources for invention. In: The Rate and Direction of Inventive Activity: Economic and Social Factors, pp. 609–625. Universities-NBER Conference Series, Princeton.

Baumol, W. & Oates, W., 1975. The Theory of Environmental Policy. Prentice-Hall, Englewood Cliffs, N.J.

Briggs, F.N. & Knowles, P.T., 1967. Introduction to Plant Breeding. Reinhold Publishing Company, Inc., New York.

Ciriacy-Wantrup, S.V., 1952. Resource Conservation. University of California Press, Berkeley.

Evensen, R.E. & Kislev, Y., 1976. A stochastic model and applied research. J. Pol. Econ. 84: 265–82.

Hey, J.D., 1979. Uncertainty in Microeconomics. New York University Press, New York.

Hirshleifer, J. & Riley, J., 1979. The analytics of uncertainty and information – An expository survey. J. Econ. Lit. 17: 1375–1421.

Kamien, M. & Schwartz, N., 1982. Market Structure and Innovation. Cambridge University Press, Cambridge.

Kamien, M. & Schwartz, N., 1972. Timing of innovations under rivalry. Econometrica 40: 43–60.

Kreps, D.M. & Wilson, R., 1982. Sequential equilibria. Econometrica 50: 863–94.

Krutilla, J., 1967. Conservation reconsidered. Am. Econ. Rev. 56: 777–86.

Mood, A.M., Graybill, F.A. & Boes, D.C., 1974. Introduction to the Theory of Statistics, 3rd ed., McGraw-Hill, New York.

Myers, N., 1978. The Sinking Ark. Pergamon Press, Oxford.

Nelson, R.R., 1959. The simple economics of basic scientific research. Pol. Econ. 67: 297–306.

Nelson, R.R., 1980. R&D, knowledge and externalities. In: C. Bliss & M. Boserup (eds.), Economic Growth and Resources, Vol. 3, Natural Resoures.

Nelson, R.R., 1981. Research on productivity growth and differences. J. Econ. Lit. 19: 1029–64.

Newberry, D. & Stiglitz, J., 1984. The Theory of Commodity Price Stabilization: A Study in the Economics of Risk. The Clarendon Press, Oxford.

Oldfield, Margery, 1984. The Value of the Conservation of Genetic Resources.

Prescott-Allen, Robert & Christine, 1980. The Contribution of Wild Plants and Animals to Industrialized Societies. Mimeo.

Reinganum, J., 1982. A dynamic game of R&D: Patent protection and competitive behavior. Econometrica 50: 671–88.

Ruff, L., 1969. Research and technological progress in a cournot economy. J. Econ. Theory 1: 397–415.

Sandmo, A., 1971. On the theory of the competitive firm under price uncertainty. Am. Econ. Rev. 61: 65–73.

Schery, R.W., 1972. Plants for Man. Prentice-Hall, Inc., Englewood Cliffs, N.J.

Shubik, M., 1982. Game Theory in the Social Sciences. M.I.T. Press, Cambridge.

Telser, L., 1982. A theory of innovation and its effects. Bell J. Econ. 13: 69–92.

Appendix A: A static model of R&D

In this appendix we analyze a purely static formulation of the R&D model and compare it to the sequential case.

In a purely static model, X and M are chosen simultaneously. Hence production X cannot depend on the realized technology Z. Consider first the case of the social optimum. Since X no longer depends on Z, (7) which defines $B(M)$, the expected benefits associated with a given stock of capital M, can be rewritten in the simpler form:

$$B(M) = G(X) - C(X)\bar{Z}(M) \tag{A.1}$$

$\bar{Z}(M)$ is the expected value of Z given M. The socially optimal level of production is given by differentiating the net benefits, $B(M) - \varphi(M)$, with respect to X.

$$g(X_{S0}) = C'(X_{S0})\,\bar{Z}(M). \tag{A.2}$$

In the static model, $X_{S0}(M)$ is only a function of M while in the sequential model a production plan, $X_{S0}(Z)$, must be calculated describing production for each realized Z. Comparing (A.2) and (6) we see that the two production specifications are related.

$$X_{S0}(M) = X_S(\bar{Z}(M)). \tag{A.3}$$

In the static case, optimal production assumes that the realized technology will be the expected technology.

The socially optimal level of capital can be found by differentiating $B-\varphi$ with respect to M, obtaining the equation

$$-C(X_{SO}(M))\frac{\mathrm{d}\bar{Z}(M)}{\mathrm{d}M} = \varphi'(M). \tag{A.4}$$

Equation (A.4) is the static analogue of (12). Integrating by parts the equation

$$\bar{Z}(M) = \int_0^{C_0} f(Z, M)\,\mathrm{d}Z,$$

and differentiating by M, permits (A.4) to be rewritten in a form similar to (12),

$$C(X_{SO}(M))\int_0^{C_0}\frac{\partial F}{\partial M}\mathrm{d}Z = \varphi'(M). \tag{A.5}$$

Comparing (A.5) and (12) reveals how the sequential model differs from the purely static one. Since

$$C(X_{SO}(Z)) = -\frac{\mathrm{d}B}{\mathrm{d}Z}(X_{SO}(Z), Z),$$

the necessary condition for the static model can be derived (using (A.3)) from the sequential condition by making the assumption:

$$B(X_{SO}(Z), Z) = B(X_{SO}(\bar{Z}(M)), \bar{Z}(M) + \frac{\mathrm{d}B}{\mathrm{d}Z}(X_{SO}(\bar{Z}(M), \bar{Z}(M))\,[Z - \bar{Z}(M)]. \tag{A.6}$$

In other words, neglecting the sequential aspect of the model is equivalent to assuming that the net benefits of a technology Z can be measured by a linear function of Z, in particular, the function obtained by expanding B in a Taylor series about the point $Z = \bar{Z}(M)$.

Consider now a static version of the industry model. In this case, a firm chooses X_i and M_i assuming that the production of the other firms, $\tilde{X}$, is fixed. Neither X_i nor $\tilde{X}$ can depend on Z. If industry output is defined as $X_N = NX_i$ and using the symmetry condition $\tilde{X} = \frac{N-1}{N}X_N$, then the industry analogue of (A.2) is:

$$g(X_N(M)) + \frac{X_N}{N}g'(X_N) - C\left(\frac{X_N}{N}\right)\bar{Z}(M) = 0. \tag{A.7}$$

Just as for the social optimum, the static N-firm production plan and the sequential production strategy are related by the equation

$$X_N(M) = X_N(\bar{Z}(M)). \tag{A.8}$$

The industry analogue to (A.4) is obtained by differentiating the expected profits of firm i, given $\tilde{X}$ and $\tilde{M}$, with respect to M_i and only then using the equilibrium condition for X_N. The result is

$$-C\left(\frac{X_N(M)}{N}\right)\frac{\mathrm{d}\bar{Z}\,(M)}{\mathrm{d}M} = \varphi'(M). \tag{A.9}$$

Rewriting $\mathrm{d}\bar{Z}(M)/\mathrm{d}M$, as in (A.5), we derive the necessary condition for M_N, the industry stock of genetic capital,

$$C\left(\frac{X_N(\bar{Z}(M))}{N}\right)\int_0^{C_0}\frac{\partial F}{\partial M}\mathrm{d}Z = \varphi'(M). \tag{A.10}$$

Equation (A.10) is the static analogue of (24). We note that the strategic effect captured by $-\frac{\partial \Pi}{\partial \tilde{X}}\frac{\mathrm{d}\tilde{X}}{\mathrm{d}Z}$ has been lost in the static formulation.

Appendix B: Linear demand and constant marginal cost

$$g(X) = g_0 - g_1X,$$
$$C(X, Z) = ZX. \tag{B.1}$$

The socially optimal production for each realized technology Z is found by setting price equal to marginal cost,

$$X_{SO}(Z) = \frac{g_0}{g_1} - \frac{Z}{g_1}. \tag{B.2}$$

The equivalent formula for the expected industry output $(X_N(Z)$ is found from solving (17),

$$X_N(Z) = \frac{N}{N+1}\left(\frac{g_0}{g_1} - \frac{Z}{g_1}\right) = \frac{N}{N+1}X_{SO}(Z). \tag{B.3}$$

In order to compare the marginal social benefits from increasing M with the marginal profits to firm i, $-\frac{\partial \Pi}{\partial \tilde{X}}\frac{\mathrm{d}X\,Y}{\mathrm{d}Z}$ must be calculated. This is accomplished by direct calculation of $\frac{\partial \Pi}{\partial \tilde{X}}$ and the implicit differentiation of equation (17) to find $\partial\tilde{X}/\partial Z$:

$$-\frac{\partial \Pi}{\partial \tilde{X}}\frac{\mathrm{d}\tilde{X}}{\mathrm{d}Z} = -\frac{X_N}{N}\frac{N-1}{N+1}. \tag{B.4}$$

Using equations (B.3) and (B.4) we note that

$$C\left(\frac{X_N}{N}\right) - \left.\frac{\partial \Pi}{\partial \tilde{X}}\frac{\partial \tilde{X}}{\partial Z}\right] = \frac{X_{SO}}{(N+1)}\left[1 - \frac{N-1}{N+1}\right]$$
$$= X_{SO}\frac{2}{(n+1)^2}. \tag{B.5}$$

Hence $B'(M)$, the marginal social benefits from increasing M, and $\prod'(M,N)$, the marginal profits to firm i from increasing M, are related by the equation:

$$\Pi'(M, N) = \frac{2}{(N+1)^2}B'(M). \tag{B.6}$$

We can now write down the analogue of equation (31) for the linear demand curve,

$$\frac{B'(M_N)}{\varphi'(M_N)} = \frac{(N+1)^2}{2}. \tag{B.7}$$

Since $(N+1)\frac{2}{2}$ is an increasing function of N, it follows via the analysis in the text, that monopoly is the market structure which produces the most R&D when the demand is linear.

Now $B'(M)$ is explicitly calculated for the linear demand curve, constant marginal costs and the special distributional assumption discussed in our example. A direct integration of equation (11) yields the result

$$B(M) = B(X_{S0}(C_0), C_0) + \frac{g_0}{g_1}\frac{C_0\alpha M}{\theta+\alpha M} - \frac{C_0}{g_1}\frac{C_0\alpha M}{\theta+2\alpha M}. \quad \text{(B.8)}$$

Next, $B(M)$ is differentiated to obtain the marginal benefits

$$B'(M) = \theta C_0\alpha\left[\frac{g_0}{g_1}\frac{1}{[\theta+\alpha M]^2} - \frac{C_0}{g_1}\frac{1}{[\theta+2\alpha M]^2}\right]. \quad \text{(B.9)}$$

For $C_0 < \frac{g_0}{2}$, $B'(M)$ is a decreasing function of M, hence B'/φ' is a decreasing function. For C_0 between $P_0/2$ and P_0, $B'(M)$ has one peak. Hence as long as the worst case technology $Z = C_0$ is not too bad, the equilibrium in M and X is certainly well behaved. For high enough C_0 the behavior of the equilibrium does however depend on the supply curve for M, $\varphi'(M)$.

CHAPTER 16

Endangered species: the economics of irreversible damage

ANTHONY C. FISHER and W. MICHAEL HANEMANN

1. Introduction

In this paper we develop some of the economic theory relevant to decisions about the protection of endangered species and their natural habitat. Two key concepts emerge from the discussion: uncertainty and irreversibility. Uncertainty is pervasive in economic life, of course. But more than the usual degree of uncertainty surrounds the potential future benefits from conserving ecosystems. Most endangered species, for example, remain as yet undiscovered in tropical moist forests that are at the same time undergoing rapid and (from the point of view of the indigenous species) destructive development. In these circumstances it is hard to predict what values – a cure for (some form of) cancer, a liquid hydrocarbon, a perennial corn, to name just three of the more interesting among current possibilities – will ultimately emerge if the species that may produce them are not lost.

Irreversibility is clearly central to thinking about endangered species or ecosystems because these represent the results of geomorphologic and biologic processes that span eons and that cannot be replicated by man. Their loss is, for all practical purposes, irreversible; and we shall argue that this, along with the uncertainty, introduces a new element into the benefit–cost calculus.

The plan of this paper is as follows. In the next section, we indicate reasons for the growing concern over species extinction. In Section 3, we show how conventional benefit–cost analysis needs to be modified to take account of uncertainty and irreversibility. In Section 4, we prove the existence of a positive 'option value' of nature preservation – the value of retaining an option to preserve or develop in the future. In Section 5, we illustrate this argument with a simple numerical example, intended to show how one would calculate option value in practice and what factors might influence its magnitude. In Section 6, we present some concluding observations.

2. The growing concern over species extinction

The conventional view of the threat to endangered species of plants and animals is that it is due to overexploitation. In some cases, this is undoubtedly correct. But the major threat is habitat modification. This can take several forms: direct conversion, as in drainage of wetlands or development of drylands for agriculture, housing, and transportation; chemical pollution, as from acid rain; and 'biological pollution', the introduction of exotic species. Of these, the most important currently appears to be direct conversion for agricultural and other development. Thus, protection of endangered species is intimately related to preservation of the natural environment.

How serious is the threatened loss of species? In the United States, over 500 species are known to have become extinct since 1600 or between one and two per year. By contrast, over a 3 000-year span

D.O. Hall, N. Myers and N.S. Margaris (eds.), Economics of ecosystems management.

 ISBN-13: 978-94-010-8928-9

during the Pleistocene period – a period of glaciation when many individual organisms died – less than 100 species were lost in North America (Opler, 1971). About two thirds of the recent losses have been in Hawaii due to the clearing of forests for cropland and the introduction of exotic species. This is significant because, despite the considerable interest in the problem in the United States, by far the greater number of species and endangered species is found in the tropical moist forests of the world. Wholesale extinctions are threatened by the clearing of forests for cropland and fuel wood. Some biologists believe that 1 000 species are disappearing worldwide each year and that this rate may reach 10 000 annually by the end of the decade (Myers, 1981). By one estimate, as many as a million of the current 5 to 10 million species could be gone by the year 2000 (Myers, 1983). Again, these numbers are highly conjectural but extinction on anything like this scale would clearly be catastrophic.[1]

Wy should anyone care if unprecedented levels of extinctions over the next few decades send uncounted species the way of the dodo and the dinosaur? Would human welfare really be much affected? The question here is, how does preservation of plant and animal populations *not now harvested* contribute the human welfare? There are at least two distinct ways in which they do this, more if we count 'non-material' welfare. An example of one is provided by the recent discovery of a wild grass related to corn in a remote, mountainous area in Mexico slated for development (Vietmeyer, 1979). The wild grass is a perennial whereas domesticated corn must be planted annually. A successful hybrid could result in substantial savings due to elimination of the need to prepare the ground and reseed each year. In a later section, we shall show how to calculate the option value of preserving a natural habitat given varying assumptions about the value of the resulting product, the value of a development project in the area, and the probability distributions of the random variables representing these values. But the example clearly illustrates one way a currently unharvested species contributes to human welfare: by conserving genetic information that may in the future be useful in some form of economic activity.

A second major way in which species are useful is as components of living ecosystems that provide the basic physical and biological supports for human life. These include maintenance of the quality of the atmosphere, control and amelioration of climate, regulation of freshwater supplies, generation and maintenance of soils, disposal of wastes, and cycling of nutrients (Ehrlich & Ehrlich, 1981). Removal of any one species can cause a system to break down because each has evolved a set of characteristics that make it a unique functional part of the system. To some extent, it may be possible to substitute for ecosystem services. For example, nitrogenous fertilizers can substitute for nitrogen-fixing organisms. But even here there are environmental problems associated with heavy use. Agriculture in the United States, already a heavy user, still derives considerably more nitrogen from natural systems (Delwiche, 1970). In general, it seems fair to say that some services of ecosystems are not substitutable at all; and for those that are, the direct and indirect (or external) costs of substitution are likely to be high.

Loss of natural populations can also adversely affect human welfare in less tangible ways. People derive pleasure from the contemplation of strikingly varied life forms such as the perhaps 15 000 different species of butterflies. Surely this pleasure would be much reduced if, for example, all butterflies looked alike. And we ought to at least note, though we can do little more here, that some of the concern for endangered species is of a religious or ethical nature which does not easily fit into our utilitarian framework.

3. Benefit–cost analysis with irreversible damages

We have discussed in an informal way what appears to us to be the leading contemporary conservation policy issue: protection of habitat for endangered species. Why should there be more concern about this than about issues involving the

[1] For a critical look at Myers' (1981) and other estimates, see Lugo & Brown (1982).

allocation of resources of comparable value? The reason is at least partly because wild lands and the natural populations they support are the results of geomorphologic and biologic processes that represent a time frame measured in eons and which, if destroyed, cannot be replaced or restored. There is thus a basic irreversibility that attends the modification of unique natural environments and indigenous species. These have indeed been referred to as 'gifts of nature' since they cannot be reproduced in all of their essential features by the efforts of man.

To economists, the important question about irreversibility is this: What are the implications for resource allocation? If the in situ resources of an environment are declining in value relative to the extractive resources, then, perhaps, irreversibility poses no special problem. An optimal program will call for conversion at a rate dictated by the changing relative values. Unfortunately, just the reverse is likely to be true, at least in the industrialized countries. Unique natural environments are in many cases likely to appreciate in value relative to goods and services they might yield if developed. Then the restriction on reversibility matters because value would be increased by going back to an earlier, less developed state.

Note that we have carefully qualified this proposition. The fact is neither economists nor ecologists can say with certainty just what the future benefits of environmental preservation will be in general or in specific cases. To return to the example cited earlier, there is a chance that conserving a Mexican mountain ecosystem will lead to the successful breeding of a perennial corn, but this cannot be known at the time a decision about the area is taken. In the remainder of this section, we present a model of the decision to develop a natural environment that takes account of both the irreversibility of development and uncertainty about the values development would preclude.

Our model is based on the original formulations of Arrow & Fisher (1974) and Henry (1974), but we adopt the more transparent notation and approach of Hanemann (1982). The decision problem is: How much of a tract of wild land should be developed in each of two periods (present and future)? We choose units of measurement such that the maximum level of development (or preservation) is just unity. Then, there are three substantive assumptions. First, development in any period is irreversible. Second, the benefits of development in the first period are known; those of development in the second period are not. These assumptions capture the essential features of our problem. A third, made more for ease in obtaining unambiguous results, is that benefits are a linear function of the level of development. Later on, we consider what happens when this assumption is relaxed.

Let us interpret the assumed structure a bit to indicate both its rationale and its limits. The benefits associated with a given level of development are the benefits of development *and* the benefits of preservation. Let the former be given, for the first period, as

$$B_{1d}(d_1) = \alpha d_1 \tag{1}$$

where d indicates development; d_1 the level of development in period 1; and α, a positive constant. Notice that $0 \leq d_1 \leq 1$.

Let the benefits of preservation be

$$B_{1p}(d_1) = \beta - \gamma d_1 \tag{2}$$

where p indicates preservation and β and γ are positive constants. Then, benefits in period 1 are

$$\begin{aligned} B_1(d_1) &= B_{1d}(d_1) + B_{1p}(d_1) \\ &= \beta + (\alpha - \gamma)d_1. \end{aligned}$$

The relationship between B_1 and d_1 is clearly linear, implying that first-period benefits will be maximized by choosing either $d_1 = 0$ (if $\alpha < \gamma$) or $d_1 = 1$ (if $\alpha > \gamma$). Thus, the problem is restricted to a choice of corner solutions. The argument would be more complicated were we to consider uncertain second-period benefits as well, but this key restriction carries over.

Second-period benefits are $B_2(d_1 + d_2, \theta)$ where d_2 is the amount of land developed in period 2 and θ is a random variable.[2] Thus, second-period benefits depend on development in periods 1 and 2 and are uncertain. Notice that $d_2 \geq 0$, and $d_1 + d_2 < 1$. We

[2] Second-period benefits can be viewed as present values. We suppress the discount factor here because it would not affect our results.

shall assume that the problem is to maximize expected benefits over both periods. This is one particular way of dealing with the uncertainty. It is not, however, as restrictive as it may seem since we have not specified that benefits are measured in money units. If, for example, benefits are measured in utility units, then our formulation is equivalent to the quite general expected utility maximization.

The remaining structural element of the problem involves the behavior of uncertainty over time. More specifically, we consider two possible cases. In the first, nothing further is learned about the value of θ by period 2 so that d_1 and d_2 are chosen in period 1. In the second, the value of θ is learned by period 2 so that it makes sense to defer a decision on d_2 to period 2. Now comes a very important assumption. It is that the learning, in case 2, does not depend on first-period development, d_1. *For the kind of uncertainty we are trying to capture,* this seems appropriate. Uncertainty is largely about the future (period 2) benefits of preservation – the value that may be discovered in some indigenous species, for example. This will be determined *not* by developing its habitat but by undertaking research into its medicinal or other properties. The research is not endogenous to our problem, but we do assume that the answer it yields, concerning the value of θ, does not depend on the development of the tract in question. Not surprisingly, this assumption importantly affects the results we shall obtain. At that point, we shall have more to say about the alternative assumption and its consequences.

Now let us write down expressions for the value to be maximized under each information structure. Where no new information is forthcoming by the second period, define $V^*(d_1)$ by

$$V^*(d_1) = B_1(d_1) + \max_{\substack{d_2 \\ 0\leq d_1 + d_2\leq 1 \\ 0\leq d_2}} \{E[B_2(d_1 + d_2, \theta)]\}. \tag{4}$$

Then, the maximum value is $V^* = V^*(d_1^*)$, where d_1^* maximizes $V^*(d_1)$ subject to $0\leq d_1\leq 1$.

Where new information is forthcoming, define $\hat{V}(d_1)$ by

$$\hat{V}(d_1) = B_1(d_1) + E\Big[\max_{\substack{d_2 \\ 0\leq d_1 + d_2\leq 1 \\ 0\leq d_2}} \{B_2(d_1 + d_2, \theta)\}\Big]. \tag{5}$$

The maximum value in this case is $\hat{V} = \hat{V}(\hat{d}_1)$ where $\hat{d}_1$ maximizes $\hat{V}(d_1)$ subject to $0\leq d_1\leq 1$.

What can we say about value-maximizing, or optimal, development in the first period in each case? Clearly, since $V^*(d_1)$ and $\hat{V}(d_1)$ are different, d_1^* and $\hat{d}_1$ will be different. A natural hypothesis is that $\hat{d}_2\leq d_1^*$ since it would seem to make sense to put off development, which is irreversible, if there is a prospect of better information about the benefits it will preclude. Put differently, if the decision-maker ignores the prospect of better information, first-period development will be too great. We can prove this result, not in general, but where the choice is between no development ($d_t = 0$) and full development ($d_t = 1$). Recall that this is precisely the choice implied by our linearity assumption.

We wish, then, to compare the alternatives of developing and preserving in each information setting. Where no information is forthcoming, we have

$$V^*(0) = B_1(0) + \max\{E[B_2(0, \theta)], E[B_2(1, \theta)]\} \tag{6}$$

and

$$V^*(1) = B_1(1) + E[B_2(1, \theta)]. \tag{7}$$

Then,

$$d_1^* = \begin{cases} 0 \text{ if } V^*(0) - V^*(1)\geq 0 \\ 1 \text{ if } V^*(0) - V^*(1)<0. \end{cases} \tag{8}$$

Where new information is forthcoming, we have

$$\hat{V}(0) = B_1(0) + E[\max\{B_2(0, \theta), B_2(1, \theta)\}] \tag{9}$$

and

$$\hat{V}(1) = B_1(1) + E[B_2(1, \theta)]. \tag{10}$$

Then

$$\hat{d}_1 = \begin{cases} 0 \text{ if } \hat{V}(0) - \hat{V}(1)\geq 0 \\ 1 \text{ if } \hat{V}(0) - \hat{V}(1)<0. \end{cases} \tag{11}$$

Notice that $V^*(1) = \hat{V}(1)$. With full development in the first period, total value over both periods must be the same since the development is locked in for the second period regardless of what is learned about the random variable θ in the first.

We have still not shown the relationship of d_1^* to $\hat{d}_1$. For this, just one more step is needed. From the convexity of the maximum operator and Jensen's inequality, it follows that

$$\hat{V}(0) - V^*(0) = E[\max\{B_2(0, \theta), B_2(1, \theta\}] - \max\{E[B_2(0, \theta)], E[B_2(1, \theta)]\} \geq 0. \quad (12)$$

Since $\hat{V}(0) \geq V^*(0)$ and $\hat{V}(1) = V^*(1)$, $\hat{d}_1 \leq d_1^*$. In practice, what this means is that optimal first-period use of the area is less likely to be full development ($d_1 = 1$) where it is possible to learn about the benefits precluded than where it is not.

To produce this result, we made a couple of assumptions that deserve further comment. First, about the linearity of the benefit function which allowed us to compare just the two alternatives of development and preservation in each information setting, Hanemann (1982) has shown that the result ($\hat{d}_1 \leq d_1^*$) does not follow in the general case where d_1 can take any value in the interval [0, 1]. On the other hand, it may be produced in a variety of special cases.

For example, suppose we partition the area into a number of separate subareas each with its own (linear) benefit function. Then, each can be ranked by suitability for development; the one for which benefits are greatest is No. 1 and so on. Suppose further that the subarea benefit functions are independent; that is, benefits of developing subarea No. 1 do not depend on whether No. 2 is developed and so on. Now proceed exactly as before for each subarea starting with No. 1. If it pays to develop in a given information setting, move to No. 2 and ask the same question. Continue until the subarea is reached for which optimal development is zero. Since $\hat{d}_{1i} \leq d_1^* i$, where i indexes the subarea, the number of subareas optimally developed will be less in the new-information case than in the no-information case. In other words, the fraction of the original area developed will be less.

Of course additional assumptions were needed to get this result. In our judgment, partitioning of an area, particularly if it is large or diverse, is plausible as is piecewise linearity. Independence of subarea benefit functions is less so. Other less restrictive assumptions about benefit functions might be devised that will yet be consistent with $d_1 \leq d_1^*$. The judgment about all such assumptions is ultimately an empirical one.

The second key assumption we made was that resolution of the uncertainty is independent of development in the first period. We have argued that this is plausible; but, again, the judgment is an empirical one. It is fairly obvious that, if resolution depends on development, and, in particular, on a positive level of development, then (some) development may be optimal *for the purpose of providing information* even where this involves an irreversible commitment of resources that turns out to be a mistake. Such a result is obtained in a recent study by Miller & Lad (forthcoming). Similarly, development for the purpose of providing information about whether the development is, in fact, irreversible may be optimal as has been shown by Viscusi & Zeckhauser (1976). The reader is free to choose among these alternative assumptions about the structure of information, but we continue to feel that ours is most plausible *for our problem*, i.e., relevant information about the properties of indigenous species will come not from developing their habitat but rather from research that, if anything, depends on preserving habitat.

4. Option value and the value of information

In the literature on nature preservation, the concept of 'option value' has played a prominent role. Beginning with the article by Weisbrod (1964), there has been a notion – advanced by some, including Weisbrod, disputed by others – that preservation carries with it a value (option value) above and beyond conventional consumer's surplus. A number of articles following Weisbrod established that option value could be identified with a risk premium: the difference between what a risk-averse consumer is willing to pay for the option of consuming (wilderness recreation, say), at a predetermined non-discriminatory price, and his expected consumer's surplus (Zeckhauser, 1969; Cicchetti & Freeman, 1971). These and other studies also exposed a difficulty; namely, that preservation, as well as development, can bring risks, e.g., floods or power failures (Schmalensee, 1972; Henry, 1974). The net option value of preserving a wilderness environment could then be negative. And, in any case, option value in this interpretation depends on risk aversion.

A different interpretation has been put forward by Arrow & Fisher (1974) and Henry (1974) and, more recently, by Conrad (1980), Hanemann (1982), and perhaps others. Unlike the first, it does not depend on risk aversion. Also, unlike the first, it is explicitly dynamic. In fact, it falls out quite naturally from the model presented in the preceding section. Option value, in this interpretation, is the gain from being able to learn about future benefits that would be precluded by development if one does not develop initially – the gain from retaining the option to preserve or develop in the future. In our terminology, this is

$$OV = \hat{V}(0) - V^*(0). \tag{13}$$

From equation (12), option value, OV, is non-negative.

We can obtain this result in a somewhat different fashion following Arrow & Fisher's (1974) original presentation. Suppose the decision-maker ignores the prospect of new information. Then he will compare $V^*(0)$ and $V^*(1)$. A correct decision – one that takes account of this prospect – can be induced, in principle, by a subsidy to preservation. The optimal subsidy will clearly be one which leads the decision-maker to compare $\hat{V}(0)$ and $\hat{V}(1)$. To solve for the subsidy, S, write

$$[V^*(0) + S] - V^*(1) = \hat{V}(0) - \hat{V}(1), \tag{14}$$

so that

$$S = [\hat{V}(0) - \hat{V}(1)] - [V^*(0) - V^*(1)]. \tag{15}$$

From equations (7) and (10), $V^*(1) = \hat{V}(1)$ and equation (15) reduces to equation (13). The subsidy, S, is just what we mean by option value – the extra value attaching to the preservation option that would be overlooked in a conventional decision, a decision that did not take account of the prospective gains from information.

It is tempting to identify this concept of option value with another one familiar in decision theory: the value of information or, more precisely, the expected value of perfect information. However, the identification is not quite correct. Option value in this interpretation is, as Hanemann (1982) shows, a *conditional* value of information, conditional on $d_1 = 0$.

The *unconditional* value of information is $\hat{V}(\hat{d}_1) - V^*(d_1^*)$ or, in other words, the gain from being able to learn about future benefits provided d_1 is optimally chosen in each case. This my mean $\hat{d}_1 = d_1^* = 0$ or it may not. In fact, two other outcomes are possible: $\hat{d}_1 = d_1^* = 1$ and $\hat{d}_1 = 0$, $d_1^* = 1$. (Note that $\hat{d}_1 = 1$, $d_1^* = 0$ is ruled out by the result that $\hat{d}_1 \leq d_1^*$.) If $\hat{d}_1 = d_1^* = 1$, the value of information is $\hat{V}(1) - V^*(1) = 0$, whereas option value is still $\hat{V}(0) - V^*(0) \geq 0$. If $\hat{d}_1 = 0$ and $d_1^* = 1$, the value of information is $\hat{V}(0) - V^*(1)$. Notice that option value is once again greater than the value of information since $\hat{V}(0) \geq \hat{V}(1) = V^*(1) \geq V^*(0)$.

To summarize, then, option value is not identical to the value of information in the development decision problem. Option value is instead a conditional value of information, conditional on a particular choice of first-period development ($d_1 = 0$) and, moreover, is equal to or greater than the (unconditional) value of information.

5. An illustration: the measurement of option value

It may be helpful to illustrate these concepts with an example. Suppose that the discounted future benefits of preserving the natural ecosystem are represented by the random variable θ_0; this can take on negative as well as positive values, since (to continue the example mentioned in the earlier sections) there is no guarantee that preserving the Mexican mountain ecosystem will, in fact, lead to the successful breeding of a perennial corn. Similarly, let the random variable θ_1 represent the discounted future benefits of agricultural or other extractive or industrial development which irreversibly destroy the natural ecosystem. This, too, can take positive or negative values since there is no guarantee that the development project will always be financially profitable.

In terms of our previous notation, $B_2(0, \theta) = \theta_0$ and $B_2(1, \theta) = \theta_1$. These benefits refer to the future. For the current period, the (expected) benefits from development are $B_1(1)$, and the benefits from preservation are $B_1(0)$. To complete the example, we need to specify the distributions of θ_0 and θ_1. In a real-world example, this would be the heart of the empirical analysis – the union of natural sciences, engineering, and economic analysis in order to quantify the benefits of development and preservation, identify the sources of uncertainty about these benefits, and elicit probability distributions for the uncertain parameters. Here we bypass these tasks and simply assume that θ_0 and θ_1 are normally distributed with means μ_0 and μ_1 and variances σ_0^2 and σ_1^2. In addition, we allow for some possible correlation between the two sets of benefits by introducing a covariance term, σ_{01}. A correlation could arise, for example, if some of the states of the world in which development is more valuable coincide with those in which preservation is more valuable; e.g., higher population levels lead to greater demands for both produced goods and environmental amenities.

Applying (6)–(10) to this model, we have

$$\hat{V}(0) = B_1(0) + E\{\max[\theta_0, \theta_1]\}$$

$$V^*(0) = B_1(0) + \max[\mu 0, \mu 1]$$

$$\hat{V}(1) = V^*(1) = B_1(1) + \mu_1$$

and

$$OV = \hat{V}(0) - V^*(0) = E\{\max[\theta_0, \theta_1]\} - \max[\mu 0, \mu 1].$$

In order to evaluate this expression we need to make use of the following property of the bivariate normal distribution due to Clark (1961):

$$E\{\max[\theta_0, \theta_1]\} = (\mu_0 - \mu_1)\,\Phi\,\frac{\mu_0 - \mu_1}{\sigma} + \mu_1 + \sigma\varphi\,\frac{\mu_0 - \mu_1}{\sigma} \quad (16)$$

where φ and Φ are, respectively, the standard normal p.d.f. and c.d.f. and

$$\sigma = (\sigma_0^2 + \sigma_1^2 - 2\sigma_{01})^{1/2}.$$

Hence,

$$\hat{V}(0) - \hat{V}(1) = B_1(0) - B_1(1) + (\mu_0 - \mu_1)\,\Phi\,\frac{\mu_0 - \mu_1}{\sigma} + \sigma\varphi\,\frac{\mu_0 - \mu_1}{\sigma} \quad (17)$$

and

$$OV = (\mu_0 - \mu)\,\Phi\,\frac{\mu_0 - \mu_1}{\sigma} + \sigma\varphi\,\frac{\mu_0 - \mu_1}{\sigma} + \mu_1 - \max[\mu_0, \mu_1], \quad (18)$$

while

$$V^*(0) - V^*(1) = B_1(0) - B_1(1) + \max[\mu_0, \mu_1] - \mu_1. \quad (19)$$

The formulas in (17)–(19) permit us to make an explicit analysis of how changes in the distributions of θ_0 and θ_1 – i.e., the changes in the underlying uncertainty concerning the future benefits of preservation and development – affect the optimal development decision and the option value. Some simple calculus reveals that

$$\frac{d[\hat{V}(0) - \hat{V}(1)]}{d(\mu_0 - \mu_1)} > 0,$$

$$\frac{d[\hat{V}(0) - \hat{V}(1)]}{d\sigma} > 0,$$

$$\frac{d[V^*(0) - V^*(1)]}{d(\mu_0 - \mu_1)} \begin{cases} > 0 \text{ if } \mu_0 - \mu_1 > 0, \\ = 0 \text{ otherwise} \end{cases}$$

$$\frac{d[V^*(0) - V^*(1)]}{d\sigma} = 0$$

and

$$\frac{dOV}{d|\mu_0 - \mu_1|} < 0, \qquad \frac{dOV}{d\sigma} > 0.$$

In order to interpret these results, observe that $E\{\theta_0 - \theta_1\} = \mu_0 - \mu_1$ while $\text{var}\{\theta_0 - \theta_1\} = \sigma^2$. Accordingly, an *increase* in $(\mu_0 - \mu_1)$, the difference between the expected future benefits of preservation and development:

- lowers $\hat{d}_1$ – i.e., strengthens the case for currently preserving the natural ecosystem if the prospect of future information is recognized;

- lowers d_1^* if $\mu_0 > \mu_1$ but otherwise leaves d_1^* unaffected;
- lowers the option value, OV, if $\mu_0 > \mu_1$ and raises it if $\mu_0 < \mu_1$.

Similarly, an increase in σ, which can come about from an increase in the uncertainty underlying the benefits of preservation (σ_0^2) or development (σ_1^2) as well as a reduction in the covariance between them (σ_{01}),

- has no effect on d_1^* but lowers $\hat{d}_1$;
- raises the option value.

These qualitative properties are illustrated numerically in Table 1. Here, we consider a case where the discounted future benefits of development are normally distributed with a mean of \$1 million, say, and a standard deviation of \$750 000. Thus, there is a 90.88% probability that the development project will have a positive payoff. In the various rows of the table, we consider several alternative distributions for the discounted future benefits of ecosystem preservation and various degrees of correlation between them and the benefits of development ($\varrho = \sigma_{01}/\sigma_0\sigma_1$). In row (a) we consider a case where the preservation benefits are normally distributed with a mean of \$590 000 and a standard deviation of \$1.5 million. In this and all the other examples, the benefits of preservation are far more uncertain than those of development and their expected value is smaller. In row (a), for example, the probability that ecosystem preservation has a positive payoff in the future is only 63.06%.

Since $\mu_0 < \mu_1$ in each case, observe from (19) that $V^*(0) - V^*(1) = B_1(0) - B_1(1)$. Therefore, from (8), $d_1^* = 0$ only if $B_1(0) - B_1(1) \geq 0$ – i.e., if one ignores the possibility of future information, the current development decision is based solely on a comparison of current preservation and development benefits. Similarly, observe from (17) and (18) that $\hat{V}(0) - \hat{V}(1) = B_1(0) - B_1(1) + OV$. Therefore, from (11), $\hat{d}_1 = 0$ only if $B_1(0) - B_1(1) + OV \geq 0$ – i.e., when the possibility of future information is recognized, the current development decision should be based on an assessment of current preservation and development benefits *plus* option value. In the situation depicted in row (a), option value amounts to \$448 600 and development should be postponed as long as $B_1(1) - B_1(0) \leq 0.4486$, i.e., current development benefits exceed current preservation benefits by less than \$448 600.

In the case just considered, there is no correlation between the future benefits of preservation and development. In row (b), we consider a case where, although the marginal distributions of these benefits are the same as before, there is a correlation of −0.5 between them – i.e., future states of the world which make ecosystem preservation more valuable also tend to make development less valuable and vice versa. In this case, the option value rises from \$448 600 to \$566 600, an increase of more than 26%, and development should be postponed as long as current development benefits exceed current preservation benefits by less than \$566 600. What is happening here, in effect, is that the case for postponing irreversible development is enhanced because, in future periods, ecosystem preservation provides something of a hedge against factors adverse to development. The situation is reversed in row (c). Here, preservation and development benefits are positively correlated with a correlation coefficient of 0.5. Therefore, future circumstances which enhance the value of preservation also enhance the value of development. Ac-

Table 1. Calculations of option value.

μ_0	μ_1	σ_0	σ_1	ϱ	$P\{\theta_0>0\}$	$P\{I_1>0\}$	OV	Row
0.5	1.0	1.5	0.75	0	0.6306	0.9088	0.4486	(a)
0.5	1.0	1.5	0.75	− 0.5	0.6306	0.9088	0.5666	(b)
0.5	1.0	1.5	0.75	0.5	0.6306	0.9088	0.3062	(c)
− 1.0	1.0	1.5	0.75	0	0.2525	0.9088	0.0955	(d)

cordingly, the advantage from postponing the development project is diminished: the option value falls from \$448 600 to \$306 200, a reduction of more than 30%.

Finally, it is instructive to compare rows (a) and (d). In row (d) the prospects for preservation are far less attractive – in addition to foregoing the benefits of development, the future costs of preserving the natural ecosystem are expected to exceed the benefits by \$1 million, and there is only a 25.25% chance that ecosystem preservation will have a positive payoff. Nevertheless, the option value is \$95 500 – a not-insignificant amount – and development should be postponed as long as current development benefits exceed current preservation benefits by less than this amount.

Although the foregoing is merely a contrived example, we hope that it will have served to illustrate the power of the analytical tools developed in Sections 3 and 4. We have already emphasized that, in practical applications involving real-world development decisions, by far the most challenging problems are the construction of the benefit functions for ecosystem preservation and industrial development, the identification of their myriad uncertainties, and the quantification of probability distributions for these uncertain components. Nevertheless, even if these empirical problems can be resolved, it still is essential to have a sound conceptual framework for combining these benefit functions and determining the optimal decision. This is provided, we believe, by the analytical framework developed above.

6. Concluding remarks: some thoughts on policy

We have shown that, in principle, a positive option value is attached to ecosystem preservation and that this option value can be calculated if one knows something about the probability distributions of relevant benefit functions. The rather abstract analysis that lies behind these results is hardly part of the equipment of economic and environmental policymakers. Yet, the results have implications for policy, and policy does seem to be moving in the direction indicated. In this concluding section, we offer a few very brief thoughts on existing policy and regulations and on future challenges.

In the United States, where the potential losses are less severe than elsewhere in the world, the Endangered Species Act of 1973 authorizes the Secretary of the Interior to designate areas of 'critical habitat' and requires all federal agencies to ensure that activities authorized, funded, or carried out by them do not jeopardize the existence of endangered species *or modify their critical habitat* (emphasis added). The bar on activities that affect habitat has become quite controversial as it seems to some a blank check for regulation. Somewhat surprisingly, in view of the controversy, the Act was reauthorized for three years with a large majority and with only minor changes by the U.S. Congress in 1982. An explanation for this anticlimactic action may be that the potential impacts of the Act on economic activity have been largely unrealized (Harrington & Fisher, 1982).

Much more serious are the impacts on economic activity that could result from a thoroughgoing attempt to prevent extinctions in the tropical moist forests that are home to most species and most endangered species. These forests are currently being converted – cleared for farming and fuel wood – at a rate that has been estimated as between 100 000 and 200 000 km^2, or between 1% and 2% of the total annually.[3] Any attempt to prevent this largely subsistence-level activity would clearly be very costly unless alternative sources of food or fuel were found for those affected. On the other hand, a continuation threatens enormous losses.

The challenge is two-fold. First, ways must be found to discriminate among areas slated for conversion so that those richest in species can be afforded some measure of protection. Such an approach would recognize that some conversion will take place. The object would be to minimize the related losses.[4] Second, ways must be found to finance the desired protection. The 1975 Convention on International Trade in Endangered

[3] The high estimate is due to Myers (1981) and the low estimate to Lugo & Brown. For further details, see Harrington & Fisher.

[4] For a discussion of some promising ways to discriminate, see Harrington & Fisher (1982).

Species, which prohibits trade in specimens of endangered species, does not address the problem of preserving large areas of habitat for the thousands and maybe millions of plant and animal species that have yet to be listed as endangered because they have yet to be discovered and separately identified. The protection of even limited areas of habitat could be very costly, as we have noted, and the poor countries where they largely occur are not likely to want to bear these costs by themselves, especially when most of the potential benefits go to agriculture, industry, and medicine in the rich countries. Perhaps, development aid could be tied to a measure of environmental protection. The World Bank is already beginning to do this, and the practice might be extended to cover habitat preservation. Aid has often been tied in the past to purchase of goods of the donor country. Recipients might prefer a tie to protection of their own potentially valuable biological resources.

References

Arrow, K.J. & Fisher, A.C., 1974. Environmental preservation, uncertainty, and irreversibility. Quart. J. Econ. 88: 00–00.

Cicchetti, C.J. & Freeman, A.M., 1971. Option demand and consumer surplus: further comment. Q. J. Econ. 85: 00–00.

Clark, C.E., 1961. The greatest of a finite set of random variables. Operations Res. 9: 145–162.

Conrad, J.M., 1980. Quasi-option value and the expected value of information. Quart. J. Econ 94: 00–00.

Delwiche, C.C., 1970. The nitrogen cycle. Sci. Am. 223: 00–00.

Ehrlich, P. & Ehrlich, A., 1981. Extinction: The Causes and Consequences of the Disappearance of Species. Random House, New York.

Hanemann, W.M., 1982. Information and the concept of option value. University of California, Deparment of Agricultural and Resource Economics, Working Paper No. 228, Berkeley.

Harrington, W. & Fisher, A.C., 1982. Endangered species. In: P.R. Portney (ed.,), Current Issues in Nat. Res. Pol. Resources for the Future, Washington, D.C.

Henry, C., 1974. Investment decisions under uncertainty: the irreversibility effect. Am. Econ. Rev. 64: 00–00.

Lugo, A. & Brown, S., 1982. Conversion of tropical moist forests: a critique. Interciencia 7: 00–00.

Miller, J.R. & Lad. F., 1984. Flexibility, learning, and irreversibility in environmental decisions: A bayesian approach. J. Environ. Econ. Manage.

Myers, N., 1981. The exhausted earth. Foreign Pol. 42: 00–00.

Myers, N., 1983. A wealth of wild species. Westview Press, Boulder, Colorado.

Opler, P.A., 1971. The parade of passing species: a survey of extinction in the U.S. The Science Teacher 44: 00–00.

Schmalensee, R., 1972. Option demand and consumer's surplus: valuing price changes under uncertainty. Am. Econ. Rev. 62: 00–00.

Vietmeyer, N.D., 1979. A wild relative may give corn perennial genes. Smithsonian 10: 00–00.

Viscusi, W.K. & Zeckhauser, R.J., 1976. Environmental policy choice under uncertainty. J. Environ. Econ. Manage. 3: 00–00.

Weisbrod, B., 1964. Collective-consumption services of individual-consumption goods. Quart. J. Econ. 78: 00–00.

Zeckhauser, R.J., 1969. Resource allocation with probabalistic individual preferences. Am. Econ. Rev. 59: 00–00.

CHAPTER 17

Endangered species and the North–South dialogue

N. MYERS

Abstract

We are almost certainly losing species at a rate of one per day, perhaps more. This rate is likely to accelerate until we could well lose one million out of Earth's 5–10 million species by the end of the century, and a good many more during the course of the next century, i.e. until humankind reaches a style of living that reflects ecological accord with the planetary habitat that we share with millions of fellow species. Unlike other forms of planetary impoverishment, such as pollution, the mass extinction of species represents an irreversible loss of unique natural resources that could serve the expanding needs of innovative agriculture, medicine, industry, energy and genetic engineering into the indefinite future. This situation possesses a politico-economic dimension, insofar as a food two thirds of all species occur in the tropics, which means, by a large, in the developing world. At the same time, those nations with the technological expertise to exploit the genetic resources of species occur in the developed world, i.e. the genetically depauperate North. As has been recognized by a few governments of the North, notably the U.S. government, this carries security implications, even affecting the strategic interests of developed nations – especiallty if nations of the South attempt, as certain of them are already planning, to implement some sort of 'cartelization' of their gene reservoirs, in order to exert political leverage with the nations of the North. Hence the need to approach the threatened species issue within a context of the North–South dialogue.

Introduction

With each day that goes by, we benefit from wild species and their genetic resources in many more ways than we may be aware of. Through their contributions to modern agriculture, medicine, industry, energy and genetic engineering, wild plants and animals already support our welfare in thousands of different forms – and this, after scientists have conducted cursory screening of only one in ten of Earth's 250 000 higher plant species, more detailed screening of only one in 100 plant species, and systematized and intensive examination of only about 5000 species altogether. As for animals, the number screened in any way at all amounts to only a very trifling fraction of the 5–10 million species that make up the planetary spectrum. Were we to undertake a comprehensive and methodical screening of many more of Earth's species, we would surely come up with many times more products than we currently enjoy in our daily lives (Myers, 1983).

Yet while we enjoy these many benefits from wild species, the extinction rate has reached unprecedented levels. Detailed documentation is still forthcoming, to demonstrate the true rate. But we can realistically estimate that we are now losing

D.O. Hall, N. Myers and N.S. Margaris (eds.), Economics of ecosystems management.
 ISBN-13: 978-94-010-8928-9

one species per day right now; and by the end of the century we could well lose one million species. By the middle of the next century, we may even withness the disappearance of at least one third of the planetary complement of species (Myers, 1979, 1984).

This mega-scale extinction of species can be viewed as one of the great 'sleeper issues' of our times, on the grounds that few resource problems are more profound in their ultimate implications, while less appreaciated by political leaders and the general public.

Fortunately there is an emergent realisation that Earth's stock of species represent some of the most valuable natural resources with which we can confront the challenges of the future. All too slowly but nonetheless steadily, we are coming to appreciate that birds and reptiles, insects and sea slugs, fishes and mammals, flowers and mosses, even fungi and bacteria, hold all manner of benefits in store for us if only we can ensure their survival. To consider the needs of agriculture, for example, we can recognize that the Green Revolution is being supplemented (not supplanted) by an equally revolutionary phenomenon, the Gene Revolution – a breakthrough in agricultural technology that may soon enable us to harvest crops from deserts, farm tomatoes in seawater, grow super-potatoes in many new localities, and enjoy entirely new crops such as a 'pomato). Genes are the hereditary materials of each species' make-up; we can now isolate and manipulate them. The sophisticated techniques of genetic engineering may even be bringing us closer to the day when every citizen on Earth will be adequately nourished. Apart from the potential of genetic engineering, we can look for many other agricultural advances, through improved versions of conventional crops and through entirely new foods developed from wild plants (much the same applies to domestic livestock). As for medicine, we can look forward to one advance after another, some of them to match the discovery of penicillin. Scientists in the medicinal field foresee more innovative advances during the last two decades of this century than during the previous two centuries – again, partly through conventional adaptations of wild genetic resources and related strategies, partly through the applications of genetic engineering. As for industry, our creative applications of the gene reservoirs of wild species may seen make our present industrial scene appear like a hangover from the Stone Age.

So much for some basic scientific considerations. Much documentation is available in the literature on the biology, ecology, biochemistry, pharmacology, pharmacognosy, and several other related fields that illuminate our applications, both actual and potential, of wild species and their genetic resources. These various aspects have been dealt with at length elsewhere, and will not be considered further here. Rather this paper will focus on certain economic and political dimensions of the topic.

Distribution of wild species and their gene pools

At least two thirds of all species occur in the tropics. Although the tropics cover only a small portion of the Earth's surface, we need not be surprised at this concentration of species. With their year-round warmth and often with year-round moisture, the tropics have served as the planet's main powerhouse of evolution. By contrast, the temperate zones and the remainder of the planet are relatively deficient in species concentrations and gene reservoirs. Indeed the developed world can be viewed as genetically depauperate. Yet it is the developed nations that, in the main, possess the technological capacity to exploit species and their genetic resources for economic advantage. This situation raises several issues salient to North–South relations, and in particular to economic questions addressed in negotiations within the North–South dialogue.

The need to preserve germplasm resources is but one of several global resource and environmental issues that emerged during the 1970s, and that are likely to receive increasing attention during the foreseeable future. These problems, which often underline differences between the developed and developing worlds, include food, energy, and pollution, as well as 'environmental commons' issues such as the oceans. Antarctica, the ozone layer, and the carbon dioxide cycle. They highlight the interdependent nature of society at large, and the need for collective action on the part of the community of nations. Moreover, many of these problems are interrelated: progress can be made on one front only by tackling several other problems simultaneously.

Characteristic of problems that affect most if not all nations, and that thus can be characterised as intrinsically international if not supranational, is the problem of biotic impoverishment. Plainly this is not merely an issue for wild life enthusiasts. The demise of a single species represents an irreversible loss of a unique natural resource. The planet is currently afflicted with various severe forms of environmental degradation, for instance desertification and pollution. But whereas these forms of degradation can generally be reversed, extinction of species cannot. When a species is eliminated, it is gone for good – and, in strictly utilitarian terms, that will frequently turn out to be for bad.

Yet herein arises a complex question of 'ownership), plus associated questions of property rights. To whom, for example, does the ayaeay belong – or 'belong'? The ayeaye, a form of primitive primate, is a species endemic to Madagascar. It thereby comes under to sovereign jurisdiction of Madagascar. At the same time, however, the ayeaye can surely be regarded as a subject for acceptable concern on the part of the rest of humankind – especially if this lemur were to prove (as has the cotton-topped marmoset among other primates) to offer a critical research model for the study of cancer. Similar considerations apply to the other 18 species of lemur in Madagascar, a special group of primates, probably with special insights to fundamental aspects of biology, evolution, physiology, and the like; and all of them threatened.

Much the same applies to other species that occur within single nations, such as the monkey-earing eagle of the Philippines and the snail darter of the United States. And the same too for species that occur in several countries, such as the cheetah that finds habitats in about 20 countries of subsaharan Africa.

In light of this situation, we can surely venture a view to the effect that the Earth's stocks of genetic variability constitute part, in some senses at least, of the common heritage of all persons now alive on Earth – plus all future inhabitants. They can hardly, in equity, be deemed to be the sole exclusive concern of nations within whose territorial boundaries they happen to exist. Each citizen of Madagascar enjoys his or her share in the planetary patrimony of species and genetic resources every time he or she takes a cop of coffee: without the wild relatives of the coffee plant, and their germplasm infusions for the coffee crop in a dozen producer countries, a Madagascar citizen would be contemplating a one-dollar cup of coffee. The main gene centre of wild coffee germplasm, Ethiopia, enjoys concesssionary supplies of grain and other relief food from North America, thanks to the exceptional productivity of the great grain belt across Canada and the United states – which continues to flourish, and even expands its bounty year by year, by virtue of wheat and corn genetic resources from Central America, the eastern Mediterranean and other centres of genetic variability. Brazil, which supplies wild rubber germplasm to Southeast Asia's rubber plantations (and, like Ethiopia, is growing increasingly leary of traditional approaches to trade in wild genetic resources) depends on unimpeded access to germplasm supplies in diverse parts of the world in order to sustain its sugarcane, soybean and other leading crops. Without access to foreign sources of fresh germplasm year by year, the nations of Europe would quickly find their aggriculture declining in output.

These illustrations are drawn from the field of agriculture, especially crop agriculture. We can draw similar parallels with regard to medicine, industry, and other major sectors of modern economies.

At the same time, les us remember that whereas the ayeaye and other lemurs may in certain strong senses belong to Madagascar, and the State of Madagascar may exercise some form of theoretical property rights in the species, this does not obviate a fundamental difficulty, that of 'common property'. No organization, and no person or group of persons, in Madagascar exercises active ownership over the lemurs. Of course the government exercises nominal ownership; but this is a far different kind of ownership from the find exercised by an individual over, say, his horse. Nor can any authority in Madagascar be readily charged with responsibility to make certain the lemurs survive – not, at any rate, with sufficient power to withstand the economic forces tending towards the lemur's destruction (principally forest depletion on the part of multitudes of small-scale farmers; Madagascar's forests are only one fifth of what they once were). True, Madagascar can assert, through its government, that it seeks to exercise ownership rights by declaring leagal protection for the ayeaye and other lemurs as national resources. But the government cannot adequately enforce this assertion, since too many citizens have a strong private interest in seeing the lemur's forest environments give way to cultivation, and the government has too little capacity to enforce its laws in every last corner of the forest habitats.

Much the same applies to the cheetah in Africa (Meyers, 1975, 1976). Neither the governments in the principal countries concerned, nor conservation bodies in those countries, nor the global community, can exercise property rights or any other form of protective authority to prevent the cheetah being depleted by stockmen who are unwilling to see their domestic herds marauded by a predator with which they share savannah grasslands. African governments can pass laws to protect 'their' cheetahs, but they do not have sufficient enforcement powers to protect scattered individual cheetahs across thousands of square kilometres of savannah Africa. So the stockman who disposes of cheetah to prevent commercial loss has good hope of getting away with his illegal action. Still more to the point, the cultivator who digs up cheetah habitat can hardly be charged with any offence at all, either legal or moral – supposing, as is generally the case with cheetah habitat, the land lies outside a park or other protected area.

Responsibility of the developed world for developing-world species

The instance of the cheetah points up an international linkage that affects the survival prospects of cheetah habitat. The stockman increasingly seeks to produce more beef for European markets, in order to meet marketplace demands of European consumers who seek beef at 'reasonable', i.e. non-inflationary prices (at least, as compared with prices for beef from domestic sources within Europe). So the stockman sees incentive to make his livestock-raising operations more extensive, encroaching into additional parts of tranditional cheetah habitat. At the same time, he often has to operate on shrinking profit margins. So he sees incentive to make his operations more intensive as well, and to eliminate any sources of 'waste' – which include, notably, wild predators that take his livestock. To this extent, marketplace pressures several thousand kilometres away, in a sector in the developed world, militate against the cheetah's survival outlook in savannah Africa.

Similar economic-ecologic linkages can be traced at international level. Between North America and Central America, there is the 'hamburger connection' (Myers, 1981). Between all main sectors of the developed world and tropical forest regions of the developing world, there are commercial timber connections that foster unduly negligent harvesting of specialist hardwoods from tropical forests (Myers, 1980, 1984). Many other such linkages operate to the disadvantage of species located in the developing world – and, let us remind ourselves, the developing world, being roughly co-extensive with the tropics, harbours at least two thirds of all Earth's species.

Insofar as everybody bears some responsibility for the species problem, wherever it is located, and insofar as everybody will suffer from mega-extinction of species(through loss of utilitarian benefits, if for no other reason), everbody shares the responsibility of relieving the problem. This postulates efforts by national and international institutions on a scale that reflects the increasingly interdependent character and needs of the global community (Myers & Myers, 1982, 1983.

Exclusive and inclusive rights and activities

In these circumstances, a central problem – and an opportunity – lies with rights and activities of nation-states that can be either exclusive or inclusive. This is a basic issue of international law, and like the concept of sovereignty, it is in a state of flux – however little that fact may be recognised by some portions of the global community.

'Exclusive' refers to those rights and activities that affect predominantly only one nation. The term is thus associated with national independence. 'Inclusive' refers to those rights and activities that have significant effects for other nations, or externality effects. Thus the term links up with interdependence. By implication, 'inclusive' refers especially to those resources that are, in whatever degree, resources of common heritage (generally resources of common property, too). Examples include species concentrations, important ecosystems, the oceans and major water bodies, marine fisheries, rare geomorphological features, and the planetary air mantle. Some of these, e.g. the planetary air mantle and the high seas, are clearly resources that belong entirely to the entire community. Yet they are sometimes subject to exploitation by individual countries with severe repercussions for all humankind. France or China explodes a nuclear device in the atmosphere, and thereby exposes people in far-away lands to radioactive follout. Virtually all industrialized nations use the open oceans as dumping grounds for huge amounts of toxic chemical wastes.

Other categories of resources, such as concentrations of species and important ecosystems, 'belong' to individual nations on the grounds that they are located within territorial jurisdictions. At the same time, they are clearly of legitimate interest to the community at large. This applies equally to major tracts of tropical forests, with their 'watershed services' for territories far beyond the forestlands themselves. A nation may exploit its forests as resources over which it believes it exercises sovereign rights; and the nation thereby engages in what is, from certain standpoints, an apparently exclusive activity. Yet the forest exploitation, if not strictly regulated, may precipitate severe spillover consequences for other nations. Deforestation in Nepal leads to disruption of river systems (flash floods, followed by reduced flow) for nations downstream, viz. India and Bangladesh, causing $2 billion of damage in mid-1978. Widespread forest exploitation in Indonesia or Brazil can lead to extinction of untold stocks of species. As economic and technological activities grow evermore advanced and complex, they trigger off growing numbers of externalities within an increasingly integrated global community. In turn, the adverse impact of these activities on common heritage/common property resources becomes more pervasive and acute.

The situation has been graphically expressed by two Yale law professors, McDougall & Schneider (1977):

'It is today widely recognized that the globe – or, more precisely, the entire earthspace environment – is an ecological unity both in a basic scientific sense and in the interdependencies of the social processes by which man uses it. All living creatures, including man, that inhabit the planet, are united with each other and with their non-living surroundings by a network of complex and interdependent natural and cultural components known as the 'planetary ecosystem' . . . It is the more specific ecological unities or interdependencies of this comprehensive ecosystem whigh make it a single shareable, and necessarily shared, resource. What is true about interdependencies in the enjoyment of the atmosphere, the oceans, the air space and outer space, drainage basins, and land masses when considered separately, is equally true of the indivisible whole which they comprise.'

This illuminates the dilemma of inclusive and exclusive rights and activities of nations, with particular regard to environmental resources. Certain resources, due to their outstanding importance for the community at large, deserve to be maintained as shared resources under inclusive competence, as free as possible of exclusive competence or dominance. These resources include the atmosphere, the climate and weather, and the oceans and major water bodies. Almost all species occur within the territories of individual nations, yet they should, in some sense, be considered as an inclusive interest of the entire community. Thus species straddle both exclusive and inclusive concepts, a situation that raises complex difficulties for their conservation.

The community of nations is faced, then, with a growing problem of resource exploitation and environmental spillovers. The problem has not been ignored by international bodies, even though they have taken few practical steps to face up to the challenge. The 1972 Stockholm Conference on the Human Environment came up with a Declaration that asserted, under Principle 21, that

'States have, in accordance with the Charter of the United Nations and the principles of international law, the sovereign right to exploit their own resources pursuant to their own environmental policies, and the responsibility to ensure that activities within their jurisdictional control do not cause damage to the environment of other states or of areas beyond the limits of national jurisdiction.'

Similarly, the recent Charter on Economic Rights and Duties of States proclaimed, under Resolution 3281 of the U.N. General Assembly, that

'The protection, preservation and enhancement of the environment for the present and future generations is the responsibility of all States. All States shall endeavour to establish their own environmental and developmental policies in conformity with such responsibility. The environmental policies of all States should enhance and not adversely affected the present and future development potential of developing countries. All States have the responsibility to ensure that activities within their jurisdictional control do not cause damage to the environment of other States or of areas beyond the limits of national jurisdiction. All States should cooperate in evolving international norms and regulations in the field of the environment.'

Significantly, both these last two statements stress the responsibility of nations to make sure that their own activities do not harm the environment of 'downstream nations'. It is clear how this responsibility applies to the environment of nations in close ecological association. How about nations further afield, with which the relationships may not be so direct and plain? If a nation is to avoid activities that may harm 'the environment of other nations', may this not mean the planetary ecosystem in the sense referred to by McDougall and Schneider? Were a tropical forest nation to cut down a sector of its forests that include the sole habitat of some primate or bird species, could this not be construed as eliminating an integral component of the planetary ecosystem, and thus inflicting a degree of environmental injury on all nations?

High-value resources: rights and interests

The argument can be carried further. A number of resources, notably oil reservoirs, are located, through geographical accident, within the territories of only a few nations. Oil is of such critical importance to the welfare of all nations that the entire global community can assert a strong legitimate interest in how it is made available. A case can even be made for 'rights of access' on the part of all needy nations, as well as 'rights of possession' on the part of the fortunate few. A similar case can be made for a range of high-value resources, such as grain reserves, fertilizer stocks and other scarce commodities.

Indeed, the time may be coming when the Earth's endowment of high-value resources will be seen as an asset to be handled for the benefit of all mankind. This will mean that the community of nations will have to reconsider what is implied by ownership, management, use, allocation and conservation of these resources. This would be a tough and virtually unprecedented challenge. Yet some step of this sort may become necessary as global society grows ever-more hungry for critical resources. As long as there seemed to be plenty of oil to meet everybody's needs for ever, there was no need for a commonly agreed way to share out Earth's bounty. Now that we are into a fundamentally different scene, the ways of the past no longer measure up.

It is within this perspective that the problem of species can be looked at afresh. The guts of the situation boils down to this: outright possession of an individual species is generally limited to a few nations (if not a single nation), whereas significant benefits can accrue to all nations (through modernized agriculture, medicine and other pragmatic purposes). Much the same applies to tropical forests: if the forests continue to decline through over-exploitation on the part of possessor nations, the adverse repercussions will fall not only on the nations in question and on neighbouring nations (through degradation of watersheds, river systems and the like), but on nations far and wide (through disruption of climatic patterns, elimination of exceptional concentrations of species, etc.). As already pointed out, the two sets of problems are inter-related. The principal cocoa-growing nations are located in West Africa, while the genetic resources on which modern cocoa plantations depend for their continued productivity are found in cocoa's original source areas, the forests of Central and South America. The principal banana-growing nations are in Central America and the West Indies, while the genetic base is in the forests of South-east Asia.

In sum, then, the challenge of conservation of species, as of tropical forests, is of special concern to the entire global community. Even though almost all species exist within national jurisdictions, their survival can hardly be viewed as a matter solely for the nations in question. This implies two forms of responsibility – highly compelling responsibility, considering what is at stake. First, any nation faces a duty to safeguard the species within its territory, on behalf of all humankind. Second, all humankind faces a duty to offer whatever support is required – finance, skills, etc. – to enable individual nations to discharge their duties.

Some scope for action

We have seen that everybody bears some responsibility for the species problem. We have also seen that everybody will suffer through mass extinction of species, if only through the utilitarian benefits foregone. This argues that everybody should share the responsibility of relieving the problem. In short, the situation postulates efforts by institutions, at both national and international levels, of a nature and scale that reflect the increasingly interdependent character and needs of the global community.

The first priority in addressing the issue of disappearing species is to establish the problem on political agendas as a strategic resource issue. Consciousness raising will entail an extensive educational campaign through the mass media, as well as forums for political leaders and policy makers. Formal policy commitments on the part of governments and international forums such as the U.N. General Assembly should then be possible. A two thirds majority endorsement of a commitment to species preservation by the United Nations would foster international participation and minimize resistance to collective initiatives.

A species trust convention could be created in recognition of the principle that species constitute 'resources of common heritage, to be maintained on behalf of the global community now and forever'. Collective responsibility for the common heritage would not mean collective rights to particular resources. Far from playing down sovereignty, this approach would emphasize it as a functional concept. Thus, a country would no longer be left to rely on its own isolated efforts to protect its species, but could expect assistance from the community at large.

Developing countries often have difficulty keeping their citizens alive, let alone guaranteeing the survival of their wild creatures. The benefits of efforts at species conservation tend to accrue to those states with the technological expertise to exploit species' genetic resources – that is, the developed countries. Most people in the developing world do not live long enough to contract cancer or heart disease. To the extent that developing countries are currently trying to safeguard their species through parks and related measures, their efforts amount to a resource handout to developed countries.

Trifling budgets

Several programmes designed to maintain the Earth's stock of species already exist. But they are limited in both scope and scale, and they need to be massively supplemented. International development organizations, such as the World Bank, the United Nations, and bilateral aid agencies, could subsidize conservation activities in developing countries throug greatly increased support for protected-area networks and similar measures to assist species. The World Bank has generally directed no attention to species as such, despite their value as highly diverse stocks of natural resources. But just recently, the Bank has come out with a forth-right assertion concerning depletion of tropical forests – a problem that the Bank views as substantial enough, and deteriorating fast enough, to warrant some mega-scale action. In order to safeguard the millions of species in tropical forests, among other rationales, the Bank is proposing an overall programme to assist the tropical forest situation in general, with a budget to be contributed by developed nations (in reflection of their stake in tropical forests' survival) to the tune of $ 1 billion a year for ten years (Spears, 1983). This parallels and independently worked-out programme of almost identical form, with the same budgetary proposals, developed by Myers (1984).

The principal United Nations agency in this area, UNESCO, operates a project on natural areas and genetic resources – essentially a clearing-house for information. In addition, its World Heritage Fund supports a handful of exceptional parks around the world. But both these activities receive trifling budgets. UNESCO further seeks to establish a global system of biosphere reserves representing the earth's 200 biotic provinces that would harbour sample communities of species. But only one third of the reserves have been set up, even though instituting and operating the rest would cost only about $ 80 million a year.

The United Nations Environment Programme is still reviewing its options in regard to threatened species, genetic resources, and wild-land habitats, with a total budget of only slightly more than $ 1 million. As for bilateral agencies, the U.S. Agency for International Development leads the field in recognizing the value of species conservation. To date, however, the agency's operational activities in this field have achieved no more than make a preliminary dent in the problem.

Although citizen conservation groups have made contributions, they have been unable to attract sufficient funding. The World Wildlife Fund, working in conjunction with its scientific arm, the International Union for Conservation of Nature and Natural Resources, has been able to mobilize only about $ 70 million since its founding in 1961.

This dismal record confirms that the species problem tends to be perceived largely in scientific terms and therefore lacks political clout. What is needed is a campaign of sufficient scope to confront the overall problem. International assistance should recognize common responsibility on the part of the community of states for a deteriorating asset of common heritage. An appropriate measure could involve creation of a trust for species to establish an international fund in support of conservation efforts in those countries where the need is greatest and local funds are in shortest supply.

It is difficult to offer more than approximate estimates of funding requirements necessary to make significant progress toward species preservation. Some preliminary calculations, of strictly exploratory and tentative sort, suggests that we could do a substantive job with a budget of $ 200 million a year (Myers, 1983).

International institutional initiatives

At least as important as the matter of funding is the question of norms and procedures with respect to resources issues. The species problem, as one of an emergent category of controversial resource issues that are intrinsically international in nature, will establish precedents for other global-commons questions, such as carbon dioxide, chlorofluoromethanes, and deep-sea mineral deposits. Fortunately, there are some precedents already available.

At least three precedents should be of help: the Convention on Wetlands of International Importance, the Convention on Conservation of Islands for Science – both of which safeguard prime habitats with their species concentrations – and the Convention on International Trade in Endangered Species (CITES), established in 1975 (largely through U.S. instigation) and now ratified by more than 60 nations. CITES seeks to cut down the vast commerce in threatened creatures and their products, and it has had some success in restraining those persons who make money at the expense of declining species. But it does nothing about the great number of people who harbour no malign intent toward wildlife, but who constitute a far greater threat through their disruption of wildlife habitats. For every one species that is endangered because of the poacher's poisoned arrow, hundreds are in trouble because of the cultivator's digging hoe. Nonetheless, CITES could serve as a model for a further convention of comprehensive scope to assist threatened species through extensive networks of protected areas and gene banks.

The role of the United States

The United States has played a splendidly pioneering role in this field during the late 1970s. At least two initiatives are worthy of mention: the measure to require AID to take explicit account of save-species considerations in its development aid programmes; and the Biological Diversity Strategy Conference convened by the State Department in 1981. Because of this fine trail-blazing record, the United States deserves special attention here. Herewith some suggestions for further initiatives on the part of the United States.

The nation could take several steps to instigate a save-species strategy on the part of the international community. In conjunction with other developed and gene-poor nations, the United States should recognize the benefits of maintaining a low political profile in its activities, broaching the issue in a manner that reinforces concensus and cooperation, while playing down politics and reducing divisiveness. It should demonstrate a spirit akin to that with which it is handling the carbon dioxide question – an issue where, as with species, a central concern lies with the distribution of benefits and costs of resource management. Thus, the best role for the United States in a save-species campaign lies with quiet diplomatic activities, plus financial contribution and technical back-up for whatever international measures emerge, leaving other parties to blow clarion calls for action.

Apart from instigating a strategy at the multilateral level, the United States could pursue a number of bilateral measures. For example, it could offer assistance to Mexico in order to protect natural environments that may harbour wild strains of modern crops. Indeed, giving species preservation greater visibility within the U.S. aid programme may increase domestic support for development assistance in general.

The means are certainly available to save many species. The funding could be made available if it were given the priority according to food, energy, and pollution. The missing element is political will – and it is in this direction that the United States can foster a growing concensus among the community of nations, in order to generate the key factor of political commitment. If nothing is done before the problem becomes all too apparent, the process of mass extinction may have generated too much momentum to be halted. As is the case with the carbon dioxide problem, the international community must act ahead of time, before it is too late. The United States has not only taken a lead in drawing attention to the threatened species problem through its 1981 Strategy Conference, but it has taken a similar initiative with respect to tropical forests, as natural resources of exceptional value and as the Earth's major repository of species. The United States can pursue further measures in support of a generalized save-species campaign.

Let us conclude this paper by returning to the North–South dimension of the issue. In the light of this context, there is a premium on a cooperative spirit for a save-species campaign. Inasmuch as most species exist in developing countries, developed countries can exercise only indirect influence over their survival prospects. This means that the developed world will derive greater benefit from careful coordination within the international community.

An approach in this spirit is all the more pertinent because developing countries may be less inclined to participate in an international campaign to safeguard the world's wealth of species as long as cooperative relations are lacking in respect to their principal concern, North–South economic issues. However, and increased emphasis by the developed countries on the preservation of species might also help induce the pirit of international cooperation necessary in addressing broader international economic, resource, and security issues. A campaign in support of species could help articulate the common interests of nations. It might even encourage governments to adopt a more collective approach to other international issues, leading to an enhanced world order.

References

McDougall, M.S. & Schneider, J., 1977. The protection of the environment and world public order. In: B. Preger, H.D. Lasswell & J. McHale (ed), World Priorities, pp. 81:114. Transaction Books, New Brunswick, New Jersey, U.S.A.

Myers, N., 1975. The cheetah Acinonyx jubatus in Africa. IUCN Monograph No. 4. International Union for Conservation of Nature and Natural Resources, Morges, Switzerland.

Myers, N., 1976. The cheetah in Africa under threat. Environ Affairs 5(4): 617–647.

Myers, N., 1979. The Sinking Ark. Pergamon Press, Oxford, U.K. and New York, U.S.A. (third printing in 1983).

Myers, N., 1980. Conversion of Tropical Moist Forests. Report to National Academy of Sciences. National Research Council, Washington D.C., U.S.A.

Myers, N., 1981. The hamburger connection: how Central America's forests become North America's hamburgers. Ambio 10(1): 3–8.

Myers, N. & Myers, D.M., 1982. From the 'duck pond' to the global commons: increasing awareness of the supranational nature of emerging environmental issues. Ambio 11(4): 195–201.

Myers, N., 1983. A Wealth of Wild Species. Westview Press, Boulder, Colorado, U.S.A.

Myers, N. & Myers, D.M., 1983. How the global community can respond to international environmental problems. Ambio 12(1): 20–26.

Myers, N., 1984. The Primary Source. W.W. Norton Publishers Inc., New York, U.S.A. (in press).

Spears, J.S., 1983. Saving the Tropical Forest Ecosystem. Office of the Forestry Adviser, The World Bank, Washington D.C., U.S.A.

CHAPTER 18

Present and future pest management

BENGT LUNDHOLM

Since agriculture started there has been a continuous battle between man and pests. After the second World War chemical pesticides came into general use. Their efficiency was obvious and their side effects seemed to be small. Some thought that man had developed the ultimate weapon against the pests. A new industry developed and is still growing, supporting powerful multinational companies.

But then *Silent Spring* was published and a lot of unexpected side effects were revealed. A very strong opinion against pesticides in general appeared. A fierce debate started, where emotions on *both* sides were very strong, and where *both* sides engaged 'objective' experts.

Pesticides turned out to have many unexpected side effects, with risks in three main areas:

1. risks in relation to human health;
2. risks related to the environment;
3. risks that the efficiency of the pesticides is decreasing.

In evaluating the risks, both the space and time dimensions have to be considered. In relation to space it has to be kept in mind that the pesticides and their metabolites can be dispersed over long distances. They may even have a global distribution and thus the effects, if any, may appear at any place, especially as we have the phenomena of biological concentration. Almost the same situation exists in relation to time, as the effects may appear long after the application.

Of importance in risk evalution, especially from our point of view, is to be aware of the enormous differences that exist between the industrialized and the developing countries, caused by different biological, economic and social conditions.

Pesticides may affect human health in two ways. The first is when labourers handle pesticides during industrial production or agricultural application. In industrial countries, where the main factories are situated, workers are exposed to pesticides, but there are in general strict health controls to eliminate the risks. In the developing countries the farmworkers are in the risk zone. According to reports from WHO there are annually 500 000 cases of toxification in the developing countries, of which 5 000 die. It has to be noted, that these cases reflect only effects appearing a short time after exposure. Thus the long-term effects such as cancer are not included.

The other health effect is caused by the residues of pesticides contaminating food and drink. Here we have an ongoing discussion related to carcinogenicy and terratogenicy. The experts from the chemical industry claim that the residues are so slight that they can be safely neglected and that, at least in the industrial countries, the consumers are protected by sufficient controls. Experts from the other side claim that control is very slack and that there exist no threshold levels for carcinogenic substances – even the smallest amount will have effects. These effects will become apparent in a large population, although the risk for the individual is small. However, the general public and thus

D.O. Hall, N. Myers and N.S. Margaris (eds.), Economics of ecosystems management.

ISBN-13: 978-94-010-8928-9

also the politicians are very concerned about these risks.

The risks in the relation to the environment are well known. I will not elaborate on the decline of raptorial birds or DDT in Antarctic seals. But from the industrial side it has been claimed that this is the main problem. In a book called 'Pesticides and Human Welfare' sponsored by seven chemical companies the following is a summing up.

> 'It has to be realised that a decision often involves weighing an advantage to human beings against those to wildlife.'
> (Pesticides and Human Welfare, 1976. D.L. Gunn & J.G.R. Stevens (eds.).

In this paper I will deal in more detail with the third risk: the decreased efficiency of the pesticides and especially insecticides. From the beginning, it was noticed that an application of insecticide could result in an increased pest population. The reason for this was that the insecticide not only killed the pests but also the natural enemies of the pests. And as the pest insects then multiplied very fast without natural enemies present, the end result was very negative from the farmer's point of view, as repeated pesticide applications were necessary. Another effect of pesticide applications was that the pest itself disappeared to be replaced instead by new pest species.

These two phenomena, resurgences of the pest and appearance of new secondary pests, are very common and the remedy was always more pesticide applications. It is important to realize that once you have started to use pesticides, you have to go on using chemicals so as not to lose your crop. You are in a vicious circle. In the report 'Man's impact on the global environment' of 1970 written by the Massachusetts Institute of Technology, pesticides were compared with narcotic drugs. 'Thus pesticides not only create the demand for future use (addiction), they also create the demand to use more pesticide more often (habituation). Our agricultural system is already locked into this process and it is now spreading to the developing countries.'

From a commercial point of view, pesticides seemed to be the ideal commodity, as they created a steadily increased demand. Even more marvelous – if herbicides were used the insect population increased and you had to apply insecticides! This was first studied by Dr. David Pimentel at Cornell University. He once passed a cornfield with certain parts heavily infested by insects and began to investigate the situation. He found that the infested parts had been treated with herbicides. The herbicides had changed the chemical composition of the plants and made them more palatable to the insects.

In this chemical warfare the pests are far from defenceless, and develop resistance against toxic substances. This resistance is genetic and renders the substances harmless. The ability to establish resistance depends on the existence of suitable genes. Even with an extremely small frequency these genes will be favoured by the very strong selection pressures caused by the chemical treatment. The time required to establish resistence is dependent on the number of generations required. If there are many pest generations during the year as in the humid tropics, resistance will develop rapidly. Another factor which not has been given sufficient notice, is that partly because of increased pesticides use, and partly because of the increased numbers of biologically active chemicals in the environment, there must be a general increase in the mutation rate. This will of course also give mutations for resistance. This phenomenon is of general biological importance and has been stressed by the Russian botanist E. Slepyan.

In the beginning, the solution to the resistance problem was to find new pesticides. The chemical industry carried out extensive research in order to handle this situation. After the Second World War a large number of pesticides were developed and put on the market. But resistance developed rapidly in response to insecticides and more slowly among herbicides and fungicides. I will deal in more detail with the insecticides. Four main groups of insecticides have been developed. We have the whole range of organochlorine compounds including DDT (introduced 1939) Lindane (1942), chlordane, dieldrin and toxaphene (between 1945 and 1948). The second group were the organophosphorous insecticides, with parathion and malathion released in 1945; the carbamate group appeared in 1953 The first of the fourth group, the synthetic pyrethroids, came out in 1978. This arsenal has, however, its limitations because of the phenomenon of cross-resistance. This has not been properly considered in the general discussion in spite of the fact that it has been known for a long time.

When an insect has become resistant against one chemical substance it is at the same time showing resistance towards other chemicals belonging to the same group. Often, it also shows some resistance to chemicals from other groups or quickly develops such a resistance.

The development of resistance can be illustrated by the situation in relation to malaria (see Table 1). The increased resistance in the malaria mosquitoes is partly the result of the unrestricted use of pesticides in tropical and subtropical countries especially in agriculture. In 1968, 38 species of the genus *Anopheles* were resistant. By 1975, 42 species were resistant: 41 had resistance to dieldrin, 24 to DDT and of these 21 species were resistant to both substances. Also revealing is the resistance in the genus *Culex*: in 1968 19 species were resistant and by 1975 the number had increased to 41, of which 35 were resistant to DDT, 26 to dieldrin, 16 to organophosphorous and 2 to carbamates.

It has correctly been claimed that pesticides and especially DDT have saved millions of lives by reducing the mosquito numbers. But now mosquitoes have become resistant and malaria is making a comeback. It has become very difficult to find new chemical weapons against malaria, as the following figures show. The World Health Organization (WHO) has a special mechanism for testing chemical formulations to be used against mosquitoes. Between 1960 and 1970 about 2 000 formulations were tested. In 1963 the number was 29 and in 1974 it was down to 7. Of the tested formulas very few are approved for use (see Table 2).

Table 1. Resistance in mosquitoes.

Anopheles:	
1968	38 species resistant
1975	42 species resistant
	41 resistant to dieldrin
	24 resistant to DDT
	21 doubly resistant
Culex:	
1968	19 species resistant
1975	41 species resistant
	35 resistant to DDT
	26 resistant to dieldrin
	16 resistant to organophosphorous
	2 resistant to carbamates

But the industry also has its problems. New safe and efficient chemicals have to be found. Every year more and more chemicals have to be screened in order to find one new pesticide. In 1955 2 000 had to be screened: in 1978 the figure was over 15 000 (Figure 1).

The cost of development has increased because of the very elaborate safety testing necessary in order to have the pesticide registered. The time to develop a new compound is now, in average, 8 years (see Fig. 2).

From an ecological point of view pesticides with a narrow spectrum are better than those with a broad spectrum, but from the commercial point of view, broad spectrum pesticides are more attractive as their market is larger. The industry also has to consider that a new pesticide, because of the resistance problem, may have short commercial life.

The big chemical companies are well aware of the difficulties and are trying new ways to keep their markets. One is to take over the seed trade and they will in the future sell seeds and pesticides as a package. The marriage between seed and pesticide will mean new problems in relation to both plant protection and plant breeding.

Table 2. WHO testing scheme.

1960–1974	ca. 2000 formulations
1963	– 200
1973	– 72
1974	– 29
1975	– 7*
	Relative costs
	DDT 1
	Malathion 3.5
	Carbamate 10
	Pyrethrin 100

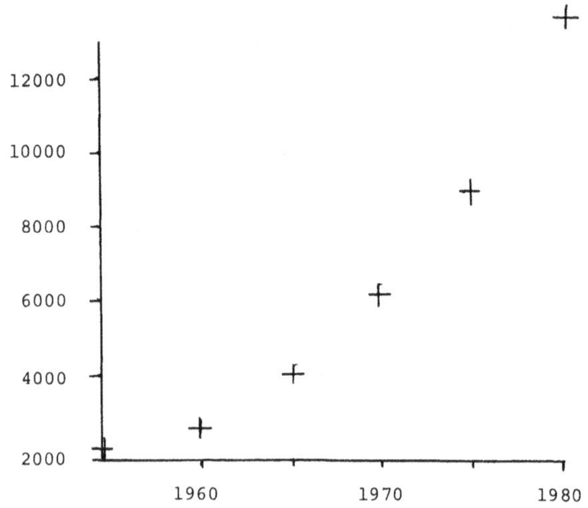

Fig. 1. Number of chemicals screened to get one registered pesticide.

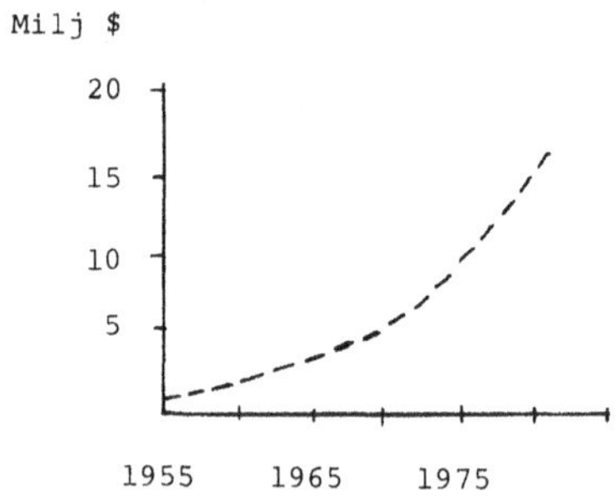

Fig. 2. Costs to develop one pesticide.

As pointed out earlier, resistance is developing most rapidly in tropical and subtropical areas. It is in these areas that we have serious problems, in spite of the fact that so far the amount of pesticides used there is rather small (Table 3). In order to increase agricultural production in the developing tropical countries pesticides use will increase. This will mean severe problems with resistance in the near future.

Table 3. Pesticide use.

North America	31%
Western Europe	22%
Asia	16%
Eastern Europe	13%
Latin America	12%
Africa	5%
China	not included

To sum up, the future of the chemical pesticides is not bright. They will turn out to be a cul-de-sac. Other alternatives to pest control have to be found.

Before chemical pesticides became dominant several other methods were used. These were more adapted to the biological conditions and may be called *biological control methods*. Some still use this term only when insect predators are used to control an insect pest. I use the term in a wider sense to include all methods which offer alternatives to chemical control. It may be the use of natural enemies, pathogens, pheromones, methods that develop resistance in the crop plant against the pest, or it may be agricultural practice to suppress the pest.

I will now give a general comparison between these biological methods and the use of pesticides. I have stressed that the chemical pesticides are well adapted to the market economy. This means that the chemical companies have very good marketing and are making good profits. The biological methods are not well suited to commercialization. As a result the chemical pesticides are well advertised, but the advertising for biological methods is minimal.

We shall now look at the different methods from the point of the farmer. An application of pesticides gives a visible result. At once a lot of dead animals! The results from biological methods are not seen at all. There are no dead animals, rather unborn once! The pesticide application may give a 100% reduction of the pest in a field for a short time. Using biological methods you may continously have a few pest specimens in your crop, but hopefully they will cause no economic damage. Using pesticides will mean annual and probably increasing costs whereas, biological methods are very cheap for the farmer. The use of biological methods will need more careful planning and also planning that considers areas outside the farmers own farm. The choice may be difficult.

The risks and safety of the two approaches are quite different. The risks of biological methods are very small in relation to the environment, even if some cases are known where careless application have backfired. These methods also have insignificant risks in relation to people handling farm commodities and also in relation to the consumers. It is thus obvious that the biological methods offer alternatives with very small risks to man and environment.

Important also are the differences in relation to research and development. It has been pointed out that R & D in relation to chemical pesticides is handled in a very efficient way by the multinational chemical corporations. Earlier, the industry supported very little research on biological control as it was not considered possible to make a profit from the results. Today, some new companies are developing biological methods, such as the use of pathogens and pheromones. But in general the research on biological control has to be supported by public funds. There is also a very marked difference in the amount of knowledge needed for the application of pesticides and biological controls as biological methods have to be adapted to the agroecosystems. So far we have limited knowledge about both the structure and functioning of these agroecosystems. We thus need a very large input of research in order to use the biological methods in a proper way. We have a large number of cases where biological control has failed. This can be attributed to a large extent to the limited knowledge about the environment where the methods were put to use. I think this background research has to be based on public funding and to a large extent will be done by the universities. I also think that neither the scientists nor the politicians are aware of what is really needed in order to get good results.

I have elaborated on the shortcomings of chemical pesticides, but now I want to give some remarks in relation to the possibilities of the different biological methods. From many points of view the method that seems most promising is to breed resistance into crop plants against pests. But even this method may fail in the long run, as it has turned out that the insects sooner or later break through the resistance mechanisms.

It is estimated that a modern cultivar of a cereal will only last between 5 and 10 years. As it takes a long time to develop a new cultivar, the plant breeders have to try to keep well ahead of the insects. It is however possible that the new type of horizontal resistance will last longer than the earlier type of resistance which was based on a single gene. But it takes a very long time to screen and develop this new type of resistance. It is probably wrong to call it 'new' as it is the original type that has been developed in nature. Notice that this old type does not give a 100% protection, but allows some insects to feed on the plant. These insects are the food or prey of their natural enemies, that is important in keeping the insect populations at a low and constant level.

Before the golden age of the pesticides the classical method of control was to find and introduce a natural enemy. Several successes have been reported, but there have also been many cases of failure. Now, new successful results have been reported, especially from Australia and the Pacific Islands. At the same time Professor Brian Beirne at Simon Frasers University in Vancover is investigating why there have been so many failures. First he points out that very often the method has been used too late as a last resort when other methods have been unsuccessful. Used at the correct time the biological control might have solved the problem. More important is, however, that the techniques in applying the method have been wrong in many cases, e.g. the number of specimens put out in most cases have been too small. In the future it is possible that our knowledge will increase and the techniques will improve and the results will be much better.

An interesting new possibility in this has been developed by Dr. Marjorie A. Hoy, of the University of California. Usually it is impossible to use natural enemies and pesticides at the same time as the pesticides will kill the natural enemies. The widespread use of pesticides has actually interfered and wrecked cases where biological control was working well. But not only pests but also their natural enemies develop resistance. Resistance is based on the genetic structure, which can be changed by breeding. Dr. Hoy has now developed carnivorous mites resistant to certain pesticides. These mites can be used in greenhouses to control herbivorous mites while at the same time pesticides are used to control other types of pests. This technique opens up important new possibilities.

Another area in which promising progress has been reported is the use of pathogens to control pests, especially the finding of new extremely efficient strains of *Bacillus thuringensis*. This will be important for both agriculture and the fight against malaria.

Field tests with pheromones have also given excellent results. These control methods will in the near future be used in cases where no chemical methods exist, for instance in connection with forest pests.

In the future we will probably have new methods based on the mechanisms that the plants use to protect themselves against different types of attacks. Ecochemistry has found a series of new defence chemicals that might prove useful for crop protection. However, some of these substances are toxic to man and have to be avoided. But it is also possible that new principles to protect crops will be found, for instance the possibility of inocculating crop plants against disease.

Even if ongoing research is indicating that biological control will offer good alternatives to the chemical pesticides, it is dangerous to believe that one special method will solve crop protection problems. Pest organisms are formidable enemies! For example, the insects: there are now between 2–3000 pest insects, but the potential number of pest species in this group is 3 million, adapted to all sorts of surroundings and climates. The total bodyweight of the insects is twelve times larger than that of the human race. Insects have a very rapid turnover rate and there are an immense number of individuals in every species.

Man can not win outright the battle with pests over the crops, but it is important not to loose ground, as has already occurred in some cases. Cotton growing has been abandoned in several areas, because of the unwise use of pesticides. If present agricultural techniques, based entirely on pesticides, are not changed, it will be very difficult – if not impossible – to protect the crops. We will have a large food disaster when the pesticides fail, and this will originate in a tropical area. It is also dangerous to build the defence of the crops on only one method even if it is well adapted to the biological situation. A new approach to pest management is needed. In recent years, scientists in both industrialised and developing countries have worked out a very promising approach called integrated pest management (IPM). This makes it possible to reduce the amounts of chemicals and at the same time, gives a better, and probably cheaper, method of controlling pests and diseases.

Integrated pest management (IPM) is based on an holistic approach to pest control, where a combination of methods is used in order to give long term and efficient protection of crops. Some of the components of IPM are:

1. 'chemical' pesticides to be restricted to avoid damage to the environment and human health and to prevent the development of resistance;
2. 'biological' pesticides – pheromones and naturally produced chemicals – to be used;
3. use of natural enemies, 'classical biological control';
4. use of pathogens – different types of micro-organisms;
5. use of 'sterile males' or other genetic methods;
6. use of plants resistant to attacks of pests and diseases;
7. use of cultural methods to avoid attack;
8. monitoring and forecasting of pest populations.

Integrated pest management can only be developed using a systems approach based on a good knowledge of the relevant agro-ecosystem and also of the possibilities of the different control methods and their interrelationships. IPM must develop through very close and well coordinated cooperation between experts representing many different specialities, as such an integrated programme can only be based on knowledge from different areas. Attempts to achieve this kind of coordination will reveal that there are many gaps in our knowledge that have to be filled. To fill these gaps research has to be organized. It is also important that this research has a suitable structure and coordination.

Most interesting is that this type of approach has been developed for large-scale crop protection systems in China. This was not motivated by a lack of confidence in pesticides, but rather by a lack of pesticides. Post war China could not afford expensive pesticides! After twenty years of trial and error there are well-functioning integrated systems. These are based on cheap labour and it has been claimed that such systems could not work in industrialized countries. However, that IPM can also function very well in other parts of the world has been shown by the so-called Huffaker project in the USA.

Carl B. Huffaker has now completed an extensive programme on integrated pest control and reported the results in the book 'New Technology of Pest Control'. It was a five-year project, with 18 universities and 250 scientists participating. Funding was provided by the National Science Foundation, the EPA, and the U.S Department of Agriculture. The programme dealt with different crops including corn, soybean, cotton, alfalfa, and apples. Some of the results were:

1. for some crops a pesticide reduction of between 30 and 50%;
2. for cotton, systems have been developed which consume less energy, less water and less fertilizer;
3. in Florida, biological control methods have been developed for citrus plantations;
4. biological methods to control pests in apple orchards;
5. resistant cultivars have been developed for cotton and soybean;
6. a new information system for farmers has been formulated.

This is a representative sample of what integrated pest control may achieve.

In the first chapter of the book called 'Rationale, Organization and Development of a National Integrated Pest Management Project' and written by C.B. Huffaker and Ray F. Smith it is pointed out what earlier had prevented a break through in this field and what is needed to establish new projects of IPM.

I quote from the book:

'The first is: the chemical industry has for too long dominated the pest control scene, and this has resulted in an almost complete departure from some of the older more ecologically based methods for pest control. A virtual army of insecticide salesmen have in many parts of the country practically replaced the traditional dependence of the farmer upon his university, their researchers, and extension farm advisors for advice on pest control.'

'The second need is that the methods of funding and managing research programs to develop improved pest control must allow for some changes.' This project was funded by the personal intervention of President Nixon.

'Thirdly, the management and research priorities of such programs must be revamped. At present, the management is automatically subject to the cross-currents, opposing view-points and – yes parochialism and personal interests or special background of the administrators of the various levels in the several universities usually involved in such coordinated programs as now exist. A program of appropriate scope and technical depth, centered on use of systems, science and modelling as a means of setting research priorities, guiding the research, evaluation of direct results on pest abundance, and on the economic and social benefits to the farmer and society, requires a strong centralized management largely independent of the domination by the administrators.'

CHAPTER 19

The economics of wetland conservation

J.K. BOWERS

This paper is concerned with one class of wetland habitat in one country, the United Kingdom. The habitat however has close parallels in much of Europe, particularly Western Europe, and appears to be under threat everywhere for broadly similar reasons. It is argued that effective conservation requires action in part, at least, at the international level, specifically within the EEC; community action is required since it is community policies that present the threat.

The habitats are the grazing marshes located on the flood plains of many English and Welsh rivers. These marshes consist mainly of permanent grassland maintained by a high water table and periodic flooding. They are devoted to cattle enterprises – the production of beef and dairying – and to a smaller extent, to the raising of sheep. Where flooding is frequent, agricultural use is confined to summer grazing perhaps with the addition of a hay crop. In other dryer parts some animals may be found throughout the year.

Where the water table is high and/or flooding is especially frequent, the grassland is often floristically rich and a wide range of neutral grassland types is to be found. The marshes include areas, usually small, which are of national and sometimes international importance for nature conservation (Ratcliffe, 1977). In other areas fertiliser application or occasional re-seeding has reduced the botanical value. Where the river systems are tidal, salinity of flood water serves to reduce species diversity and hence botanical value of the grassland.

In many cases these marshes contain net-works of dykes or drainage ditches (serving a subsidiary function of wet hedges.) Where not saline nor subject to fertiliser pollution these form additional valuable habitats for aquatic plants, insects, and amphibians and contain national and some European rarities. Finally the marshes are important breeding grounds for wading birds, particularly, redshank (*Tringa totanus*) snipe (*Capella gallinago*), curlew (*Numenius arquata*) and lapwing (*Vanellus vanellus*) with smaller numbers of black-tailed godwit (*Limosa limosa*) and ruff (*Philomelus pugnax*) and during winter floods hold vast numbers of wintering wildfowl. The population numbers of several duck species and of Bewicks Swan (*Cygnus columbianus bewickii*) on these marshes are of international importance.

The main blocks of such marshes in England and Wales are the Somerset Levels and Moors, which constitute the flood plain of the river Parret and its tributories, and the Norfolk Broads being the valleys of the rivers Yare, Bure and Waveney. Smaller but important sites are the river Derwent in Yorkshire, the Ouse and Nene washes in Cambridgeshire, the rivers of the Hampshire Basin (Avon, Test, Itchen and Stour), the Arun in Sussex, the Soar in Leicestershire and parts of the Severn. An important set of grazing marshes lie on the rivers draining into Lough Neagh in Ulster. Conservation of these pose essentially similar problems, differing only in the institutional framework of decision-making.

D.O. Hall, N. Myers and N.S. Margaris (eds.), Economics of ecosystems management.
 ISBN-13: 978-94-010-8928-9

The threat to these habitats is from land drainage for agricultural improvement, either the conversion to arable, mainly cereals and, increasingly, oilseed rape; or pasture improvement: reseeding with commercial grass species permitting higher stocking densities. Piecemeal drainage has been continuing for centuries and the surviving grazing marsh is a small fragment of what previously existed. Further piecemeal reclamation is currently constrained in part by the capacity of the existing river systems to move flood water. Hence progress in land drainage requires large-scale work to increase the capacity of the system by widening and deepening the rivers, or cutting additional channels or by the provision of hydraulic barriers to exclude surge tides plus the raising of river banks to prevent over-topping. Comprehensive schemes exist and are in various stages from formulation to implementation for almost all of the river systems listed above. The approval of the Soar Navigation Bill by the House of Lords in July 1983, with the strong objections of the Conservationists denied a hearing, must mean the destruction of the Soar marshes within a very few years. Progress on the rest is restrained by public expenditure limits and by continuing opposition from conservation bodies. As argued below, economic viability is not a constraint on the implementation of arterial drainage schemes.

Arterial capacity is only one factor, albeit an important one, explaining the survival to the present of these marshes. Another is undoubtedly the structure of land holdings. Common grazing is rare, but fragmentation of holdings arising from the past enclosure of common grazing appears fairly widespread, so that consolidation of holdings is required for agricultural improvements. Thus in the case of the main arterial scheme for the Norfolk Broads, the Yare Barrier, it has been shown that, despite the contrary assumption underlying the scheme, flood risk is not a constraint on conversion to arable over much of the benefit area (Bowers, 1983). Confirmation of this view comes from the fact that ancilliary schemes involving subsections of the marshes have been proposed and in one case carried out without the main scheme on which they were supposed to be dependent. Anticipatory conversion of marsh land before implementation of an arterial scheme seems to be quite common.

In fact only in extreme cases does flood risk or height of water table pose a total block on conversion of these marshes. Drainage of individual fields or blocks of fields is an investment decision, the returns on which are the difference between gross margins on existing and proposed agricultural enterprises, with adjustment for differences in farm specific costs and for risk of flood loss. The Common Agricultural Policy of the EEC contain a built in bias in favour of intensification. In addition the structure of support and price trends favour cereals against livestock enterprises. The level of flood risk at which it is worth the farmers' while to drain and convert his marshland is thus an endogenous variable whose value has shown a tendency to fall over time.

This is not say of course that arterial drainage works are irrelevant to the problem. They are a factor additional to general support policies and serve to increase the rate of agricultural change of these wetlands and in some cases make possible change which, at current agricultural prices and with current technology would not otherwise occur. Let us consider the two factors separately.

Drainage of marshland is one aspect among many of the environmental damage resulting from trends in modern farming. These trends may be classified under four headings: increasing intensity of land use, increasing chemical and capital intensity, specialisation of production and a shift from grassland to arable. While a convenient aid to exposition these processes are inter-related in complex ways. The forms of environmental damage resulting are manifold: loss of permanent pasture, loss of hedgerows and field margins, grubbing out of small woodlands, drainage of upland bogs as well as lowland marshes, filling in of farm ponds, pollution of water courses from fertiliser run-off, depletion of bird of prey populations from persistant pesticide residues, loss of footpaths and access, loss of landscape diversity, damage and destruction of field monuments and archeological remains.[1] The results in terms of loss of floristic and

[1] For a survey see Bowers & Cheshire (1983), Chapter 2.

faunal diversity is devastating (Nature Conservancy Council, 1977). Taken to its logical conclusion modern farming is an environment suitable only for agricultural crops and some highly specialised and adaptable agricultural pests. Since the vast majority of land in the United Kingdom is devoted to agriculture the implications for the environment of these agricultural trends is thus extremely serious.

The standard analysis of the process is that it is a public goods problem. Economic pressures are forcing farmers to change their technique, to improve productivity and increase output. Good environment is a joint product with conventional agricultural output but it is a public good. Hence no account is taken of it in deciding either on output level, product mix or production technique. Policy prescriptions follow from the diagnosis. This story needs modification in a number of respects.

First it is not clear whether the public goods problem characterises the behaviour of the individual farmers or the Ministry of Agriculture. Production technique is determined within a price system where capital inputs are subsidised – more so than in manufacturing or service industries – while land and labour are not. Furthermore switches of technique in the direction of increased intensity of land-use – i.e. the replacement of land and labour by capital, termed farm improvements, are subsidised. Changes to increase environmental outputs conversely would not typically attract grant aid. These substantial price incentives are bolstered by a large volume of propaganda – farm advice – in favour of the trends we have identified, this in turn backed by state-supported applied research in intensive agriculture. Clearly the authorities believe that left to themselves farmers would do less environmental damage.

Second if increasing output appears as an economic necessity to the individual farmer, it looks not to be so for the state and still less for the European Community. Most of the agricultural products are in surplus within the EEC and the problem is not that of increasing production so much as disposing of what is already produced. The difference between private and social valuation of agricultural output is large: effective protection rates for the U.K. in some major products are indicated in Table 1. The level of protection appears to be on a rising trend both within the EEC (i.e., for the U.K., since 1972) and in comparison with pre-EEC regimes (Black & Bowers, 1981).

Despite this, official pressure to increase output as rapidly as possible continues as it has done for more than forty years (Bowers & Cheshire, 1983). The official explanations are rationalisations rather than reasons. The argument that farm improvements are required to make farmers more 'efficient' looks strange when viewed against the figures for protection levels. There is little doubt that U.K. farmers are overcapitalised and over-specialised. The directions of change that policy induces are unlikely to realise economies of scale. For a start the bulk of support and subsidy is directed at increased specialisation and intensification *within* the existing farm structure. Support for farm amalgamations, for increasing the size of units, is small in comparison with that for intensification. The result is more a shift of factor proportions than an increase in scale. Despite the high rates of growth of output per man and per hectare, overall productivity growth has been lower or at best no better than that of the internationally poor performance of U.K. manufacturing (Bowers & Cheshire, 1983, Ch. 6).

Maximising current output growth does not seem a sensible strategy either for safeguarding future food supplies or for conserving the productivity of British farm land. Balance of payments arguments are short term expedients and cannot be used to defend a more-or-less permanent commit-

Table 1. Effective protection rate for agricultural products in the United Kingdom.

Product	%
Wheat	126
Barley	160
Sugar	*
Beef	727
Lamb	38
Milk	*

* At world prices value added at the margin is negative.
Source: Black & Bowers (1981).

ment to expansion of output. Surpluses may form a convenient and useful source of food aid but the policies that create them by disrupting world markets in agricultural produce have arguably done more damage to the economies of Third World countries than the surpluses have repaired.

A common justification for agricultural protection is income support. This provides no reason *per se* for making subsidies output linked and still less for encouraging intensification. Indeed if the objective is supplementation of income a case could be made for relating payments to output of the public good – environment – which would otherwise be undersupplied. For the U.K. at least agricultural subsidies are regressive (Josling, 1976), at current levels of urban unemployment probably strongly so.

The true explanation of the policy of output maximisation via intensification and specialisation is to be seen within the Bureaucracy of the Ministry of Agriculture: it has defined its goals in these terms; there have been no effective pressures to identify departmental goals with the national interest. There have been many pressures however re-enforcing the 'Departmental View' notably from agricultural supply industries, farm chemical and machinery manufacturers; from land owners who see the subsidy system capitalised in rising land values and from the farmers themselves who, having invested in increased intensity, in response to official propaganda and incentives, have demanded policies to validate their decisions.

The arguments above relate to the objectives of national policies. The C.A.P. is the outcome of negotiation between national governments and reflects the aims of its members. Many of the specifically national policies such as farm improvment grants are in accordance with EEC directives and nothing additional needs to be said except in two respects. First the shift from pasture to arable is a direct result of the system of support within the CAP, whereby, unlike the pre-EEC U.K. guaranteed price system, the annual price increases tend to increase the profitability of arable enterprises c.f. livestock enterprises. This arises because cost of grain inputs into livestock enterprises rises *paru passu* with producer prices for grain. To off-set its effect livestock prices would have to increase typically by about 40% more than cereal prices. They do not do so partly because, given the problems of negotiation, a more or less uniform across-the-board increase is almost inevitable, and until the present because the authorities have tried to eliminate livestock surpluses by encouraging a switch to grain.[2]

The second, specifically European problem is the prisoners' dilemma inherent in a system of common financing of agricultural expenditure where structural policies to raise national output still lie in the hands of member governments. This point brings us specifically back to our problem of the Wetlands. Arterial drainage of these areas leads to increased agricultural output. The cost of this output in the form of additional subsidies and the disposal of the resulting surpluses is met by community funds. The U.K., by draining the Wetlands, is thus able to alter the balance of receipts and disbursements to the CAP in its favour. This argument is of course applicable to all countries of the EEC and so far as I can tell is being generally applied. Each country seeks to alter the balance of funds in its favour; the result, since almost all the products are in surplus and can only be disposed of at a loss which must be financed from community funds, is that all are worse off. Unless the EEC takes control of public investment to raise agricultural output: land drainage, land reclamation, irrigation etc., the surpluses and hence the cost of the CAP rise even faster than the system of high support prices and farm improvement grants dictates. Furthermore the volume of environmental damage is increased thereby also.

Not all of the cost of additional output e.g. in the U.K. is met by non-U.K. sources. In fact, depending on the product and how it is disposed of, up to 40% can be met by U.K. consumers and taxpayers (Bowers, 1982). The U.K. Ministry of Agriculture however behaves as though all were met from abroad.

Arterial drainage works in England and Wales are the responsibility of Regional Water Authori-

[2] They are now encouraging switching to oilseed rape to reduce cereal surpluses.

ties, (the system is somewhat different in Scotland and Northern Ireland). These statutory bodies are responsible for water supply and sewerage disposal as well as land drainage and flood protection. The land drainage divisions are however largely separate and under the control of Land Drainage Committees comprised mainly of farmers and landowners appointed by the Ministry of Agriculture (MAFF). As a result of a determined fight by MAFF in conjunction with the farmers unions and other agricultural pressure groups at the time of the creation of the Water Authorities land drainage remains within its ambit whilst other Authority functions are under the control of the Department of the Environment.

Arterial drainage works are grant-aided by MAFF, the rest of the money being met from local taxation (rate precepts). Powers exist to levy additional charges on farmers who benefit, but these are never used. Comprehensive surveys of drainage 'problems' within their regions ordered by MAFF, provide the W.A's with a portfolio of possible schemes. Allocation of priorities and embodiment into 5 year rolling capital works programmes is the responsibility of the Land Drainage Committees.

In order to attract grant aid, arterial schemes must be subject to a cost–benefit analysis, carried out by the W.A's under MAFF guidelines. A benefit–cost ratio of greater than unity when benefit and costs are discounted at 5% is a normal requirement for grant aid. Until a Public Inquiry into a scheme at Amberley Wild Brooks on the river Arun in Sussex in 1977 when the cost–benefit analysis was severely criticised by conservationists, these studies were made readily available. Now they are treated as secret by the Water Authorities. A study of a number of CBA's (Bowers, 1983) shows that there are good reasons for keeping them concealed from objectors.

While in all cases the benefit–cost ratios were claimed to be greater than unity the studies contained a number of technical defects all of which served to raise the value of discounted benefits. The true ratios were in fact substantially below unity. The main defects were as follows.

1. Agricultural outputs were valued at farm-gate prices with no attempt made to adjust for the considerable volume of subsidy implicit in them. This might lead one to suppose that what was being conducted was some sort of financial appraisal rather than a proper social cost–benefit analysis. This is not in fact the case since capital expenditure both by the W.A. and the farmers was valued gross of grant. In their evidence to the House of Lords Committee on the Soar Navigation Bill, Severn-Trent Water Authority described the exercise as a 'quasi-financial appraisal', whatever that may mean. MAFF have now admitted that their guidelines are in error in this regard, inevitably since they are in conflict with H.M. Treasury guidelines on the conduct of CBA in the Public Sector. Rationalisations along the lines that all the subsidy paid is a transfer from other EEC countries are therefore unnecessary.
2. The yields used for anticipated, post-scheme outputs are theoretical maxima for the soil-type and location and not estimates of what is likely to be realised. Frequently too current prescheme yields are excessively low. In the case of the Soar the pattern of assumed yields made current cropping patterns incomprehensible.
3. The rates of 'take-up' of benefits i.e. the rate at which pasture will be improved are optimistically set and are unrelated to the level of flood protection planned. There is in fact no economic modelling of the interaction between cropping and flood risk. The latter is arbitrarily set usually at levels which are excessive in relation to scheme objectives. The former, as with current yields is usually determined by asking potential beneficiaries what they currently achieve and what their plans are. There is thus a substantial 'free-rider' effect built into the methodology which is not recognised nor corrected for.
4. Appraisal period is not related to structure life but is treated as a variable to be varied at will to achieve the desired level of benefits.
5. No adjustment is made for anticipated post-scheme flood losses.
6. Intangibles e.g. environmental damage are ignored. Following the Wildlife and Countryside Act 1981, the W.A's have an obligation to take

account of the environmental implications of their activities. They have interpreted this to mean that they must consult with the statutory conservation bodies and perhaps make minor modifications to their schemes to meet these bodies' wishes. Having done this they are then entitled to ignore remaining environmental disbenefits in scheme appraisal.

There is some evidence that W.A's are modifying their practices in response to continual criticism. But this is unlikely to be more than window-dressing since the basic problem remains. The W.A's see a satisfactory benefit–cost ratio as a formal requirement that they must satisfy if they are to get their grants and be allowed to do what they wish. They accordingly adjust their numbers to give the required outcome. Since the number of parameters is large and they are constrained by no proper economic model of the system they are ostensibly seeking to modify, they have adequate freedom to achieve their goal. Placing constraints on some of the parameters causes them to look optimistically at others. There are a number which we have not mentioned, e.g. the coverage of schemes can be adjusted to include unrelated urban flood protection schemes with high CBA ratios (Bowers & Black, 1983). The W.A's are not interested in the balance of social advantage of their schemes; their land drainage activity has its own internal logic. The pressure for them to inflate benefit–cost ratios is in fact increasing as in order to maintain their capital works programmes and keep their labour forces employed they are forced to look at increasingly marginal schemes. Severn-Trent Water Authority in fact argued before their Lordships (House of Lords, 1983) that, alone among public sector investment, land drainage employment should be shadow priced at substantially below current wage rates!

Cost–benefit analysis is a guide to a decision-maker faced with a number of competing projects between which he is indifferent. Water Authorities have a single project, or a small number that they do not regard as competing, to which they are firmly committed. The independent check on their enthusiasm should be administered by MAFF. We have already noted that it is by MAFF guidelines that the appraisals are carried out. MAFF are not interested in social benefit either. They have a volume of funds for grant-aiding land drainage and are interested only in the maximum amount of increase in agricultural output or the maximum benefit that they can give to farmers in the form of land drainage and flood protection from that expenditure. The 'quasi-financial' appraisal probably suits their purpose admirably. The final arbiter of the public interest in this situation should be the Treasury. Treasury approval is only required for schemes above a minimum level of expenditure and this level is readily avoided by phasing of schemes and treating each phase as a new scheme. Furthermore the Treasury is largely dependent on Maff for judgement of scheme viability. Criticism of MAFF and current constraints on public expenditure may improve the efficacy of this final check.

By way of conclusion we suggest some policy implications. Arterial drainage investment would be controlled, to the benefit of the environment, if responsibility for CBA were taken out of the hands of the Water Authorities and responsibility for grant-aid were taken out of the hands of MAFF. Much would be gained by removing the aura of secrecy surrounding the process and making the appraisals public documents. A requirement for a public inquiry where conservation interests are involved would be beneficial. The pressure for environmentally damaging works such as land drainage comes in part from a desire to alter the balance between disbursements and receipts within the CAP. This factor adds to the tendency for surpluses and expenditure to escalate which is a consequence of the high level of support prices. Its control should be a concern of those who wish to limit CAP expenditure as well as those who wish to protect the environment. While such changes should serve to limit the rate of destruction of surviving wetlands, proper conservation of these and other aspects of the farmed environment, needs a change in the basic structure of agricultural support policies and a divorce between income support and output growth. Without this, large-scale environmental damage is inevitable, and site safeguard measures, seeking islands of good habitat within a sea of 'productive' agricultural land, is an expedient of despair.This issue, however, is beyond the limited scope of the present paper.

References

Black, C.J. & Bowers, J.K. 1981. The Level of Protection of U.K. Agriculture. School of Economic Studies Discussion Paper No. 99, University of Leeds. 41 pp.

Bowers, J.K. 1982. Who pays the cost of U.K. agricultural expansion? School of Economic Studies Discussion Paper No. 110, University of Leeds. 11 pp.

Bowers, J.K. 1983. Cost–benefit analysis of Wetland Drainage. Environ and Planning A, 15: 227–235.

Bowers, J.K. & Black, C.J. 1983 . Investment Appraisal and the Environment: the Soar Valley Improvement Scheme. School of Economic Studies Discussion Paper No. 129, University of Leeds. 43 pp.

Bowers, J.K. & Cheshire, P.C. 1983. Agriculture Land Use and the Countryside: An Economic Analysis. Methuen. 173 pp.

House of Lords, 1983. Minutes of Evidence taken before the Committee on the Severn-Trent Water Authority Bill, 12th July and 13th July, pp. 67 and 77.

Josling, T.E. & Hamway, D. 1976. Income transfer effects of the Common Agricultural Policy. In: P. Davey, T.E. Josling & S. McFarquer (eds.), Agriculture and the State. London. 230 pp.

Nature Conservancy Council, 1977. Nature Conservation and Agriculture. H.M.S.O., London. 38 pp.

Ratcliffe, D.A. (ed.), 1977. A Nature Conservation Review, Vol. 2. Cambridge University Press, London. pp. 320.

Treasury, H.M. 1982. Investment Appraisal in the Public Sector. H.M.S.O., London. 40 pp.

CHAPTER 20

Planning guidelines for urban forest management

RUSSELL A. BEATTY

Introduction

The livability of the world's cities has been profoundly affected by explosive population growth and migration to urban centers. Worldwide population is expected to increase by 50% by the year 2000 and major shifting from rural areas to cities will continue (Barney et al., 1982). In the United States, over 90% of the population is expected to be living in and around cities by the year 2000.

The quality of the urban environment, especially in large conurbations, tends to deteriorate in proportion to increased size. Although the issue of urban quality is complex, one key influence is the amount, quality, and distribution of urban vegetation and green open space (Driver & Rosenthal, 1978; Wingo, 1978).

For example, in Athens the oxygen level has dropped from 21% to 17% (perilously close to the 16% danger point at which health is threatened). The decline has been attributed largely to the loss of nearly 4000 acres of pine forest through indiscriminate cutting, to a forest fire in 1982, and to a reduction in the amount of urban green space from 7% to 0.2% (Schance, 1983).

In the United States, *urban forestry* has emerged as a concept for the planning and management of urban trees and forests (Shafer & Moeller, 1979). Through the research efforts of the U.S. Forest Service we have learned that large parks and open spaces affect city climate (Geiger, 1950; Federer, 1971; Heisler & Herrington, 1976), and that trees can improve urban microclimates (Herrington, 1978), increase property values (Payne, 1973, 1978), improve air quality (Smith & Dochinger, 1976), and influence wildlife habitats (De Graaf & Thomas, 1976).

To date, urban forestry has not been well integrated into the design and planning of most towns and cities in the United States. Trees have been planted for shade and beauty generally without regard for ecological, sociological and economic benefits or consideration of future management. The focus has been almost exclusively on individual street and park trees. Tree master plans have generally addressed only street trees. What is really needed is a massive infusion of trees and the planting of green open spaces based upon a well-conceived urban forestry plan.

There are a few examples where urban forestry has been integrated with city planning strategies to improve the health and liveability of cities.

In Nanjing (Nanking), China, one of the five 'furnace cities' of the Yangtse Valley, the average summer temperature has dropped 5 degrees F (1.8° C), from 90 degrees F (32.2° C) to 85 degrees F (29.4° C) since 1949. A further reduction is anticipated and is directly attributed to the influence of trees planted during that period. Some thirty-four million trees have been planted in the vicinity of Nanjing since 1949 for the specific purposes of reducing summer temperatures, purifying the air, regulating climate, and beautifying the city.

Tree planting includes the afforestation of de-

D.O. Hall, N. Myers and N.S. Margaris (eds.), Economics of ecosystems management.

 ISBN-13: 978-94-010-8928-9

nuded hillsides, planting coniferous windbreaks, plantings established to control water and wind erosion of soils, planting triple rows on either side of railroads, and lining streets with deciduous trees, primarily Ginkgos and Sycamores (Bartenstein, 1981; Schumann, 1981).

With an estimated population of 1.5 million, the quantity of trees planted in Nanjing comes to twenty-three trees per person planted over a period of thirty-two years.

In Stuttgart, West Germany, urban planting is being integrated with land use planning and the control of urban development. Fingers of green open space which penetrate the city center are planted to convey flows of cooling night air and to reduce daytime temperatures (Loessner, 1978).

Since 1976, the 'Dayton (Ohio) Climate Project' has been developed to use the environment's natural cooling, cleansing, and flushing actions to reduce air and noise pollution, combat energy waste, and improve the appearance of that city (Bartenstein, 1981).

These examples illustrate significant large-scale efforts to utilize vegetation to improve the health, appearance, and livability of these cities. Space for trees was deliberately planned and provided for because importance was placed on the value of trees in improving urban conditions. But these are exceptions. In the U.S. (and, I suspect, in other countries as well) the emphasis is still on planting and caring for street trees rather than integrating urban forest planning with overall city planning strategies. A wide gap still exists between city planners and urban tree and forest managers. The consideration of the city, its people and its vegetation as an ecosystem is a foreign concept to most urban planners. And urban foresters have not been a particularly forceful influence in city planning. Although sound urban forest planning is no panacea for urban ills, it can be considered a type of holistic healing for urban areas.

The intent of this paper is to review the issues facing urban tree managers in the U.S., to suggest the attributes of an urban forestry planning process, and to suggest a model for urban forestry planning.

Municipal tree survey

In a survey of the municipal tree programs in the United States, I identified a number of problems that argue for comprehensive urban forestry planning (Beatty & Heckman, 1981):

- a serious shortage of funding coincides with low administrative priorities for trees;
- there are increased demands by citizens but tree management tends to be crisis oriented;
- tree management departments tend to have a low public visibility vis-à-vis progressive urban forestry;
- official public advocates for trees and urban forestry are generally lacking;
- a limited recognition of the values of trees and their potential roles exists in most cities; tree selection and use is generally unrelated to a broad range of benefits of trees and forests;
- increasing environmental stress is relative to city size.

Inadequate funding

A lack of adequate funding is a major obstacle to systematic tree management and the promotion of more active urban forestry programs. This situation is unlikely to improve as municipal budgets continue to suffer economic constraints, escalating inflation, and resource shortages. In the United States in 1976, the average yearly cost for each city resident for municipal tree management was $1.63 in 1976 (Ottman & Kielbaso, 1976). Today I estimate the cost to be about $3.00. This represents a mere 0.56% of the median total municipal budget in the United States. City administrators and elected officials generally remain unconvinced of the value of a comprehensive urban forestry program. They tend to consider tree management as a low prioirity activity and deferrable as compared with other municipal programs. This is partly due to inadequate education about the benefits of trees in the urban environment and their economic, social and biological values to the city. The resulting deferred maintenance seriously increases future liability and management costs through neglect (Callahan & Bunger, 1976).

Increased citizen demands

Nearly two thirds of the cities of the United States report increased demand for tree planting, and maintenance. This is probably a result of increased awareness of the benefits of trees in cities (Zube, 1973) and increased growth of suburban cities, especially in the South and West.

Low visibility

Municipal tree departments tend to have low visibility and relatively poor community relations. This is especially true in inner city neighborhoods where high rates of vandalism frequently occur. Reasons for vandalism are complex but they may include frustration from unemployment, animosity towards civic authority and poor public relations at the street level by the municipality.

A municipal tree department is frequently perceived as a crisis-oriented agency. Some cities work entirely from citizen requests, responding to daily crises rather than systematically managing the city's tree resources (Beatty & Heckman, 1980; Callahan & Bunger, 1976). As Bartenstein (1983) suggests the 'caretaker' role of tree managers has been relegated to that of 'mortician'.

Lack of public advocates

Only 30% of the cities surveyed had a tree commission or similar public body. Tree commissions can be influential advocates for a city's urban forestry program. As a governmental commission, they can help procure increased funding, improve public relations, develop educational materials on the values of trees and urban forests, and coordinate tree management activities among various city departments and among such other public agencies as water districts and public utilities which have large land holdings.

Dispersal of tree management responsibilities

The responsibility for tree management is often housed in municipal departments which have competing and often conflicting responsibilities for various components of the urban forest. A park and recreation department is the most common city agency responsible for tree planting and management (Beatty & Heckman, 1980; Ottman & Kielbaso, 1976) but public works and planning departments are also responsible for tree planting and management activities. Such other public agencies as utility companies, cemetery boards, transit authorities, and flood control districts are also responsible for tree planting and care on their lands. Their efforts are usually unrelated to a municipal tree management program even though such lands comprise a significant portion of the urban forest.

Larger cities tend to have more specialized professional tree care departments, whereas smaller city tree programs are frequently managed by someone whose job or background has no relationship to forestry (city manager, city engineer, public works director) (Ottman & Kielbaso, 1976).

Limited recognition of benefits of urban trees

Until recently, tree planting strategies have rarely been considered an important component of land use planning by city planning departments. Notable exceptions include Dayton, Ohio (Bartenstein, 1980) and several German cities – Frankfurt, Nuremberg, and Stuttgart (Schabel, 1980).

The use of trees to combat air pollution, glare and reflection, attenuate noise, control wind, or modify overall city climate is rarely a part of municipal tree planting strategies despite the availability of supporting literature and schematic guidelines (Robinette, 1972). In fact, nearly half of U.S. cities have begun to emphasize planting small ornamental street trees because of high maintenance costs associated with larger trees (Ottman & Kielbaso, 1976). By contrast, a Midwest study found that residents actually prefer large trees (Kalmbach & Kielbaso, 1979).

Environmental stress

The urban environment is generally deteriorating as a habitat for trees (NE For. Exper. Sta., 1977). Stress from the urban environment reduces the vigor of many trees and increases their susceptibility to disease and pest infestation (Bernatsky, 1978).

In large cities (over 250 000 pop.) the major stresses include soil compaction, water and nutrient deficiencies, air pollution and vandalism (Beatty & Heckman, 1980). Physiological problems are related to the lack of adequate space for trees in dense urban areas. Although trees are desired and included in development plans, their space and soil needs are not well understood by civil engineers responsible for site plans and street design. Trees are forced to grow under constant stress – too much or too little sunlight, compacted soil, underground and overhead utility lines, air pollution, channelized winds and toxic runoff.

All of these factors indicate the need to consider a more comprehensive approach to the planning, design, and management of urban tree plantings and forests.

Attributes of urban forest planning

There are a number of significant attributes that should characterize future urban forestry planning.

Comprehensive in scope

First, the process should be *comprehensive* to include all trees, forests, and potential land that is public or quasi-public in nature. The focus of attention on street trees should be broadened to include the major components of the urban forest. A typology of urban forest components can be classified in three categories of ownership:

1. *Municipal lands*: streets, roads, parking lots, parks, plazas, playgrounds, cemeteries, golf courses, sports facilities, historic sites.
2. *Agency lands*: freeways and transit corridors, parking lots, regional parks, watersheds, reservoirs, creeks and drainage easements, utility easements and schools.
3. *Private lands* (which serve the public): churches, schools and other institutions, parking lots, shopping centers, planned unit developments, vacant lots.

Even though a city is not directly responsible for the trees and forests on all these various lands, it can help coordinate the planting and management of those lands to achieve specific goals established in an urban forestry plan. Policies and guidelines that consider the whole urban forest can be developed not only for the city staff to follow on its own lands, but to suggest how other large land holders can contribute to the city's long-range goals. Such assistance may include lists of tree species recommended for various planting situations, vegetation management prescriptions, and the planning of new plantings to achieve specific benefits such as microclimate improvement, as in Dayton, Stuttgart, and Nanjing.

Broad range of benefits

Second, the urban forest planning process should incorporate the *broadest range of benefits* from trees and forests. The traditional ornamental or purely decorative approach should be discarded in favor of planning and design that integrates functional benefits with visual relationships. As Bartenstein (1982) suggests, urban forestry should no longer be considered as an 'amenity service' but included among the 'essential services' along with such health and welfare functions as fire protection.

These benefits can be categorized as (1) aesthetic, (2) microclimate improvement, (3) energy conservation, and (4) environmental improvement.

1. *Aesthetic or visual quality*: visual fitness of trees is best related to land use, size and scale of landscape space, year-round effects of foliage, and form of trees relative to intended use (broad canopy as shade tree, vertical tree for emphasis, etc.). Avoid the 'beautification syndrome' in which only short-term effects are emphasized.

2. *Microclimate modification*: massive, large-scale plantings as in Nanjing, Stuttgart, and Dayton can be designed to reduce summer air temperatures in the inner-city, to improve the microcli-

mate of residential districts, and to integrate tree planting and urban forest management with land-use planning, channel cooling breezes, and reduce the urban heat-island effect.

3. *Energy conservation*: the temperature in tree-shaded buildings on hot days can be reduced by 15–20 degrees F (3.6–7.2° C), and a net energy savings of 10–20% is possible through reduced needs for air conditioning (Deering, 1956). By integrating urban forestry with an energy conservation program, a community can make more efficient use of trees that are generally planted as a requirement for new development. Guidelines for replanting trees in older districts, especially residential districts, can also be developed by the city. In addition, wood and wood wastes could be used to generate energy in a variety of ways: fuel wood can be used in homes and mill and urban wood wastes can be used in electrical cogeneration plants, in the production of gas or simply composted and returned to the soil as mulch or soil amendment as is being done in Oakland, California.

4. *Environmental improvement*: this comprehensive term goes beyond microclimate modification and includes a host of benefits that when combined should result in overall improvement of the urban environment – physical, functional, and psychological improvements for city dwellers. An urban forestry plan might include such objectives as *air quality improvement* (planting belts of trees downwind of sources of noxious air); *noise abatement* (wide belts of dense vegetation buffers along freeways); *enhancement and management of wildlife habitation* (recognize beneficial and unique species to be favored; discourage undesirable species); and *reapportionment of urban land* to gain more tree-planted spaces. The latter might include:

- planting on vacant lots
- required planting in parking lots (1 tree/5 vehicles, for example);
- use of school properties for afforestation of natural plant communities oriented towards both nature study and environmental improvement;
- closing and/or redesigning residential streets based on the 'Woonerf'[1] model from the Netherlands (Bartenstein, 1981).

Environmental suitability

Third, urban forest planning must address the complexity of environmental factors required for the sustained growth of individual trees and remnant forests. As cities become more congested and less suitable as habitats for trees, more precise knowledge of the characteristics of the urban environment will help assure adequate, if not optimal, tree growth. The planning process should include a systematic analysis of climate (and microclimatic variations), soil conditions, ground water characteristics, air quality (and sources of pollution) and other factors that influence plant growth. Tree selection and management can then be tailored to the city's environment – its potential as well as constraints.

Social and community focus

Fourth, urban forest planning should focus on the *needs of the people of the community*. These are the people who pay the bill for community services. They must be actively involved in the planning process. Urban forestry plans will have a chance of succeeding if the urban dwellers understand the importance of urban forestry relative to the physical, economic and social welfare of the community (Bartenstein, 1983). Therefore, urban foresters and planners must enlist the support of the citizens as advocates of urban forestry. This can be done in a number of ways. The planning process should foster *community participation* and promote the *education* of citizens on the values and benefits of urban forestry.

Community participation can occur in a variety of ways at three basic levels: (1) official lay groups such as an Urban Forestry Board, (2) neighborhood or community activist groups, and (3) private individuals or corporations.

As an *official civic body*, a Tree Commission or Urban Forestry Board can initiate the urban forestry planning process by establishing policy goals

[1] 'Woonerf' is a residential precinct where pedestrians have the right-of-way over vehicles. Streets are redesigned to accommodate bosques of trees, play areas, consolidated parking, minimum vehicular space, reduced speeds, and the use of pedestrian-oriented materials (pavers, bollards, etc.)

and insuring that urban forestry be integrated into overall city planning. This group of lay persons serves as an intermediary between the city officials (including the urban forestry staff) and the public. Their work should create greater visibility for the tree program and help stimulate public interest. The commission can help fight for better municipal funding and seek financial support from the private sector for specific planting projects. It can act as a sounding board for citizens complaints and suggestions. And it can serve an educational function by interpreting the urban forestry plan for the public and city officials.

Thus, an urban forestry commission should be operational, not simply advisory. Its membership should include citizens who have demonstrated expertise and knowledge about trees and who are committed to civic improvement through sound, farsighted urban forest planning.

In addition to the 'official' involvement of citizens through an urban forestry board, urban forest planning should encourage the involvement of neighborhood groups and community activists groups. The success of a progressive urban forestry program depends on support at the grass roots level. Advocacy by the voters – the common citizens – is essential. Such cities as Oakland, California are designing their urban forestry strategies towards neighborhood associations by providing direct support – educational information, tree selection and planting guidelines, and some financial assistance. In San Francisco an activist group called 'Friends of the Urban Forest' has conducted tree inventories and initiated a number of tree planting projects.

Community participation in tree planting projects in Oakland, California has brought a dramatic reduction in vandalism of street trees by fostering citizen cooperation and an understanding of the value of trees in neighborhoods. The results of neighborhood tree planting projects provide an immediate, tangible influence on the neighborhood, a feeling of direct involvement in city government, and increased self-esteem of the participants (Cole, 1979; Ames, 1980).

And, finally, community participants should include the business community from philanthropic individuals to large corporations. Because of the paucity of funding for most urban forestry programs, a broad base of financial support from the business community can help make up the deficit. In many cases the business community has much to gain in terms of improving a city's economic vitality and in good public relations. Corporate sponsorship helps poorer neighborhoods realize that someone is interested in them and helps generate a real sense of community. Corporations benefit by demonstrating their corporate responsibility in being good citizens of the community (Ames, 1980). Succinct tree planting projects might be identified for civic organizations or for corporations to sponsor such as an 'adopt-a-block' program.

Along with regulating emission standards governmental agencies should require polluting industries to make substantial contributions to urban forestry programs for both tree planting and long-term management. Along with better city planning and industry regulation, such participation can be a major step in improving the air quality of industrial cities.

If the urban forestry program is dependent upon these types of community participation, *education* of the public must be an integral part of an urban forestry program. In the development of an urban forestry plan, educational materials can be published for distribution to private citizens, neighborhood groups, designers, corporate officials and anyone who might be involved in implementing an urban forestry program. Such materials might include interpretation of the goals of the program, information on the specific objectives (improve air quality, visual benefits, etc.), explanation of the environmental characteristics and constraints for the city, lists of trees recommended for specific planting and recommended maintenance techniques.

These materials can be distilled from a comprehensive urban forestry plan prepared by the city staff and consultants (ecologists, foresters, horticulturists, and landscape architects).

Beyond the educational benefits, improved visibility of municipal forestry and tree management should result in a broader and stronger constituency and a better funding base.

Policy goals

These attributes and elements of urban forest planning form the basis for a progressive urban forestry program in which a succinct plan can be developed. Such a plan will vary in its form and degree of complexity depending on the city size, its environmental constraints, political context, and a host of other factors. It may be a set of planting and management guidelines such as was developed for Lafayette, California (Beatty, 1977). For larger cities, the plan may comprise various documents from an inventory and analysis of the city's vegetation and environmental characteristics to specific planting and management prescriptions.

Time does not permit a full discussion of the various components of an urban forestry plan document. However, an important initial step for any city to undertake is the establishment of succinct policy goals. This process helps give direction and establishes a conceptual framework for the preparation and implementation of an urban forestry plan.

Such policy goals might include (1) economic survival, (2) neighborhood vitality, (3) preservation of city character, and (4) urban health and liveability.

1. *Economic survival* of any community is no doubt a number one priority in city management and future planning. Too often the link between economics and urban forestry is not made because of the decorative, and therefore luxury, notion of trees. A well-planted community, where trees and remnant woodlands are carefully managed and integrated into the city fabric at all scales, tends to be an attractive city. The beauty of a city enhances the opportunities to attract outside investors, which in turn stimulates the economy. Property values are increased by well-planted streets, parks, and public places and the preservation of native trees and woodlands. An improved city image results in increased civic pride and fosters self-improvement. In short, a healthy, well-planned, and properly maintained urban forest can be an indicator of economic well-being.

2. *Neighborhood vitality* is a similar goal but one that influences smaller portions of the city. Careful evaluation of the city's neighborhoods can identify areas where old trees should be preserved and renewed to maintain an area's character and thus its livability. Such a survey can reveal which neighborhoods both need and desire improvement through better tree management or new plantings to offer visual and functional benefits (microclimate, energy conservation, etc.). In some neighborhoods, inhabitants may not want trees, and the reduction or elimination or over-mature, hazardous, or messy species could improve the neighborhood's vitality. Close cooperation and participation with citizens is therefore essential in attempting to improve neighborhood vitality.

3. The *maintenance of a city's unique character* is closely related to the health and extent of its trees and woodlands. Such cities as Savannah, Ga., Palo Alto, Ca., and Sacramento, Ca., are known for their distinctive individual characters, which are due largely to the infusion of trees, parks, and woodlands into the city fabric. Character may develop by accident or by design. Nevertheless, it is fragile and once lost is difficult to regain. Implementation of this policy goal requires the combined skills of the landscape architect, city planner, and urban forester. A visual inventory is the essential basis for a management plan to insure that the city's character is preserved and enhanced and not inadvertantly lost.

4. *Urban health and liveability* is a broad umbrella which includes energy conservation, microclimate improvement, biomass utilization, and maintenance of the physical and psychological liveability of a city. Each city has its own unique needs. We know, in principle, the many benefits urban vegetation has on improving the urban environment. We do not know much about the cost-benefits ratios. In cities where the climatological benefits of the urban forest is included in land use planning, it will be valuable to assess their effects on the cities' energy budgets.

Conclusion

In conclusion there is no doubt that we must improve the livability of the world's cities. In this technological age there is a tendency to look for some 'break-through' before the invited disaster can complete itself. The solution to improving the quality of the urban environment cannot be solved with some technological breakthrough. A significant part of the solution depends upon highly skillful planning, design and careful management of urban forests which are not dependent upon expensive and resource consumptive technologies. In the past urban tree managers have focused on planting and caring for trees in isolation from the needs of the urban dwellers and from the overall city planning. The emphasis has been on the needs and difficulties of trees in modern cities rather than on the larger and more important issue of the needs and issues related to improving the habitat for people (Bartenstein, 1982). Thus the urban ecosystem necessarily must include the design and management of open space and vegetation for the benefits of people.

The economics of such planning and management are complex and not always tangible. We can estimate urban forest management costs monetarily but this is only a part of the equation. It is more difficult to evaluate the long-term human and environmental benefits and values of good urban forestry. But as Wendell Berry (1981) points out, the word 'economy' should not mean only 'money economy'. We should look to the original intent of the word which means 'order or health of the household'. In doing so we must judge economic health by the health of our human and natural household – the urban habitat where most of us live.

References

Ames, R.G., 1980. The sociology of urban tree planting. J. Arboric 6(5): 120–123.

Barney, G.O. (project dir.), 1982. The Global 2000 Report to the President, Vol I. Penguin Books, New York. 766 pp.

Bartenstein, F., 1980. The future of urban forestry. In: Breaking Ground in Urban Forestry – 1, The Pinchot Institute for Conservation Studies, USDA Forest Service. 18 pp.

Bartenstein, F., 1981. Re-greening in China. The International Dayton Line 1 (7): 5.

Bartenstein, F., 1982. Meeting urban and community needs through urban forestry. In: Proceedings of the Second National Urban Forestry Conference. Am. Forestry Assoc., Washington, D.C. pp 21–26.

Beatty, R.A., 1977. Trees for Lafayette – the master tree plan, Lafayette, California. Northern Calif. Chapt./Am. Soc. Landscape Architects Publ. Group. Oakland, Calif. 64 pp.

Beatty, R.A. & Heckman, C., 1981. Survey of municipal tree systems in the United States. Urban Ecol. 5: 81–102.

Bernatsky, A., 1978. Tree Ecology and Preservation. Elsevier, Amsterdam. 357 pp.

Berry, W., 1981. The Gift of Good Land. North Point Press, San Francisco. 281 pp.

Cole, D.W., 1979. Oakland urban forestry experiment: a cooperative approach. J. Forestry 77(7): 417–419.

Deering, R.B., 1956. Effect of living shade on house temperatures. J. Forestry 54: 399–400.

De Graaf, R.M. & Thomas, J.W., 1976. Wildlife habitat in or near human settlements. In: Trees and Forests for Human Settlements, pp. 54–59. Centre for Urban Forestry Studies, Univ. of Toronto, Canada.

Driver, B.L. & Rosenthal, D., 1978. Social benefits of urban forests and related green spaces in cities. In: Proc. of the National Urban Forestry Conf. USDA Forest Service and SUNY. Washington, D.C., pp. 98–113.

Federer, C.A., 1971. Effects of trees in modifying urban microclimate. In: Trees and Forests in an Urbanizing Environment, pp. 23–28. Cooperative Extension Serv., Univ. of Mass., USDA and County Extens. Serv.

Geiger, R., 1950. The Climate near the Ground. Harvard Univ. Press, Cambridge, Mass. 482 pp.

Heisler, G.M. & Herrington, L.P., 1976. Selection of trees for modifying metropolitan climates. In: Better Trees for Metropolitan Landscapes, pp. 31–38. Northeast For. Exper. Stn., Upper Darby, Pa.

Herrington, L.P., 1978. Urban vegetation and microclimate. In: Proc. of the National Urban Forestry Conf. USDA Forest Service and SUNY. Washington, D.C., pp. 256–266.

Kalmbach, K. & Kielbaso, J.J., 1979. Resident attitudes toward selected characteristics of street tree plantings. J. Arboric 5(6): 124–129.

Northeastern Forest Experiment Station, 1977. Proc. of the Conf. on Metropolitan Physical Environment. N.E. Forest Exper. Stn., Upper Darby, Pa. 447 pp.

Ottman, K.A. & Kielbaso, J.J., 1976. Managing municipal trees. Urban Data Service Report 11/76. International City Management Association, Washington, D.C. 15 pp.

Payne, B., 1973. The twenty-nine tree home improvement plan. Nat. Hist. 82 (9): 74–75.

Payne, B., 1978. Urban land values, trees and urban forest management. In: Proc. of the National Urban Forestry Conf. USDA Forest Service and SUNY, Washington, D.C., pp. 177–182.

Robinette, G.O., 1972. Plants, people and environmental quality. USDA, National Park Service, Washington, D.C. 142 pp.
Schabel, H.G., 1980. Urban forestry in the Federal Republic of Germany. J. Arboric. 6(11): 281–286.
Schanche, D.A., 1983. 'The cloud' chokes Athens. In: San Francisco Chronicle, May 29, 1983.
Shafer, E.L. & Moeller, G.H., 1979. Urban forestry: its scope and complexity. J. Forestry 5(9): 206–209.
Smith, W.H. & Dochinger, L.S., 1975. Air pollution and metropolitan woody vegetation. Pinchot Institute Consortium for Environ. Forestry Studies, Yale University Printing Service. 74 pp.
Wingo, L., 1978. Assessing the benefits of urban public open space. In: Proc. of the National Urban Forestry Conf. USDA Forest Service and SUNY. Washington, D.C., pp. 29–44.
Zube, E.H., 1973. The natural history of urban trees. In: The Metro Forest, A Natural History, Special Supplement, 82(9).

CHAPTER 21

The contribution made by vegetation buffer zones to improved air quality in urban areas

MARTIN MADDERS and MARGARET LAWRENCE

1. Introduction

There is much experimental evidence available illustrating the effects of air pollution on sensitive indigenous plant species (Saunders & Wood, 1974), indeed, certain species of lichen are so sensitive that their presence or absence from suitable substrates may be used in bioassay of levels of air pollution. However, plant communities of more robust species may play a significant positive role in relation to air quality in urban areas. In the U.K. our experience in the use of vegetation buffer zones as a sink to air pollutants, may not be so extensive as in other parts of Europe, but in the development of new towns such as Runcorn and Milton Keynes, the role played by mature plant communities as an air hygiene measure receives serious consideration by town planners and landscape architects.

2. The detrimental effects of air pollution

Although air pollution is undoubtedly aesthetically offensive, its more important detrimental effects are increased morbidity of the urban populations, and the damage caused to the fabric and vegetation of our urban landscapes.

It has been established that general air pollution, and more specifically smoke and sulphur dioxide, are found in sufficient concentrations in European cities to lead to exacerbation of chronic respiratory diseases (Wood & Lawrence, 1980). Pollution indices have been positively linked with morbidity and mortality from lung cancer, pneumonia, chest diseases in children, cardiovascular diseases, respiratory tuberculosis and stomach cancer. In 1972 the World Health Organisation set a long-term goal of 40 μgm of air below which it was believed unlikely that smoke would produce any ill effects, with a similar goal of 60 μgm of air being the target level for sulphur dioxide (WHO, 1972). Wood & Lawrence (1980) found statistical relationships which suggested that if sulphur dioxide levels prevailing over the whole of Greater Manchester could be reduced to the level of the cleanest district at the time of survey, the total number of deaths could be reduced by up to 2000 per year, with bronchitis and emphysema deaths down by about 450 per annum. It has been suggested that particulate lead pollution may retard mental development in children, while asbestos released from disintegrating clutch and brake linings has been considered as a causal factor in lung cancer.

It has also been suggested that the reduction in sunlight leading to lower ultra-violet levels as a result of pollution can be a further stress factor affecting physiological processes in humans. Horbert & Kirchgeorg (1982) further consider that the psychological effects as a result of a decrease in light and colour contrasts, as well as those resulting from shortened range of vision, should not be underestimated.

The polluted air of towns causes a widespread soiling of buildings and necessitates a more fre-

D.O. Hall, N. Myers and N.S. Margaris (eds.), Economics of ecosystems management.

quent washing of windows, paintwork and laundry. Acidulated rainwater resulting from sulphur pollution can cause considerable damage to all types of building materials. Meanwhile sulphur and nitrogen oxides and other airborne pollutants accelerate metal corrosion and even eat into the galvanised protection of bridges and electricity pylons. Within buildings synthetic dyes fade more quickly and fabrics and wallpapers are weakened and need more frequent replacing in polluted areas. Air pollution may also cause severe economic losses to farmers by depressing plant growth rates and hence lowering crop yields, and horticulturalists may be limited to a restricted range of tolerant species in polluted locations. Complete estimates of pollution-damage costs are difficult, but Table 1 is an attempt to estimate some of the costs in the U.K. for 1976 but does not for example include reduced yields in forestry (Saunders, 1976).

3. The effects of vegetation on air pollution concentrations

The evolution of plant leaves as efficient gas-exchange systems for respiration and photosynthesis, and their structure as large pore-bearing surfaces with an internal structure for enhancing diffusion of water-soluble gases, means that they can also effectively remove pollutant gases from the air. Any vegetated area has at least three different sites at which gas molecules may be absorbed: (i) plant cuticles, (ii) the linings of the stomata and lenticels, and, (iii) the soil surface. Each of these sites has a different affinity for a variety of gases, and surface water, whether rain or dew, especially in winter, must also be considered as a sink for air pollutants (Fowler, 1981).

Table 1. Air pollution costs.

Category	Cost (in £ millions)
Painting	6
Laundry	164
Building cleaning	2
Window cleaning	5
Corrosion	42
Textile damage	33
Agriculture	195
Health	540

Any plant community has a large surface area for its volume which has the effect of increasing turbulence in air passing through it. Turbulence affects polluted air in at least 4 ways: (i) it mixes and dilutes it with less polluted air, (ii) smoke, other particulates and aerosols are deposited on leaf surfaces, (iii) gravity settling of particulates occurs in less turbulent air on the lee side of the vegetation, and, (iv) toxic gases are brought into close contact with the gaseous exchange systems of the leaves and soil surface.

Quantitative work on removal rates by vegetation indicates that plants can be surprisingly effective sinks for air pollution. Scorer (1972) establishes that trees are good absorbers of sulphur dioxide, and that a hedge of *Crataegus* sp. may remove as much as fifty percent of sulphur dioxide from polluted air blowing through it, while Kalyushnyi (1952) concludes that parks with trees have an ambient sulphur dioxide level which is some twenty-five percent lower than in city streets nearby. Keller (1979) has quantified a sixty percent drop in airborne lead on the downwind side of a hedge, and an eighty-five percent reduction behind a shelterbelt of trees, while Warren (1973) concluded that a roadside hedge can reduce traffic-generated airborne lead by forty percent on the leeward side. Raad (1965) quotes almost complete dust interception by a thirty metre belt of trees, and even a single row of trees lining a street may reduce airborne particulates by twenty-five percent.

Based on field and laboratory evidence, Keller (1979), Raad (1965), Davis (1975), and Kuttler (1982) have estimated the pollution sink potential of one hectare of vegetation (see Table 2).

The soil surface around plants also acts as an effective pollution sink, leading to the active absorption of very significant quantities of gaseous pollutants through microbial and chemical reactions. It has been reported that carbon monoxide, sulphur dioxide, ozone, nitrogen oxides and hydrocarbons are all absorbed (Desanto et al., 1976; see also Table 3).

In choosing plant species for effective use in air

Table 2. Vegetation sink potentials.

Type of cover	Pollutant removed	Removal rate (tonnes/ha/yr)	Reference
Forest (unspecified)	SO_2	0.29	Davis
Mixed woodland	CO	2.2	Keller
Mixed woodland	NOx	0.35	Keller
Lime trees	Dust	40	Raad
Beech trees	Dust	60	Raad
Grass	Dust	0.31	Kuttler
Clover	Dust	0.77	Kuttler
Soil surface	Dust	0.13	Kuttler

Table 3. Ability of soils to absorb air pollutants.

Pollutant	Removal rate (kg/ha/yr)
Ammonia	48.18
CO	1752
Hydrocarbons	0.43
NOx	17.5

pollution sink zones, it is necessary to take into account the fact that some plants (for example conifers) which would be most effective due to their large surface area, rapid growth and evergreen habit, are also particularly sensitive to common pollutants such as oxides of sulphur and nitrogen. This is unfortunate because generally it is the evergreen species which could make the most effective contribution to pollution control. However, a degree of nursing protection can be achieved by the use of a surrounding deciduous tree barrier. In general, the air pollution tolerance of conifers is increased where the plants are not simultaneously subjected to additional environmental stress such as poor soil, wind exposure and competition for light. Even in winter the effectiveness of a deciduous shelter belt does not fall to nil and according to Jensen (1954) the pollution uptake may only be reduced by some forty percent of the efficiency of the mid-growing season. This may be because in winter even without leaves, trees present to polluted air a large surface area which is often wet due to rain or condensation.

4. The use of plants to ameliorate air pollution

There are three main categories of potential uses of planting to absorb air pollution: those to absorb pollution from area sources (e.g. an urban agglomeration, or a large industrial area); those for ameliorating emissions from a specific point source (e.g. a single factory); and linear sources (e.g. transport routes).

4.1 *Large area sources of pollution*

The use of vegetated parkland zones within urban areas has been suggested by several authors in the past as a means of absorbing pollution, ameliorating the 'urban heat island' and to provide recreational facilities (Bernatzky, 1982). Designs for such zones have ranged in nature from concentric urban green belts to numerous equi-spaced, smaller scale parkland areas. Bach (1971) has suggested that residential areas should be enveloped by buffers of vegetation which would function to protect homes from air pollution and to a lesser extent from noise, whilst offering additional space for recreation. This kind of buffering has been used in the plans for the new town outside Nuremberg in West Germany. Recently, an American research team has made favourable comparisons in cost between conventional means of emission control, and control by vegetation pollution sinks (Desanto et al., 1976).

4.2 *Point sources of pollution*

The impact of pollution from point sources may be reduced by the establishment of vegetation between the pollution source and nearby sensitive land uses which might otherwise be damaged by the emissions. It has been reported that a thirty metre depth of plantation between industrial and residential areas can give effective dust interception and can significantly reduce the concentrations of gaseous pollutants. To achieve the best possible removal rates for particulates, gases and odours, the vegetation should be planted as close as possible to the area to be protected. This use of vegetated areas as air pollution sinks has been practised in several countries in Europe. In Romania for example about a fifth of the total urban area is parkland for recreation and pollution control, while the establishment of large urban green zones for air-hygiene has been incorporated in legislation since 1954 in Czechoslovakia. In the USSR 'sanitary protection zones' have long been a feature around polluting industries, and the dimensions of these areas which vary according to the type of industry, are laid down in legislation. For example, cement-works require a sanitary protection zone at least 1 km in width, while printing works require 100 metres. The size of these zones may be doubled where special risks exist e.g. when a hospital lies downwind. In the Netherlands, private landowners, encouraged by the ministry of Agriculture, have established extensive woodland around the large oil refineries near Rotterdam. In India it is understood that there are plans to establish a great woodland buffer zone to help protect the Taj Mahal from damage by air pollution. Perhaps there are monuments in Greece too which might benefit from the same treatment.

4.3 *Linear sources of pollution*

Traffic using roads is a significant source of toxic gaseous and particulate air pollution, which is often released in close proximity to our homes, schools, hospitals and shopping complexes, work places and other sensitive land uses such as sports fields. In many places roadside planting has been sponsored for aesthetic reasons, albeit adding the benefits of an effective sink for the toxins. However, both the Netherlands and the USSR have been using functional vegetated buffer zones specifically to reduce pollution levels. In Amsterdam for example, roadside planting is the joint responsibility of the Landscape Planning Department of the Forestry Service and the Ministry of Transport Highways Authority.

In Berlin there are plans to create a new Autobahn expressway through the Tiergarten district which would introduce to the area an extra 110 000 vehicles each day with consequent decline in air quality. By careful landscape design, it is planned to create a wind-permeable region of vegetation around the new road system through which air polluted by traffic would diffuse before reaching the city.

Although the following section deals specifically with air pollution sinks for roadside locations in Northern Europe, the same principles can easily be adapted for use around pollution-sensitive land uses (e.g. housing estates, schools, and hospitals) or around pollution point sources (e.g. factories, incinerators, or industrial estates).

5. A suggested North-European roadside shelterbelt

5.1 *Type of planting and choice of species*

An air pollution sink alongside a road is essentially similar in planting design and maintenance to Caborn's (1965) wind shelterbelts. The profile of the belt should present a tall vertical and rough leading edge, there should be no large gaps for the wind to pass through, and they should be permeable to encourage air turbulence and mixing within the shelterbelt. It is important that the planting should not be so dense nor the profile so smooth that the vegetation assumes a solid aerodynamic shape over which polluted air will flow and remain unchanged in composition. It is, of course, necessary for the species to be effective interceptors of the ambient air pollutants, as well as being tolerant of them. It is also important that the chosen species

Plate 1. Urban schools and playgrounds are often fully exposed to air pollution from road traffic.

Diagram 1. A suggested North-European shelterbelt matrix.

A. Lombardy poplar, *Populus nigra* 'Italica' Japanese Larch, *Larix kaempferi* (both quick-growing)	A	B	C	B	A	B
	B	C	B	C	B	C
	C	B	A	B	C	B
	B	C	B	C	B	C
B. Hawthorne, *Crataegus monogyna,* Whitebeam, *Sorbus aria,* Holly, *Ilex aquifolium* (shade tolerators)	A	B	C	B	A	B
	B	C	B	C	B	C
	C	B	A	B	C	B
	B	C	B	C	B	C
C. Lime, *Tilia X europaea,* Oak, *Quercus* sp., Ash, *Fraxinus excelsior* (permanent species to take over eventually from 'A'	←			15 meters		→

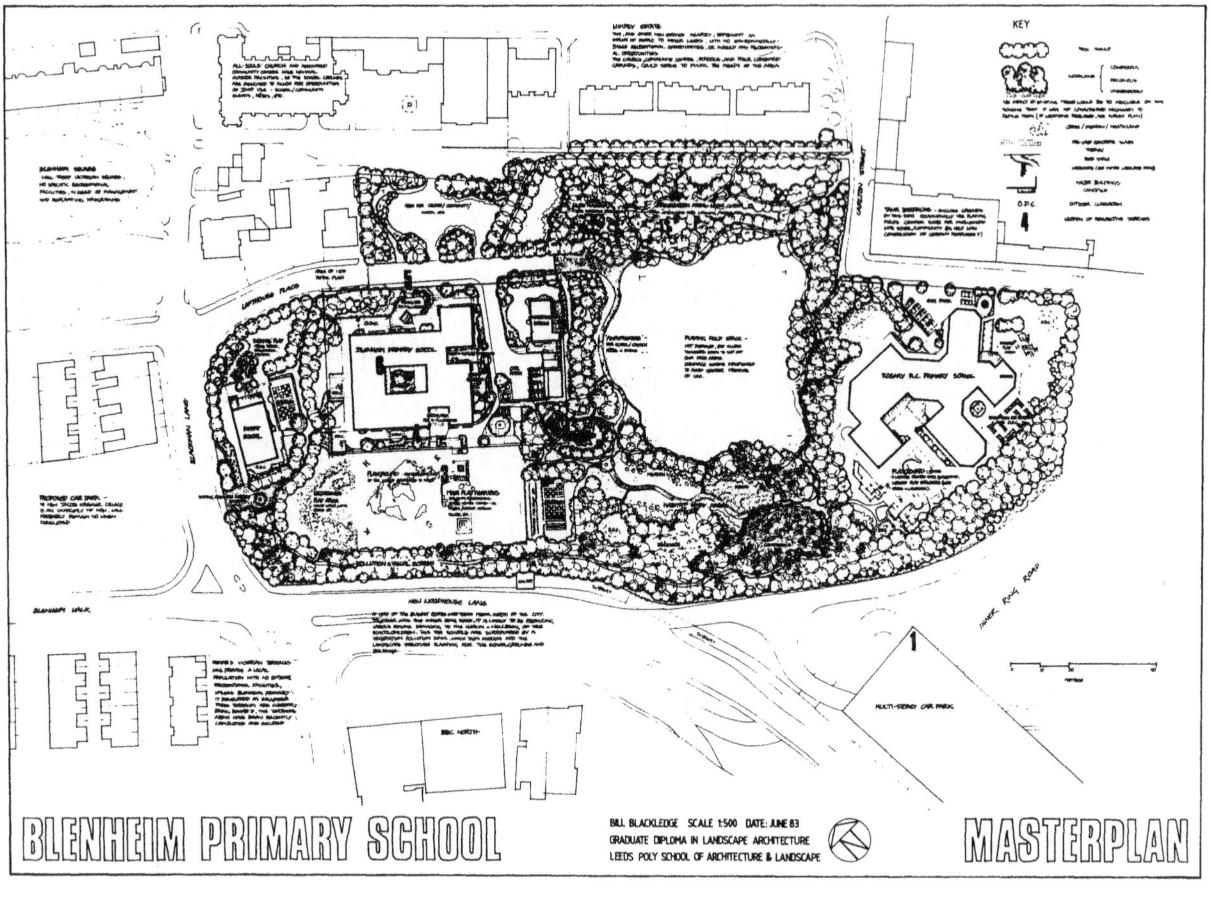

Plate 2. This design by W. Blackledge of Leeds Polytechnic incorporates air pollution buffer zone planting within the grounds of a local Primary School which is almost surrounded by busy roads.

are tolerant of such adverse local factors as salt spray and increased dryness of air near roads (Madders, 1980).

Larch is selected for rapid growth even in poor soils, for its large surface area of deciduous needles and because it is more tolerant to air pollution than most other conifers. If the area is not too heavily polluted, then *Pinus nigra* may substitute for *Larix* and improve shelterbelt performance in winter. *Populus nigra* is resistant to air pollution, establishes quickly, and has a large surface area of fluttering leaves which enhances air turbulence and mixing. *Crataegus* and *Sorbus* are tolerant to shade and to air pollution and provide a shrub layer between tree canopy and ground through which air must filter. The use of *Corylus* at this level might be considered if coppicing is to be part of the maintenance regime. *Ilex* is slow growing and shade tolerant, and because it retains its leaves, it reduces the risk of low gaps in the shelterbelt in winter. *Quercus* are slow growing, ecologically dominant, canopy-forming species which need little maintenance, while the large surface area of *Fraxinus* leaves makes it particularly efficient at dust interception. *Tilia,* especially when propagated from cuttings produces extensive basal branching which may substitute for some shrub planting.

5.2 *Maintenance*

Establishment of the shelterbelt would follow normal forestry practice, and then between twelve and twenty years, thinning becomes necessary to give a spacing for the sub-canopy species of about one sixth of their height. It is suggested by Caborn that half the *Populus* and *Larix* be removed on a 15-year cycle to be replaced by saplings. In Poland, shade-tolerant shrubs are replaced on a fifteen-year cycle to prevent the development of low-level gaps in the shelterbelt. A general guide to maintenance recommended by Caborn is that for maximum efficiency there should be an overall porosity of forty percent, which is when a viewer standing close to one side of the belt can detect movement on the other side, but not recognise the outline of the moving object.

6. Conclusions

Air pollution in European cities causes illness in humans, damage to the urban fabric, and reduced crop yields. Buffer zones of vegetation carefully designed within the city may make a significant contribution to improving air quality. Their existence should not absolve industry from the need to reduce emission levels, but may act as a second line of defence in the event of equipment failure, as well as providing some protection from explosion shockwaves. Careful design of air pollution buffer zone planting can give advantages beyond the purely functional, giving identity and qualities to surroundings which can transform urban spaces for humans to exist into places which are a joy to live, work and play in.

Acknowledgements

The authors are responsible for the views presented, but are very grateful to C.M. Wood (University of Manchester), T. Keller (Swiss Federal Institute of Forestry Research), and Keith Rawling (Chief Forester, Leeds Parks Dept.)

References

Bach, W., 1971. Seven steps to better living on the urban heat island. Landscape Architecture 66: 136–141.

Bernatzky, A., 1982. The contribution of trees and green spaces to a town climate. Energy and Buildings 5: 1–10.

Caborn, J.M., 1965. Shelterbelts and Windbreaks. Faber, London.

Davis, D.D., 1975. A literature review regarding the ability of woody plants to remove sulphur dioxide and photochemical oxidants from the atomsphere. Report P5J10 397-J Part 11. US Environmental Protection Agency.

Desanto, R.E. et al., 1976, 1977. Open Space as an Air Resource Management Measure. US Environmental Protection Agency. EPA 450/3-76-028a and b.

Fowler, D., 1981. Dry deposition of airborne pollutants on forests. Institute of Terrestrial Ecology, Forest and Woodland Ecology Symposium Report No. 8, pp. 62–64.

Horbert, M. & Kirohgeorg, A., 1982. Climatic and air hygienic aspects in the planning of inter-city open spaces. Energy and Buildings 5: 11–22.

Jensen, M., 1954. Shelter Effect. Danish Techn. Press. Copenhagen, 265 pp.

Kalyushnyi, D.N., 1952. The sanitary clasification of industrial plants and sanitary clearance zones. Levine, B. (translator) USSR Literature on air pollution and related occupational diseases, Vol. 4, pp. 165–178. Distributed by US Department of Commerce, office of technical services.

Keller, T., 1979. The possibilities of using plants to alleviate the effects of the motor vehicle. TRRL Symposium Report 513 DOE/DT.

Kuttler, W., 1982. The role of evergreen vegetation in industrial agglomeration areas. Energy and Buildings 5: 23–30.

Madders, M., 1980. Traffic pollution and roadside plants. Landscape Research News 5: 18–19.

Raad, A., 1965. Green spaces and air pollution. J. Arboric. Assoc. 1: 235–244.

Saunders, P.J.W., 1976. The Estimation of Pollution Damage. Manchester University Press.

Saunders, P.J.W. & Wood, C.M., 1974. Plants and air pollution. Landscape Design 105: 28–30.

Scorer, R.S., 1972. Air Pollution. Pergammon Press, London.

Warren, J.L., 1973. Green spaces for air pollution control. North Carolina State University. School of Forestry Resources. Rayleigh. Technical Report 50.

WHO, 1972. Air Quality Criteria and Guides for Urban Air Pollutants. TR-506, World Health Organisation, Geneva.

Wood, C.M. & Lawrence, M.I.G., 1980. Air pollution and health in Greater Manchester. Planning and Environment, A, 12: 1427–39.

CHAPTER 22

Conservation versus exploitation in a developing country

MYRTO PYROVETSI

Developing countries like Greece are struggling under the pressure of progress and of rapid economic development. Under this economic pressure, land and resources are used and developed so as to yield owners the maximum profit. The economic ethic is the motivating power for the decision makers and planners who deal with the development processes, as it most certainly is for the landowner and public. As a result of following this principle, the well being and diversity of a country's natural resources is sacrificed. Incomes and monetary costs are now the ultimate criteria determining human actions, to the degree that the natural world has no significance beyond that of resources ripe for exploitation. However, there comes a time when resources are depleted; then the country must face the result of overexploitation and mismanagement, and the ecological consequences of this type of development become apparent. Ecological damage causes economic and social damage; it is obvious that development cannot be achieved by great leaps, especially when these jumps are made without consideration of the ecological consequences. All evidence indicates that development without ecological consideration is not as positive as it was once thought to be; on the contrary, in the long run, it has its heavy price which includes environmental degradation, resource depletion and biological death. At that stage, the luxury of conservation becomes a necessity. The new goal, that shall someday become an ethic for the developing countries, is to integrate conservation into the developmental processes of the country. In other words, an ethic by which development will occur in harmony with the environment, will conserve resources and employ technologies that do not abuse natural ecosystems. The question that arises then is 'how far are we from this new environmental ethic? How can we make it part of our lives?'

Today, national parks are perceived as areas that can play a very important role in reaching this goal. The park's relatively intact ecosystems provide the basis from which man might better understand and maintain the biosphere. Thus, they can be used as models, their example providing us with a sounder basis for decision making in achieving ecologically sound development. A developing country can use its national parks in order to integrate the objectives of conservation with those of development (Miller, 1978), demonstrate them and adopt a new ethic for the whole nation. I consider the above to be the first step toward a sustainable form of development for developing countries.

Since 1872, when the first National Park was established, the level of comprehension and appreciation of the concept and the worth of national parks continues to increase worldwide. An increasing number of countries are taking positive action to safeguard representative ecosystems through the national park system. Today, there are more then 1500 national parks or equivalent protected areas in more than 135 countries. The effectiveness and continued protection of a national park varies

D.O. Hall, N. Myers and N.S. Margaris (eds.), Economics of ecosystems management.
 ISBN-13: 978-94-010-8928-9

among different cultures, legal traditions and constitutional systems. The classic concept of 'maintaining representative samples of the nation's major biotic units as functioning ecosystems in perpetuity', as it was defined in New Delhi (IUCN, 1969) is the same; however, different social, political, economic and developmental systems impose different goals and expectations. The question is whether a developing nation is able to resist to the economic pressures for exploitation in a national park and thereby to accept the new ecologically 'sustainable development' ethic.

In this paper, I will attempt to analyze: firstly, the way in which traditional economic values of developed areas outcompete park values; secondly, how local exploitation and adjacent development threaten basic conservation values, and finally how an area of great biological value can equally become a victim of conservation when administrative decisions are irresponsible.

An administration's actions may sometimes become a source of confusion and conflict to the public and may have a negative impact on conservation. This often happens when actions are not based on the conservation ethic or where conservation is not implicit in the strategies applied.

As an example of the above I will try to analyze administrative activities in Lake Mikra Prespa National Park, Greece.

The area is located in the northwest corner of Greece (40°45′ N 21°06′ E) where the borders of Yugoslavia, Albania and Greece meet (Map 1). The entire region of 19 470 ha lies between high mountains (2177 m); the surface of the lake is at an elevation of about 853 m. A narrow strip of alluvial land, in the north, separates Lake Mikra Prespa from Lake Megali Prespa which belongs for the most part to Yugoslavia. Mikra Prespa is approximately 5 450 ha in area, of which 800 ha are in Albania. It is a carstic lake, naturally eutrophic, with a maximum depth of 7.7 m (Kusuris, 1983). A canal in the western part of the isthmus joins the two lakes and facilitates a continuous flow of water from Mikra to Megali Prespa, due to the difference in altitude between the lakes. The shores of Mikra Prespa are in most part flat, of silt, and form an extensive wetland. The wetland consists of broad reedbeds, wet grasslands and marshes. In its northern part the reedbeds surround an inner small lake, Lake Viro, which is a significant breeding habitat for pelicans; the broad wall of phragmites (*Phragmites communis*) provide the pelicans with isolation and protection from human disturbance.

Among the impressive number of permanently destroyed wetlands of Europe Lake Mikra Prespa remains one of the last refuges for its rich fauna. The International Union for Conservation of Nature and Natural Resources (IUCN) has recognized Lake Mikra Prespa as one of the unique wetlands of Europe, the protection of which is necessary. One of the primary values of the wetland is that it functions as a genetic reservoir for endangered water-birds. Five species which are present here are listed on the 'official list' of endangered species of the Council of Europe; these are: the White Pelican (*Pelecanus onocrotalus*), the Pygmy Cormorant (*Phalacrocorax pygmaeus*), the Great White Egret (*Egretta alba*), the Glossy Ibis (*Plegadis falcinellus*) and the Spoonbill (*Platalea leucorodia*). One species is listed vulnerable on a world basis by IUCN Red Data Book (IUCN, 1972); it is the Dalmatian Pelican (*Pelecanus crispus*). All the above threatened species breed in Prespa. In addition to these, other species such as Little Egret (*Egretta garzetta*), Grey Heron (*Ardea cinerea*), Purple heron (*Ardea purpurea*), Squacco heron (*Ardeola ralloides*), Little bittern (*Ixobrychus minutus*), Night heron (*Nycticorax nycticorax*), Cormorant (*Phalacrocorax carbo*), Great Crested Grebe (*Podiceps cristatus*), as well as a rich waterfowl fauna maintain breeding populations on the same lake. Equally significant is the native flora containing plant species which are rare in other parts of Greece. A small forest of juniper (*Juniperus foetidissima*) is the only one of its kind existing in the country.

The fauna and flora, the geomorphological characteristics, the archaeological sites (interesting ancient and byzantine ruins), the local traditional architecture and the natural beauty of its landscape led the Greek Government to declare it a National Park in 1974. The Park contains nine small and very old villages inhabited by approximately 1200 people, mostly of old age.

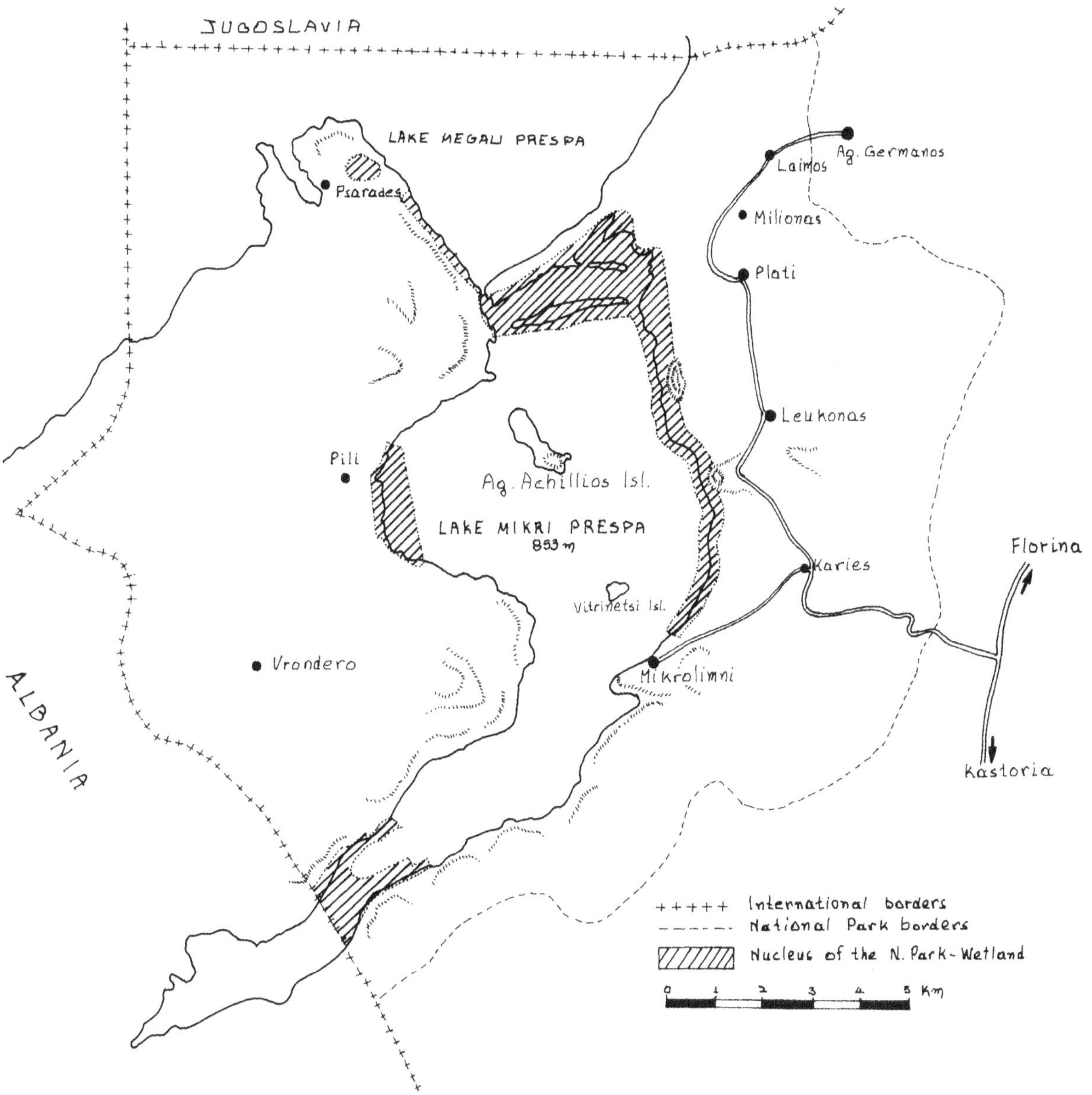

Briefly, the history of the area reveals how things usually evolve in a developing country. Until the late 1960s Prespa was lucky to remain free of the industrialization and threats of development from which most other natural areas in Greece suffer, because of its isolated geographical location and for related political reasons. Local people lived wholly within the ecosystem and in relative harmony with it. Farming, fishing, grazing and tree felling were practiced in a sustainable manner, without any external influence or pressure. However, in the mid-1960s a government agency informed the villagers that the fish in the lake would increase if pelicans were exterminated in the area. 'Pelicans eat your fish-Hunt them' was the order. The administration even encouraged people to kill the pelicans with a bounty of 50 drachmas ($0.50) per adult and 5 drachmas (5 cents) per destroyed egg, which was a good payment for that time. Although it may seem incredible, this continued in the area for some years. Hundreds of pelicans were killed during this time, mainly by hunters who were attracted to the area by the opportunity to gain money.

It should be noted here that Lake Mikra Prespa is the only breeding area for the two species of Pelecanidae in Greece (Crivelli, 1980). Since the drainage of Lake Malik, Albania, in 1938 and Crna Reka, Yugoslavia, where breeding had been reported in the past (Makatsch, 1950), Prespa is the only breeding place in the broader area. In Europe, the Dalmatian Pelican breeds in Lake Skadar (Yugoslavia), Lake Srebana (Bulgaria), with very few pairs, and in Danube Delta, Romania (Crivelli, 1980). The White Pelican is found breeding only in Mikra Prespa and in the Danube Delta.

The order to exterminate pelicans from the area was the first experience these local people had with the 'newly coming civilization'. A few years later, the same agency, under the pressure of IUCN, declared that pelicans were strictly protected and designated the whole area as a national park.

At present, a large part of the Park's resources are being consumed at an alarming rate. Many problems have been created in the name of development and under the National Park protection. I will analyze some of these problems, under the various levels of conservation performance.

At the planning level

The only plans that exist for the area are the so-called 'Five-Year Economic Development Plans'. There is a complete lack of environmental planning and rational use allocation. Present planning ignores the fact that each area of land has its own characteristics which make it more or less suitable for a particular use. For example: the wetland, which is the nucleus of the park, is the breeding and feeding habitat of the water-birds. In its reeds, *Salix* trees and low bushes, the cormorants, herons, spoonbills, ibis and grebes build their nests and breed. In the wet grasslands, which are periodically flooded, the herons, spoonbills and ibis find a rich water fauna upon which to feed. The whole area, extending as a zone 200 meters wide from the lake, has been put by Presidential Decree under strict protection from any human distrurbance or interference. Yet, in this same area a certain agency (whose authority is conservation) planned and implemented a high cost drainage system. With this system they transformed a big portion of the highly productive wetland into farmland. The unique wet grasslands with the high water table, so critical as feeding habitats, disappeared. As a consequence of the above, heron numbers declined; i.e. 100–200 pairs of Little Egret were reported breeding in Prespa in 1968 (Brosselin & Molinier, 1968) but in 1978 this birds' population fell to 17 pairs (Crivelli, 1980). The Glossy Ibis was reported being present with 20–30 adults in 1968 (Brosselin & Molinier, 1968) while in 1982 and in 1983 the species has disappeared completely from Prespa (Pyrovetsi, unpubl. data).

Continuing its economically orientated activities the same agency built an equally expensive irrigation system for the newly created farms. However, the whole project, despite its high economic cost, proved to be inefficient economically. Today, from the 859 ha planned to be irrigated, 685 ha are wheat farms, having no need of irrigation, while crops which require irrigation occupy 174 ha (Ministry of Agriculture, 1982).

This gives an efficiency of 20,25% for the whole irrigation project. Furthermore, there are plans to expand the draining and irrigation projects in order to increase the agricultural land.

It is obvious that the authorities do not consider the impact of their proposed activity on the functions and characteristics which were identified previously for this particular wetland. Thus, the use of the land is not allocated on the basis of ecosystem evaluation and environmental assessment. The latter would ensure that ecological, social, physical and economic criteria were used for decision-making. In this case, economics, not environmental sense, dictated the land management decisions and, even then, the outcome showed that a poor economic evaluation was done. I believe that the single most important wetland characteristic we should strive to protect in Prespa is the frequency and duration of standing water, a characteristic which was completely lost with the draining. Realistically speaking, wetlands drained are usually wetlands lost. Evidently, uses that depend on irreplaceble ecosystem characteristics, such as the wet grasslands should have priority over other uses.

Sometimes exploitation of an ecosystem may be incompatible not only with other uses of that ecosystem but also with the functioning of other ecosystems, as well. In Mikra Prespa the new farms are in close proximity with the reedbeds where the birds breed; agricultural activities in these farms cause great disturbance to the nesting activities and have a negative impact on the breeding.

At the policy level

A well-defined, generally agreed policy of conservation performance is missing. Instead, the desire for an economic return is reflected in every action of the various agencies. The previous example is indicative of this. The conflict between conservation and exploitation is most often weighted towards economic benefit. This is often a result of intensive competition among the various agencies for higher productivity. In other words, policy goals are concerned mainly with productivity and only incidentally with conservation. Similar economic goals are reflected in the policies formed for forestry, agriculture and fisheries in Mikra Prespa National Park. The first priority ought to be conservation policies at more than one level of government; these should be clear, effective, and relatively harmonius. The absence of local citizens' participation in the elaboration of these policies is a serious fault.

At the legislative and administrative level

The designation of the area as a National Park was followed by the enactment of a series of laws and regulations. However, these laws, whatever their quality, failed to be implemented, sometimes as a result of the law being so stringent. The local people had to flout them in order to survive in the area. Land owners felt abused as they were denied by these laws their constitutional right to cultivate their own land. There is no law, or decision to compensate owners for losses of property rights which is a matter of simple justice. The problem becomes move apparent if we consider that the people in the area have the lowest income in the country and must struggle for their survival. Furthermore, the law implies a degree of public understanding and cooperation that simply does not exist.

Example: grazing of cattle is prohibited on the wetland because of the disturbance it causes to the biologically significant bird activities (breeding). The Administration even fenced the area to prevent grazing pressure by the cattle. Today, there are openings in the fence, made by the villagers, so that the cattle can go through to the heart of the wetland and graze very close to the waterfowl nests. This disturbance creates a varity of negative effects to the nesting and feeding activities of the waterbirds.

The problem as a whole is closely related with the fact that the government did not give any option of alternative occupation to the public when it issued the orders against certain land uses. As an alternative, people should be either compensated or given land to farm elsewhere. For the cattle grazing, the government should facilitate the use of the nearby mountain rangeland which provides an excellent quality of grass in high quantities for productive cattle-raising. Instead of organizing and resolving difficulties, the law created much confusion and hostile feelings and today very few conservation regulations are effective in the area. For local people park regulations are only one of the many inconveniences they have to face as part of their struggle for survival; if they obey them, it is only from fear of getting caught.

Another serious organizational problem is the lack of coordination among agencies; sometimes orders are issued that make the same agency responsible for both the exploitation and the protection of a resource.

An example is that hunting is generally prohibited within the National Park. Nevertheless, during the hunting season it is permitted to hunt certain game species. The same agency issued both conflicting laws. The result is that hunters hunt indiscriminately in the area, being covered by the vague law and the lack of Park guards.

Another example is the operation of a marble quarry and a tin factory on the lakeshore, within the water-birds' reproductive biotopes; both industrial installations are operating under the jurisdiction of the Ministry of Industry. The Ministry of Agriculture had previously denied licenses to the quarry and the factory. In this case, jurisdictional conflicts between agencies prevent the law from being implemented. As a first step toward the resolution of the conflict a special organization is needed to promote and coordinate conservation measures. The same organization would bridge gaps and reduce conflicts and duplication.

At the management level

The National Park is not a self-regulating system since it is subject to human activities such as cultivation, pasturage, settlement etc. within its boundaries. Therefore, the significance of its management is becoming more important: to minimize the problems caused by human disturbance and exploitation of its resources as well as to maintain and promote the biological and ecological diversity of the area. This kind of management of the Park's resources should be based on a flow of continuous research, assessment and monitoring. At present, there is a complete absence of scientific activity which might lead to a rational management plan. On the contrary, a high proportion of serious management mistakes is observed.

For example, recently an agency introduced pheasants in the fields close to the wetland, within the nucleus of the Park. Although the pheasants are not expected to survive in these habitats, it seemed that this action ignored the fundamental principle about the introduction of exotic species (Petrides, 1968). While there may be biological objections concerning the pheasants' habitat requirements which are not met in the area, there are also more cogent objections concerning the pheasants' impact on the fragile wetland ecosystems; their impact on the indigenous endangered bird species may be even more serious. If the policy of the national park management is to conserve natural communities such introduction of alien plant or animal species should be avoided as a matter of principle.

Equally important is the status of predators at Prespa National Park. Predator control is still in effect there. Wolves and foxes are killed, under the bounty system, for the damage they sometimes cause to the livestock. Indigenous predators are an integral and important element of natural ecosystems, living in quantitative balance with their prey populations (Lamprey, 1973). The attitudes of pastoral and agricultural people toward predators are unlikely to change, but the area as a national park should play a part in saving the wolf and fox from extinction in northern Greece.

Wildlife management is closely associated with management of the vegetation which supports it. In Prespa's wetland the extensive reedbeds remain intact, without any periodic burning or cutting of the reeds in order to improve them as a wildlife habitat. It might be considered as a conservation measure, however the continuous decomposition of the reeds causes an anoxic environment within a meter water depth and has a negative impact on the quality of the water, the lake condition and on its fisheries. Along with this, there is no watershed management which results in further degradation of the lake.

Information and public support level

There is a lack of any information or educational program which would increase awareness for the benefits of conservation. Ultimately, ecosystems and species are being destroyed because people do not know that it is in their interest not to harm them. Until local people understand why they should safeguard ecosystem and species, they will not do so and will continue to be hostile toward the park idea. Another facet of the same problem is the absence of any public participation. This is particularly important in rural areas such as Prespa, and in their development. Without the active involvement of people no conservation plan can be achieved (Pyrovetsi et al., 1983). Similarly, communication and the exchange of information between the government and local people concerning their expectations and problems, is essential.

All the conservation problems analyzed before are directly related with the lack of rural development; probably this is the most serious conservation problem in the area. All the villagers are preoccupied with the day-to-day struggle for existence; when the source of their next meal is a major worry, principles such as ecosystem preservation or resource conservation have little relevance. Rural development is a *must* for Prespa. Short-term measures to ensure human well-being should go hand-in-hand with long-term measures which will safeguard its natural, cultural, and aesthetic resources and improve the quality of life. Reconciliation of man with nature is no longer merely desirable at Prespa, it has become a necessity.

Today, I would like to focus attention on Prespa and to ask the government, the public, and all of you attending this Symposium, to overcome any apathy or resistance and take all measures needed in order to save Prespa and similar natural areas in Greece.

References

Brosselin, M. & Molinier, A., 1968. Visite au lac de Mikra Prespa (Grèce). Bureau MAR, duplicated report, 20 pp.

Crivelli, A., 1980. The importance of Greece for the conservation of two species of pelicans: *Pelecanus onocrotalus* L. and *Pelecanus crispus* Bruch. Newsletter of the Hellenic Society for the Protection of Nature 'NATURE' 23/24: 35–39.

IUCN, 1969. Pro. X General Assembly, New Delhi. IUCN Bull. n.s. Supplementary paper 27. Morges, Swiss.

IUCN, 1972. IUCN Red Data Book (Mammals, reptiles, birds). Morges, Swiss.

Kusuris, Th. & Diapulis, A., 1983. Lake Mikra Prespa. Technical Study, Institute of Oceanographic and Fisheries Research, Athens.

Lamprey, H., 1973. Management of flora and fauna in national parks. Trends: Jan.–Mar., 1973.

Makatsch, W., 1950. Die Vogelwelt Macedoniens.

Ministry of Agriculture, 1982. Report on the Draining and Irrigation Systems of Prespa. Kozani.

Miller, K., 1978. Planning national parks for ecodevelopment, Vol. I, II. Instituto de la Casa Fotográfica y Ciencias de la Naturaleza Centro Iberamericano de Cooperación, Spain.

Petrides, G., 1968. Problems in species' introductions. IUCN Bulletin 2(7): 70–71.

Pyrovetsi, M. et al., 1983. Prespa. Ministry of Youth and Sports, duplicated report. 95 pp.

CHAPTER 23

Environmental impact assessment and its contribution to ecosystem management

JULIE E. MARTIN

Introduction

In this paper environmental impact assessment (EIA) is taken to mean the systematic examination of the likely environmental consequences of proposed projects, programmes, plans and policies. The results of the assessment – which are assembled in a document known as an environmental impact statement (EIS) – are intended to provide decision-makers with a balanced appraisal of the environmental, social and health implications of alternative courses of action. When an EIS has been prepared it is used by decision-makers as a contribution to the information base upon which a decision is made. In this way, EIA can assist in formulation and evaluation of environmentally sound development proposals. As such it is a tool for ecosystem management.

The EIA approach to evaluation of development actions evolved in the early 1970s, in response to a number of stimuli. A prime factor was the growing scale, pace and associated repercussions of resource development schemes such as large dams, motorways, and nuclear power plants, which were undertaken in post-war years. Unforeseen harmful impacts occurred which reduced predicted benefits. Also important was the upsurge in environmental activism as the public became increasingly aware of the environmental consequences of development actions. Finally, there was considerable evidence of inadequacy in existing appraisal techniques. Projects were assessed mainly on grounds of technical and economic feasibility, and potential environmental, social and health impacts were rarely examined explicitly or rigorously. Even when considered, the assessment usually took the form of cost–benefit analysis, which attempts to express all impacts in terms of resource costs valued in monetary terms. Many environmental, social and health impacts do not readily lend themselves to economic analysis. They may be difficult to quantify, as in the case of disturbance of the special cultural and social patterns of native peoples. They may also be long term and indirect. For example, the Aswan High Dam had deleterious secondary effects such as a decrease in agricultural productivity. In addition, there was a failure to consider the policy context in which proposals were put forward. A mechanism for addressing more fundamental questions, regarding for instance the need for a development, possible alternatives, and appropriate levels of safety and environmental protection, was sought.

As a result of these various stimuli the influential US National Environmental Protection Act (NEPA) was passed in 1969, making EIA of major federal actions a procedural requirement. Since the passing of the Act many other countries have adopted, or are in the process of adopting, some form of EIA, and there has been great controversy regarding its role and potential. The paper will outline the main principles of EIA, and will then examine a number of major areas of debate. The contribution of EIA to ecosystem management will

D.O. Hall, N. Myers and N.S. Margaris (eds.), Economics of ecosystems management.

 ISBN-13: 978-94-010-8928-9

be illustrated using the example of integrated development of the coastal zone. Finally, the advantages and disadvantages of EIA as a mechanism for maintaining environmental quality will be summarised.

Principles of EIA

EIA aims at logical and rational decision-making. It is generally agreed that EIA should be concerned with the identification, measurement, interpretation and communication of environmental impacts of the proposed action. Attempts should be made to reduce potential adverse impacts and increase likely benefits through the identification and assessment of alternative sites and/or processes. The 'no go' alternative should also be assessed. Public involvement has an important part to play in determining issues of significance, providing local information, and helping to identify alternatives. EIA is intended to be an 'objective', technical and predictive exercise, with no decision-making component. The results of the assessment are presented in the EIS document as a discussion of the beneficial and adverse impacts considered to be relevant to the project, plan or policy. This report is one component of the information on which the decision-maker ultimately makes a choice. At this stage other factors – particularly economic and political factors – may influence the outcome of the decision. Ideally, a final decision can be made with due regard being paid to the likely consequences of a particular course of action. To be effective EIA should be implemented at an early stage of project planning and design. It must be an integral component of the design of projects, rather than something utilised after the design phase is complete. Preferably, EIA should be part of an incremental decision-making process which has a number of decision points, and proposals when implemented should be subject to monitoring and auditing programmes. In this way there can be a continuous feedback between EIA findings, project design and locations.

Conceptually, it is useful to distinguish between EIA methods and EIA techniques. Methods are structured mechanisms for the identification of impacts and the organisation of results. Generally, all methods share a common objective of ensuring that as many potential relevant impacts as possible are identified, measured and described, although some go further and incorporate means whereby impacts from different project designs can be evaluated and compared. These latter methods not only identify impacts, but also quantify, weight and aggregate them. Techniques are used to predict future states of specific environmental parameters. Techniques for EIA can be grouped into a number of subject areas such as hazard, noise, transport, air pollution, ecology, landscape character and visual impact. Information and data obtained using techniques can be organised, presented, and in some cases, evaluated, according to the quidelines of a particular method.

Commonly-used methods include checklists, matrices, manuals and networks. It is not the purpose of this paper to look at methods in any detail, as these have been extensively documented elsewhere (see, for example, Bisset, 1983a; Clark, 1983; Canter, 1983). However, it may be useful to look briefly at selected recent developments which help demonstrate the capabilities of EIA methods for application to ecosystem management.

Systems diagrams

A method which began to appear in the EIA literature in 1977–78 is the systems diagram. The theoretical basis for this method resides in the work of Odum (1971) in the field of ecological energetics. In fact Odum first suggested the use of these diagrams in EIA in 1972 (Odum, 1972).

A systems diagram consists of a chart showing environmental and sometimes socio-economic components linked together by solid lines. These lines indicate the direction and amount of energy flow between the components. Systems diagrams are based on the assumption that energy flow can be used as a common denominator to measure the impacts of a development action on the system likely to be changed. Thus systems diagrams enable comparative measurements of the magnitude of different impacts to be made in a common unit.

Activities associated with a project likely to cause impacts may be included in the systems diagram.

The use of systems diagrams in EIA has usually focused on ecological impacts only although Lavine *et al.* (1978) have argued that systems diagrams can provide a bridge between economic and environmental systems. This can be achieved by transferring all types of energy flow in the natural and man-made (economic) environments into a common unit. They can then be converted into money terms by the use of known or calculated energy/money ratios for different national or regional economies.

Simulation modelling

This method is based on the work of Holling (1978) and is often known as Adaptive Environmental Assessment and Management (AEAM). AEAM developed in response to a number of perceived weaknesses in EIA practice. First, EISs were becoming increasingly lengthy (to ensure that results would not be challenged as uncomprehensive). Second, it was felt that EISs were deficient in impact prediction. Third, it was considered that communication between EIA personnel and decision-makers was poor, and that EIA was having insufficient influence in decision-making.

To overcome these alleged weaknesses of EIA, AEAM uses small workshops of scientists, decision-makers and computer modelling experts to construct a simulation model of the system(s) likely to be affected by a development. The key component of AEAM is the workshop in which participants have to reach a consensus on the important features of the relationships which characterise the system(s) studied. Likely outcomes (impacts) resulting from the introduction of exogenous factors (development projects or resource management strategies) can be seen quickly by operating the model under different assumptions. Constructing the model shows up data deficiencies, thus indicating priority areas for research.

AEAM has been applied mainly for the management of resources such as forests, fisheries and water resources. The philosophy behind AEAM has had considerable influence on EIA thinking and practice. For example, it has focused attention on the need for cost-effectiveness, concentration on key issues, and improved accessibility of results to non-experts.

'Sound ecological principles'

Bisset (1983b) has commented on 'the new approach to EIA', which is an increasingly scientific one. A UK study (PADC, 1983) has shown that EISs usually offer vague generalisations about possible impacts which are difficult, if not impossible, to test in a rigorous manner. Such predictions are of little value to decision-makers because of their ambiguous nature. Also, without testable predictions it is impossible to use projects as 'natural experiments' from which environmental impact information could be obtained for use in future EIAs. Since the late 1970s a considerable number of publications have appeared attempting to show a new way forward for EIA. There has been an emphasis on the application of the classical scientific method, and improved analysis of the ecological component. Examples of work in this field are a US study examining ways of improving the scientific content and methodology of EIA (Advanced Studies in Science, Technology and Public Policy, 1982), and a major Canadian study to formulate 'guidelines' to improve the ecological contribution to EIA (Beanlands & Duinker, 1982).

Major areas of debate

The contribution of EIA to ecosystem management depends not only on the availability of appropriate, effective methods, but also on the outcome of debate concerning its scope and application. The following topical issues are therefore of relevance.

The use of EIA at policy level and in forward planning

In principle, EIA procedures should apply to all actions likely to have a significant environmental effect. A comprehensive EIA system would include the appraisal of policies, plans, programmes and projects. Lee & Wood (1978) have called this a 'tiered' EIA structure. Higher order policy or plan evaluations would be conducted first, at national or regional level; lower order programme or project evaluations would then be implemented locally. There are a number of advantages to such a tiered approach. It enables the really major questions of need, safety, environmental protection and compensation to be decided in general terms so that subsequent proposals are more likely to be practicable, consistent, and not unnecessarily constrained. It facilitates optimum site selection and full consideration of alternatives, which cannot usually be achieved at project level. It also allows more time for collection and analysis of environmental data, and eliminates repetition, as a higher order EIA may obviate the need for numerous similar project EIAs.

Unfortunately comparatively few policy or plan level EIAs have yet been attempted, although the evaluation of higher order decisions is widely believed to be very important. O'Riordan (1981) for example, has expressed the view that 'unless an environmental instinct is implanted at the policy determining level . . . EIA will tend to be a cosmetic exercise'. Clark *et al.* (1981) have commented on the difficulties encountered when both project and policy issues are considered at the same time, as at the public inquiry into the nuclear fuel reprocessing plant at Windscale in the UK. Why should policy and plan level EIA be so poorly developed? Foster (1983), who has reviewed some practical and research approaches to policy and plan level EIA, has identified a number of difficulties in using EIA at these levels. The include the lack of a site-specific environment to study, imprecise knowledge of the future, freedom to establish goals and objectives, and lack of suitable methods. Another factor is undoubtedly the reluctance of governments to open the decision-making process to public scrutiny.

Although various mechanisms have been proposed for policy-level EIA, for instance by means of investigatory parliamentary committees backed by research secretariats (O'Riordan & Sewell, 1981), or by independent assessment bodies (Clark *et al.*, 1981), it appears unlikely that they will be implemented in the near future. In the meantime, limited progress is being made towards the use of EIA methods and techniques in strategic land use planning. For example, in Scotland, to establish a policy context for the appraisal of petrochemical developments, the Scottish Development Department (1979) has compiled background information and generated and reviewed alternative strategies for siting of petrochemical development.

The use of EIA to identify social distribution of costs and benefits

Conflict between local and national or other interests in relation to a development proposal is a common problem. It cannot be resolved by EIA but EIA can sometimes help to clarify the issues at stake, before a decision is taken, usually on the basis of political factors. In addition to identifying, predicting and discussing individual impacts and how they affect particular environmental components and sectors of a human population, EIA can provide an overview of the distribution of impacts in a cumulative and spatial sense. This is especially important for impacts which affect the quality of human life, whether they affect people individually or collectively.

One of the best examples of an area in which EIA could encourage a greater degree of social justice is pollution control. A recent UK study (Miller & Wood, 1983) criticises existing UK pollution control practice because it is based on the principle of 'best practicable means' and does not specifically require consideration of the incremental effects of pollutants, nor permit application of quantitative environmental quality standards. With reference to a number of detailed case studies of air, water, land and noise pollution it concludes that had environmental impact assessment been required, serious pollution impacts which left local populations with a net detriment and no means of compensation might have been avoided.

Use of EIA in this way to identify social distribution of costs and benefits is in itself a political issue. It is quite likely that one of the reasons why national and local politicians view the idea of EIA with caution is that they fear it could reveal what environmentalists press for and some powerful interest groups wish to keep dormant, namely the maldistribution of environmental impact amongst the population at large.

Financial aspects of EIA

A major criticism which has been directed at EIA is that it causes considerable costs and delays. Initially, EIAs may be expensive to implement, particularly in areas where little is known about existing environmental and social conditions. Design changes produced as a result of EIA findings may also increase capital costs, but can be argued that savings to local, regional and national economies arising from the avoidance of deleterious impacts and from the maximisation of beneficial impacts will outweigh the costs of an EIA system in the long term. The cost of an EIA system will decline once procedures and techniques have been established and assessment personnel have become accustomed to their tasks.

The costs of EIA are commensurate with the complexity and significance of the problem and the level of detail required. In many countries, the cost is borne by the proponent of the development, whilst in others it is borne by the authorising agency. In those countries with experience in EIA, the costs vary between 0.5% and 2% of the project value. It is misleading, however, to regard the 'actual costs' of EIAs as being saved if an EIA is not undertaken, because much of the information required will have to be collected by some means for submission of planning applications or other purposes. Moreover, it can be argued that thorough investigation of impacts at an early stage of project planning may save money by speeding up the process of implementing a proposal, because inflationary costs arise from delays. Project proponents may also benefit from improved project design and siting which may obviate the need for expensive ameliorative action such as the introduction of pollution control equipment or payment of compensation. British Gas, A UK industry which undertakes in-house environmental appraisal, claims an overall cost saving as a result (Dean & Graham, 1978).

The financial benefits to the public from the implementation of EIA have usually not been determined, because it is difficult to assign monetary values to such benefits. Many of the environmental amenities that would otherwise have been degraded or destroyed have a unique value, which over time will far outweigh EIA costs. Many cases show that the use of EIA has allowed the choice of an option that is both environmentally and economically superior to the original choice – the Alaskan Pipeline being a good example.

The draft EEC directive on EIA

In the member states of the European Economic Community (EEC) the EIA debate has focused on the proposal by the European Commission that environmental assessment should be mandatory for certain types of proposed project. The aim is to incorporate into the laws and administrative practices of the member states a number of common principles for environmental assessment, and to eliminate or reduce disparities between measures in force in different countries. The draft directive would impose full environmental assessment on the member states for certain types of proposal, the principal of which are specified mining, energy, large industrial and civil engineering projects. For a wide range of other activities, including aspects of agriculture and manufacturing industry, oil and gas extraction, and reservoirs, the decision to require a full or simplified assessment would rest with the member states, who must base their decisions on criteria and thresholds subject to examination by the European Commission. The directive sets out the procedures to be followed by the various parties involved. The developer (whether public or private sector) is to prepare a file of information on the effects of the project on the environment. The competent authority in the member state is to seek the views of other authorities, agencies and the public on the proposed project, and also to prepare and assessment report on which to base its decision as to whether or not the project should proceed.

The draft directive has been examined at expert and minister level in the Council of Ministers since 1980, and has been the subject of considerable disagreement and controversy, Within the UK attitudes are clearly divided. British central government has traditionally resisted mandatory rules on the grounds that they are inflexible, and has argued that the current informal use of EIA in the UK, as a supplement to the existing land use planning system, is practical, pragmatic, and preferable. It is feared that detailed project appraisal will contribute to costly delays in project implementation and increase public spending by collecting unnecessary information. These are seen as undesirable attributes at a time when encouraging industrial development and controlling public spending are given high priority. Another objection is that environmental assessment of major agricultural and forestry projects is proposed. These are not at present subject to planning controls in the UK, and strong interest groups oppose the extension of controls to these types of projects.

Elsewhere in the UK attitudes toward the draft directive are more favourable. For example, it is generally supported by the House of Lords Select Committee on the European Communities, the Royal Town Planning Institute, and consultants who have practical experience of conducting environmental assessments. Those in favour of the directive believe that mandatory environmental assessment would be beneficial even in countries with a well-developed decision-making procedure, because it would require the *systematic* examination of environmental factors. They also believe that it need not increase project appraisal costs, for reasons already discussed.

It now seems likely that the directive will be adopted, in some form, before the end of 1983. However, there are fears in some quarters that it will be watered down and weakened before it becomes law (Clark, 1982).

The application of EIA to integrated development of the coastal zone

The coastal zone has been widely recognised – for example through UNEP's Regional Seas Programme – as one of the habitats most threatened by development pressures. Coastal resources for settlement, food production, commercial and industrial development, tourism, and nature conservation are finite, because the coast itself is finite. Coastal areas are experiencing increasing pressure on available space, and serious conflicts between competing uses. These problems are frequently aggravated by peculiarities of ownership, anomalous boundaries, archaic legislation, and confusion as to responsibilities. Moreover, coastal ecosystems are characterised by a high degree of complexity because they are at the junction of two environments, the land and the sea.

EIA methods can provide a useful tool for the analysis of complex coastal ecosystems, and the identification of likely impacts of development actions on such ecosystems. The systems diagram and simulation modelling approaches, discussed above, are particularly well suited to analysis of coastal ecosystems because of their ability to trace the primary, secondary and tertiary consequences of changes in ecosystem attributes. An example of coastal application of the systems diagram approach is the Activity Assessment Routine (Texas General Land Office, 1978) developed for estimating the environmental effects of development activities on the Texas coastal region; while AEAM (simulation modelling) has been applied to assess the impacts of industrial development around Mobile Bay, Alabama (Environment Canada, 1982). The use of EIA methods for management of coastal and marine environments has been encouraged by UNEP Regional Seas Programme.

As well as contributing to the evaluation of coastal developments at the project level, there is considerable scope for the beneficial use of EIA at plan and policy level, for the preparation and evaluation of coastal planning and management strategies. Attention has been drawn to the need for 'integrated' planning and management of coastal areas, for example in the EEC (Commission of the European Communities, 1978). It is felt that because so many other national and international policies – on energy, transport, industry and fisheries – overlap and even conflict with each other in the coastal zone, specific coastal policy is needed. There is also growing awareness that the coastline is a resource of national importance, while decisions relating to the planning and development of the coastline have at least until recently been made almost exclusively at the local level. Local councils and small communities are unable to stand up against strong and often economically advantageous developments. The result is disjointed development and serious conflicts of use. These problems could be reduced by means of an integrated plan or policy for users of the coastal zone as a whole, which would enable development pressures to be approached in a rational and comprehensive way.

There are as yet few examples of comprehensive coastal development strategies, although in some countries worthwhile progress has been made. For instance, the Netherlands has excellent, farsighted regional coastal plans. Brittany has a regional coastal planning strategy, which combines physical and socio-economic objectives in the belief that physical and economic planning can no longer be divorced. In Scotland, the coastal planning guidelines (Scotland Development Department, 1974) issued in 1974 divided the coastline into 'preferred development zones' and 'preferred conservation zones', a sensible modest basis for protecting coastal amenity while expediting oil and gas development.

An interesting recent experiment in comprehensive coastal zone planning is the Integrated Development Programme for the Western Isles of Scotland. Funded jointly by the Department of Agriculture and Fisheries for Scotland and the EEC, the programme attempts to alleviate a wide range of social and economic problems, such as unemployment and out-migration, by means of grants for agricultural improvement, fish farming, tourism, craft industries and other activities. It is to be the subject of a programme EIA.

Conclusion: the advantages and disadvantages of EIA

EIA contributes to ecosystem management by providing a mechanism by which information regarding proposed resource developments and their likely effects can be systematically collected, structured, and analysed.

Potentially, it can assist in the clarification of national priorities in all aspects of environmental management, can open up governmental activities to more detailed public participation, and can encourage greater equity in the distribution of resource costs and benefits. Unfortunately these roles are not always well-developed is practice, partly because EIA has been applied mainly at the project level rather than at plan or policy level.

There is evidence that EIA can benefit local, regional and national economies in the short and long term by enabling negative environmental impacts to be avoided and positive impacts maximised. For example, EIA can assist in the selection of the most suitable site(s) for a particular form of development or in the identification of alternative forms of development to achieve similar aims. EIA can also reduce developers' costs by speeding up the process of project implementation and improving project design so that the need for expensive mitigation measures can be avoided. The use of EIA has increased governmental awareness and responsiveness to environmental concerns in some countries.

Finally, disadvantages attributed to EIA arise largely from its abuse. It has at times been used in an encyclopaedic fashion, which may confuse rather than clarify issues; or superficially, to justify decisions already made. Its true function should be to make decision-making more efficient and more just.

References

Advanced Studies in Science, Technology and Public Policy, 1982. A Study of Ways to Improve the Scientific Content and Methodology of Environmental Impact Analysis. Advanced Studies in Science, Technology and Public Policy, Bloomington, Indiana.

Beanlands, G.E. & Duinker, P.N., 1982. An Ecological Framework for Environmental Impact Assessment in Canada. Institute for Resource and Environmental Studies, Halifax, Nova Scotia.

Bisset, R., 1983a. Introduction to methods for environmental impact assessment. In: Environmental Impact Assessment, PADC Environmental Impact Assessment and Planning Unit (eds.). Martinus Nijhoff, The Hague.

Bisset, R., 1983b. An overview of recent EIA methods. Paper

presented at the Symposium on Environmental Impact Assessment, Chania, Crete, 10–17 April 1983.
Canter, L.W., 1983. Methods for environmental impact assessment: theory and application (emphasis on weighting-scaling checklists and networks). In: Environmental Impact Assessment, PADC Environmental Impact Assessment and Planning Unit (eds.). Martinus Nijhoff, The Hague.
Clark, B.D., 1982. The EEC draft directive on environmental assessment. Scottish Planning Law and Practice 5: 6–7.
Clark, B.D., 1983. EIA manuals: general objectives and the PADC manual. In: Environmental Impact Assessment, PADC Environmental impact Assessment and Planning Unit (eds.). Martinus Nijhoff, the Hague.
Clark, B.D. et al., 1981. The British experience. In: O'Riordan, T. & Sewell, W.R.D. (eds.), Sewell Project Appraisal and Policy Review. Wiley, Chichester.
Commission of the European Communities, 1978. Integrated Management of the Coastal Areas of the European Community. ENV/465/78-EN, Commission of the European Communities, Brussels.
Dean, F.E. & Graham, G., 1978. The application of environmental impact analysis in the Britisch gas industry. Paper presented at the UNECE Symposium on the Gas Industry and the Environment, Minsk.
Environment Canada, 1982. Review and Evaluation of Adaptive Environmental Assessment and Management. Environment Canada, Vanvouver.
Foster, B., 1983. Land-use policy and plan-making: the role of EIA in forward planning. Paper presented at the Symposium on Environmental Impact Assessment, Chania, Crete, 10–17 April 1983.
Holling, C.A. (ed), 1978. Adaptive Environmental Assessment and Management. Wiley, Chichester.
Lavine, M.J. et al., 1978. Bridging the gap between economic and environmental concerns in EIA. EIA Rev. 1(2).
Lee, N. & Wood, C., 1978. EIA – a European perspective. Built Environment 4(2): 101–110.
Miller, C. & Wood, C., 1983. Planning and Pollution: An Examination of the Role of Land Use Planning in the Protection of Environmental Quality. Clarendon Press, Oxford.
Odum, H.T., 1972. Environment, Power and Society, Wiley Interscience, New York.
Odum, H.T., 1972. Use of energy diagrams for environmental impact statements. In: Tools of Coastal Management, Proceedings of the Conference. Washington, D.C., 14–15 February 1972. Marine Technology Society, Washington, D.C.
O'Riordan, T., 1981. Environmentalism, 2nd ed. Pion, London.
O'Riordan, T. & Sewell, W.R.D., 1981. From project appriasal to policy review. In: O'Riordan, T. & Sewell, W.R.D. (eds.), Project Appraisal and Policy Review. Wiley, Chichester.
PADC Environmental Impact Assessment and Planning Unit, 1983. Post-Development Audits to Test the Effectiveness of Environmental Impact Prediction Methods and Techniques. PADC Environmental Impact Assessment and Planning Unit, Aberdeen.
Scottish Development Department, 1974 North Sea Oil and Gas: Coastal Planning Guidelines. H.M.S.O., Edinburgh.
Scottish Development Department, 1977. National Planning Guidelines for Petrochemical Developments. Scottish Development Department, Edinburgh.
Texas General Land Office, 1978. Activity Assessment Routine: Ecological Systems Component (User's Manual). Texas General Land Office, Austin, Texas.

CHAPTER 24

Socio-economic and environmental side effects of technological change: the case of olive oil mills in rural Greece

LEONIDAS P. LOULOUDIS

Dedicated to H.N. Parliaros †

Introduction

The notion of modernization has been always related to the application of scientific innovation to the production process (Sugar, 1980). The effects of modernization cannot be restricted within the economic domain, but are diffused throughout society. It is widely held, however, that domination over nature by man forms the basis for the continuous rationalization of society. The resulting emphasis of the primacy of productive forces over the development of social relations may be attributed to this belief. Hence, the political and ideological role of technology through time, has been considered as neutral.

Nevertheless, according to recent theoretical approaches technology is viewed as the vehicle of the relations of production which implies that it posesses an endogenous political character (Sarikas, 1982). Therefore, technological change is a process which reproduces the relations of dominance/exploitation within the existing social division of labour. There is strong evidence that increasing state intervention in determining future technological research directives, serves this purpose. When technology has been defined as such, the resulting consequences from the technological application to the process of production cannot be questioned merely as being the results of 'inappropriate practices', which in turn can be changed within the context of the existing structures. Rather, they should be addressed as problems rooted deeply within these structures (Buttel, 1980).

In this paper we shall attempt to illustrate to what extent this theoretical approach of technology may be useful for a fruitful discussion on the appropriateness of the 'conventional' technological ensemble, before proceeding to seek the proper solution to certain environmental problems encountered in rural areas.

1. Olive oil mills waste: type and origin of the problem

Social and economic value of olive oil

Liquid wastes of olive oil mills have during the last ten years created a serious environmental problem which has forced some plants to suspend their operation.

This happens occasionally, mainly when nearby inhabitants protest and request suspension, and reflects two major problems.

First: There is no adequate legislative framework for the control of pollution, which for the moment is covered by the existing public hygiene regulations (Kostopoulos, 1980), also there is a lack of government personnel to supervise that mills

D.O. Hall, N. Myers and N.S. Margaris (eds.), Economics of ecosystems management.

 ISBN-13: 978-94-010-8928-9

conform to existing laws.

Second: State agencies are unable to recommend methods of confronting the problem. Government and private studies made in 1981 and 1983 were confined mainly to research proposals rather than to the proposal of immediate solutions (TEE Div. of S. Grete, 1981; Working Group of the Ministries of Agriculture and Environment, 1983).

But the suspension of olive oil mill operation is no more than a temporary solution considering the economic and social value of olive oil. Greece is the third largest producer in the world (with 235 000 tons in 1981–82) after Spain and Italy. Olive oil contributes 14% and 9% to plant and entire agricultural production respectively. 350 000 rural families are occupied with its production filling 76% of their working time, especially during winter months, when other agricultural work recedes (Xylogiannis, 1982). These families also consume half of the total production, which has an annual increase of 3%. During 1976–1977 Greece had the largest per capita consumption in the world (18.95 kg). For these reasons, although the operation of olive oil mills creates a serious environmental problem, the suspension of their operation creates simultaneously a social and economic problem.

Technological change and environmental effects (for the period 1965–83)

In order to evaluate the environmental effects due to technological change during the recent two decades, qualitative and quantitative aspects of pollution are distinguished, which should not be confused, especially as regards the olive oil mills effluents (Shnaiberg, 1980).

a. Qualitative aspect of pollution. Olive oil mill liquid waste is the by-product of the olive fruit processing to extract olive oil and consists of vegetative liquids which are 30–40% of the fruit weight, increased by water added during the operation. Chemically these vegetative liquids are constituted by (Fiestas Ros de Ursinos, 1958):

Organic matter	15.0%
Inorganic salts	1.8%
Water	83.2%

The organic matter contains total sugars 2–8%, nitrogen compounds (X6.25) 1.2–2.4%, organic acids 0.5–1.5%, polyalcohols 1.0–1.5%, colloids, pectins, tannins 1.0–1.5%, fatty components 0.03–1.0%, along with various glucosides and oleocyanine (Roncero, 1965). Organic matter, the component most responsible for the pollution effect, varies, according to olive variety, agricultural area, cultivation techniques, time of harvesting, plant diseases, and the technology of extraction, from 10 000–100 000 mg/l BOD_5. But the high content of organic matter is but one of the problems caused by olive oil wastewater. The other one is their toxicity to hydrofauna (Fiestas Ros de Ursinos, 1966) and higher plants. Our experiments have shown that this toxicity is not specialized, at least to higher plants (Louloudis, unpublished data). The 'direct toxicity' is attributed to volatile fatty acids and phenolic compounds, and is of great importance as decomposition is very difficult, remaining even after aerobic or anaerobic treatment, which nevertheless decrease the organic load at percentages larger than 50%. The effects of 'direct toxicity' are especially serious for Greece where 78% of these wastes are disposed in various ways into the soil.

b. Quantitative aspect of pollution. To understand the quantitative aspect of pollution we have to refer to the two main methods of extraction. The old, classic method has as its basic equipment the hydraulic press, while the newer centrifugal method utilizes the decanter. The rest of the equipment in both cases remains almost the same. There are quite large differences in the processing water flow diagrams between the two methods. Water consumption in centrifugal processing systems is 0.5 m^3/hr for the decanter and 0.4 m^3/h during the centrifugal separation, while in the classic method 0.5 m^3/hr for the separator (Working Group, 1983). Finally the total volume of effluent in the classic method is smaller than that corresponding to the decanter by a ratio 1:1.9–1:3.2. But there is also a

Table 1. Production method and polluting change.

Method	pH	BOD_5 mg/l	COD mg/l
Classic	4.0–4.9	30–100 000	59 000–162 000
Decanter	5.2	10– 20 000	39 000– 78 000

difference in the quality of polluting charge as shown in Table 1.

The above data show that in spite of the decrease of pH and polluting charge, in the decanter method, the charge remains high (10–20 000 mg/l BOD_5) compared with urban wastewater (300–400 mg/l BOD_5). Furthermore the 'direct toxicity' continues to be present and active. Thus we can conclude that the pollution effect quantitatively increases with plant modernization.

But the quantitative pollution effect depends mainly on the size or the productivity of olive oil mills which may be classified as shown in Table 2.

With an average of effluent of 105% of the initial olive fruit weight then the relative effluent production is as shown in Table 3.

The observation that an increase in the size of production also increases the volume of effluent is of interest, in relation to the trend in the merging and modernization of existing mills or the establishment of new ones. Between 1966–1975 the number of plants decreased by 33.7% with a parallel trend to replace animal energy powered presses with electric powered units. The Table 4 is indicative.

During the 1981–82 period these plants were reduced to 3469 while during 1982–83 the total was 2369. It is to be noted that this replacement trend in favour of electric powered units goes hand in hand with the installation of decanter systems. The number of such mills increases every year at a rate of 130–150 new units.

There are several reasons to impose such a decrease in the number of plants and therefore their gradual relocation in a smaller dispersal throughout the country. The main reason is economic. The total production capacity for 1981 reached 16057 kg/8h which exceeds the need of the (actual production) in only one month of operation at capacity (20 kg/day) (Xylogiannis, 1982).

Consequently the observed trends are not aimed at increasing the total potential production. The main target is an increase in average capacity which

Table 2. Production capacity.

mg/hr	kg/day	Class
from 800	19 200	small
from 800–1400	19 200–33 600	medium
from 1400–2500	33 600–60 000	large

Source: TEES. Crete Division.

Table 3. Effluent production (m^3).

Class	Yearly[a]	Daily	Hourly
Small plants	36 750	20.16	0.84
Medium plants	57 750	35.28	1.47
Large plants	95 000	62.88	2.62

Source TEES. Crete Division.

[a] Yearly production refers to 3–4 months operation. Sometimes 2 or 10 months.

Table 4. Technological change in olive oil mills.

Type of energy used	1965		1975		Increase/decrease
	Number	%	Number	%	
Water-Animal	1350	25	16	0.5	− 98.8
Mechanical	3145	60	481	13.5	− 84.7
Electric power	810	15	3018	86.0	+272.5
Total	5305	100	3515	100.0	− 33.7

Source: Ministry of Agriculture.

is 4.62 t/8h. This increase is due to the need to decrease the production cost resulting from the installation of decanter systems. The latter not only give a better quality product, according to their proponents, but reduce the number of man hours due to their higher degree of automation (Anastasiou, 1981). A decanter system plant of a total 90 HP demand, employs two workers, while the corresponding plant of the classic type, with the same production capacity needs 8HP energy and employs 8 workers. It is estimated that a wide change in technology of olive oil mill equipment may reduce labour needs by 60–80%. On the other hand the high cost of modern plant equipment and the operation costs (energy and water) require an increase of production capacity to at least 12 tons/8hr to be economically acceptable. This may be done either by merging smaller units, or by the establishment of new units sufficient to reach the above capacity. However this concentration of production in few large units increases qualitative/quantitative aspects of pollution as we have already indicated at the beginning of this chapter.

2. Technological change in olive oil mills and social organization (period 1750–1950)

Evolution of technology and the social background

The case of olive oil mills offers, we think, a remarkable empirical model illustrating the dynamic interaction of social function and technology.

It has been mentioned that the modernization of equipment reduced the number of plants by 33.7% within a decade, while the total number of electric powered plants increased by 272.5%. This trend in technological change, desspite the fact that it is damaging the environment by pollution, is considered the only solution, 'a deus ex machina', to overcome the lack of labour in rural areas. However an historical retrospection of the technology of olive oil mills shows that change in technology is incorporated in to structural changes in the rural community, the most direct consequence of which is the reduction of the rural population. It is worth noting that technological change is accelerated only when the corresponding structural changes in the whole social framework are not propelled by the interests of a so-called enlightened minority, but is based on an established ideological, political and economic national reality. Whithin this reality then, current technological solutions are the only possible, accepted and urgent answer.

Olive oil mill technology went through the three stages or complexes of the Geddes and Mumford (1963) evolution scheme. We can roughly say that the eotechnic stage starts in the Neolithic Age, the paleotechnic in about 1750 and the neotechnic after 1950. We have already said that the three systems coexist in the 20th century.

Two periods are of great interest: the one between 1750 and 1950, and the one after 1950. We have already mentioned the rapid changes which occurred in the 1950s, with the neotechnic type of plants prevailing after 1960. A more important period, however, is the one between 1750 and 1950 when eotechnic and paleotechnic types coexisted or formed complexes up until 1917. Long after 1775, when a paleotechnic improvement of the typical eotechnic equipment took place, that is the wooden screw press which was widely found up to the beginning of the 19th century, the most primitive wooden press (called spindle, 'adrachti', 'sfondili') still remained in use. For instance written and spoken evidence indicates that this equipment was in use in Corfu (Sordinas, 1971) as late as 1917, in Fourni (S. Argolis) (Gabrielides, 1976) till 1810, and in Tsakonia, Arcadia (Kostakis, 1963) till 1840. This shows the relevant stagnation which characterizes the promechanical period and the prevailing mentality. That is, a direct conformation to the existing situation (Vernan, 1975), which explains why when some improvements are introduced these are not adopted so fast as to eliminate the most primitive equipment of the eotechnical period or even more improved types. The same happened when between 1865–1890 an improvement of the late paleotechnic equipment was introduced (all-iron hydraulic presses). Also we observe that while the eotechnical complex dominates up to the 18th century, and in some areas of the country its implements are in existence up to the first quarter of the 20th century, the paleotechnic equipment is the most widely used for less than half a century. On the contrary, the neotechnic systems tend to replace all earlier systems or complexes within a period of less than half a century.

At this point there are, I think, some open questions in relation to the various rates of technological change diffusion, after their introduction in a certain social formation concerning their ultimate success or failure. An example is the improvement of the typical workshops of the eotechnic period, or the appearance of the early paleotechnic types, related to the growth of Greek communities abroad and the ensuing increase in product demand, especially of wine and oil during the 18th century (Evelpidis, 1950). Is it a mere coincidence that the time when all-steel hydraulic presses were introducted (1870–1890), is also the time of the first industrialization stage of the country. Olive oil mills constituted by that time (1875) the sixth largest sector of Greek industry as regards to invested capital (Tsoukalas, 1981). Another related question concerns the demands put forward by the association of engine (diesel) driven hydraulic press owners, during the olive oil convention of 1936: they asked that all horse driven mills, 'this shame of the 20th century' be closed by the State (Kakouros, 1938). The question arises as to what extent this demand is connected with the expansion of the vegetative oil and fats market during the previous decade (Kakouros, 1938) and the second attempt to industrialize Greece (1920–1930). Finally, to what extent this gradual adoption of neotechnic systems is due to the state policy investment in basic public works infrastructure (roads, electricity, transportation etc.) implemented with international capital after 1950.

If the above questions give a framework for discussion concerning the introduction of technological novelties, the question regarding the rate of acceptance or rejection of new technology in the rural areas remains still unanswered. It is my belief that this question cannot be answered outside the context of the rural structure itself. The international crisis in the marketing of raisins allowed the survival of precapitalist relations of production during the whole of the 19th century. The restricted financing possibilities to the industry and small manufacturers was further reduced, even after the end of the crisis in 1892 (Papagiannakis, 1983). Thus in rural society, life will continue to be determined by the traditional factors leading to a 'general way of thinking segregated from and closed to the meaning of enterprise, profit and the value of time . . .' (Andreadis, 1966). This way of thinking considers the sharing of the oil as a 'religious ceremony', the barter of products – 'I give you oil, give me butter' – quite natural, and the operation of the workshop as a good, companioniable affair during the winter (Kostakis, 1963). This way of thinking, which does not favor modernization, is based on a demographic reality permitting the wider possible geographical distribution of the workshops. At Fourni as an example, with an average of 200 inhabitants between 1696–1850 there were six workshops, while the Venetian census of 1770–1776 showed the existence of 990 workshops in Corfu, half the number of those operating today over the whole of Greece. The exodus of the rural population had not yet begun, and the small labour-intensive units were an integral part of the community structure and needs. Thus the brief industrial awaking of the last quarter of the 19th century, failed to affect these conditions. Likewise the more favorable to change second industrialization attempt in the 1920s receded quickly due to (a) structural inefficiencies in industrial as well as small manufacturing production and (b) the enforcement of the self-consuming models in rural areas, where a new increase in population was noted after the decline of the first immigration wave abroad. As a result even the primitive hand-operated mills were useful in areas with small production and were even profitable if operated by the owner himself and his family, or in inaccessible villages. Transportation of relatively heavy equipment to modernize these mills was impossible as stated by Filippopoulos (1938). The 'Eleourgical Review' complains of the 'illegal competition' of horse driven community workshops in alignment with the progressive industrial spirit of the same era (Epitheorisi, 1938).

Under the above conditions it was not strange to see in the 1958 statistics that out of a total of 8459 mills, 373 hand operated still survived while 3950 were horse driven and the balance of 4136 used mechanical energy (Boudouris, 1957; Papazoglov, 1954). But since the 1950s large infrastructure projects on a national scale have started which have permited social structural changes, one of these was the decrease by $^1/_3$ of the rural labour force during the ensuing 20 years (Louloudis, 1983). This new demographic reality facilitated the spread of technological change towards the neotechnic systems previously referred to.

Preindustrial conditions and environment

As shown so far the production conditions in olive oil mills up to the first quarter of the 20th century may be characterized as labour intensive and of small size. We will try here to give more information regarding the above assumption and to relate the rise of production with possible deleterious effects to the environment.

The amount of labour required per unit of production was in line with the calling of 'tsoūrmo', the 'gang', consisting of 6–10 workers. In the transportation of the olives to the plant, women were also involved, and 3 mules or donkeys were also required. The number of men was larger when the spindle – o ergátis – was hand operated by 8 men in the most primitive workshops (Florakis, 1971). Although workshops were operating continuously day and night, their size could be estimated roughly by the hourly output of the crushing mill and that of the press. A horse driven monolith crusher, commonly used during that period, crushed a batch of 80–100 'okádes' per hour or a total of 640–800 'okádes' in 8 hrs (Florakis, 1971). A wooden screw press needed 6–10 hr to press the daily output, a process which delayed the whole work. Therefore the output in these eotechnical complexes could not exceed 1 ton/8 hr, that is $^1/_3$ to $^1/_2$ the output of a medium sized plant of the late paleotechnic stage. The existing reports of that period do not indicate that the effluent created any special problem. This may be due to the low production capacity but was mainly due to the small quantities of olive fruits processed resulting from the large number of mills per geographical area. There is no indication in the litterature of any measures taken to treat the effluents. They were either disposed outside the mill, in adjacent uncultivated fields, or in a small pit. On the other hand, the other byproducts of the process were collected and used. The sediment of the oil – the sludge – was used to make soap or in the process of tanning to soften certain skins, and the olive press cake as a home or plant fuel and/or as a pig feed. The magazine 'Eliniki Georgia' in 1885 mentions the marketing of dried olive press cake (Gennadios, 1887). Only during the late paleotechnic stage, does Sarakomenos (1930) propose the construction of special reservoirs to collect wastewater 'amorgi'. These reservoirs were to have a capacity of 10.5 and 40.5 m^3 for two plants of 100 000 and 500 000 'okades' of olive processing capacity and Sarakomenos states that the effluents did not exceed 50% of the olive fruits weight. The effluent was to rest in the reservoir for 10 days. After this period obviously recognizing its fertilizing potential, the treated effluent was transferred in a covered cess pool mixed with leaves, manure etc. in a succession of thin layers of lime. Apparently this was a composting process, perhaps one of the first references in Greek literature concerning effluent treatment on a large scale, showing that the author was conscious of the environmental problem and of the importance of the recycling process, towards a more rational exploitation of natural resources. After half a century, the Sarakomenos proposals are again be considered by his co-patriots.

Conclusions

In Section 1, we analyzed the qualitative and quantitative aspects of olive oil wastewater. In the Section 2 we made an historical review of the dymanic interaction of technological change and social organization. In the light of this approach it is becoming, we hope, apparent that the environmental problems cited are not merely the results of 'inappropriate processing practices' that can be changed within the context of the present structure of the agricultural sector. Indeed these problems are clearly rooted within that structure, sharply reducing the possibilities for rural environmential improvement without major social changes in that sector of the economy, agriculture. For instance three of the available methods for waste treatment, are a priori not applicable considering the size of the plant or the volume of the effluents: the classic method of the septic tank (Fao, 1956), the method of condensation by natural evaporation (lagooning) (Fiestas Ros de Ursinos, 1977) or the direct use of diluted waste for irrigation-fertilization (Fiestas Ros de Ursinos, 1960). Of the remaining methods, the most important, being tested at present, are the extraction of valuable pigments (anthocyanins) (Tanchen et al., 1980), the production of single cell protein as a feedstuff (Fiestas Ros de Ursinos, 1958), and the anaerobic-aerobic treatment (Ntalis et al., 1982) for the production of methane and reduction of organic charge to acceptable levels; all encounter – beyond technical problems – certain economic restrictions, including the installation cost (Fiestas Ros de Ursinos, 1967), seasonality of plant operation (4 months out of 12) (Imrie & Righelato, 1976), lack of skilled personnel, and a sufficient and profitable market (Harrison et al., 1982) (See also Forage & Righelato (1978) for the production of microbial protein with 'low technology processes'.). Sine qua non economic criterion to overcome these restrictions is the large quantity of wastewater needed for treatment. Effluent treatment could be achieved by the application of these methods in the larger olive oil production units, or alternatively the processing of effluents could take place in a large central unit serving all the plants of a certain area where these are concentrated with a simultaneous total exploitation of the manufactured products, i.e. biomass and/or biogas. As an example, in Puglia, South Italy, a pilot plant of the anaerobic process was built in 1980 to serve the 2000 units of the area, which represents 40% of the total Italian production (Ntalis, pers. comm.).

In such large-scale production several problems exist, which are difficult to solve, as evidenced by the fact that although efforts started during the 1950s, mainly at the Instituto de la Grasa y sus Derivados in Seville, Spain, there is not as yet a definite, scientifically sound method solving the problem as a whole. This is especially true in the Mediterranean oil-producing countries, as indicated in the last report of the Greek Ministry of Agriculture.

However the environmental problems, we believe, provide the stimulus for the elaboration of new strategies of change. This change should treat jointly agricultural structures and environmental questions, in an holistic perspective of agricultural development. Towards this direction the conditions under which small units shall remain in their present geographical location or be abolished should be examined, as a first priority for the following reasons: first, to avoid a further worsening of rural environmental and energy problems; second, to encourage local food production and consumption so as to enhance decentralization of the food supply system; and third, to disperse the control of food processing industries to the local communities, thus increasing the employment prospects of the rural sector. This is an urgent task, in view of the unemployment problem in urban areas and the shortage of labor observed in certain rural regions.

References

Anastasiou, P., 1981. The Olive Oil Industry and Production in Greece. Modern Agricultural Technology 5: 36–39 (in Greek).

Andreadis, S., 1966. Social Discontinuity and Problems of Economic Development in Greece. Kenoniologiki Skepsi 1: 30–31 (in Greek).

Boudouris, E., 1957. The Olive Oil Processing. In: Encyclopedia Drandakis: 811–812.

Buttel, F.H., 1980. Agricultural Structure and Rural Ecology: Towards a Political Economy of Rural Development. In: Sociologia Ruralis 20 (1/2): 44–59.

Epitheorisi, E., 1938. Proposed Measures for the Protection of Greek Olive Oil Production. Elaiki Epitheorisi: 13–15 (in Greek).

Evelpidis, H., 1950. Economic and Social History of Greece. Papazisis (Athens): 31–33 (in Greek).

FAO, 1956. Report on Olive Oil Processing in Rural Mills.

Fiestas Ros de Ursinos, J.A., 1958. Estudio del Alpechin para su Aprovechamiento Industrial. Obtensin de levaduras Alimenticias. Grases y Aceites 9 (5): 249–258.

Fiestas Ros de Ursinos, J.A. & Albi Romero, M.A., 1960. Estudio del Alpechin para su Aprovechamiento Industrial. Ensayos Efectuados para su Posible Utilization como Fertilizante. Grasas y Aceites 11 (3): 123–124.

Fiestas Ros de Ursinos, J.A., 1966. Estudio del Alpechin para su Aprovechamiento Industrial. VII. Instalacion Comercial para la Obtencion de Levaduras Pienso. Grasas y Aceites 17 (2): 41–47.

Fiestas Ros de Ursinos, J.A., 1967. Estudio del Alpechin para su Aprovechamiento Industrial. Grasas y Aceites 18 (2): 45–46.

Fiestas Ros de Ursinos, J.A., 1977. Depuracion de Aguas Residuales en las Industrias de Aceitunas y Aceites de Oliva. Grasas y Aceites 28 (2): 113–121.

Filippopoulos, G.K., 1938. The Olive Oil. N. Apatsidis, Athens, p. 25 (in Greek).

Florakis, E.A., 1971. Tinos popular civilization, pp. 110–112. Hellenic Book, Athens (in Greek).

Righelato, R.C., 1978. Microbial Protein from Carbohydrate Wastes. In: M.J. Bull (ed.), Progress in Industrial Microbiology, Vol. 14, p. 59.

Gabrielides, N.E., 1976. A Study in the Cultural Ecology of an Olive – Growing Community. PhD thesis, Indiana University.

Gennadios, P., 1887. Eliniki Georgia. Athens) Vol 3. No C' (March) (in Greek).

Georgakopoulos, N.A., 1926. The Olive Oil, pp. 60–62. M.I. Saliveros, Athens (in Greek).

Harrison, B. et al., 1982. Why High Technology Facilities Fail. Biocycle (September–October): 22–25.

Imrie, F.K.E. & Righelato, R.C., 1976. Production of Microbial Protein from Carbohydrate Wastes in Developing Countries. In: Birch (ed,), Food From Waste, p. 79. Applied Science Publishers.

Kakouros, K., 1938. Speech on the Olive Oil Processing Convention (8-11/8/36). Elaiki Epitheorisi 1 (1): 15–16 (in Greek).

Kostakis, O.P., 1963. The Olive Oil Tree and the Olive Oil in Tsakonia. Laografia 21: 392–393 (in Greek).

Kostopoulos, B.O., 1980. Problems of Law Protection from Environmental Pollution. Athens, (in Greek).

Louloudis, L., 1983. Agriculture and Environment. Ecologia ke Perivalon. (January–February): 72–76.

Mumford, L., 1963. Technics and Civilization. Harcourt, Brace and World, New York.

Ntalis, D., Hartman, L. & Anagnostidis, K., 1982. Anaerobic Treatment of Olive Processing Wastes. In: Second (Wageningen, Holland) and Fourth (Broni, Italy) Specialised Meeting of EC on Anaerobic Digestion. Rand D Programme – Recycling of Urban and Industrial Waste (RUW – 056).

Papagiannakis, L., 1983. Different Stages and the Characteristics of Greek Industrialization. Economicos Tachydromos 12 (1507): 62 (in Greek).

Papazoglou, Ch., 1954. Progress in Olive Oil Production in Mediterannean Area And Greece. I elia 2–3: 567–574 (in Greek).

Roncero, V.A., 1965. Quimica del Olivo los Componente Organicos, 3e Parte. Grasas y Aceites 16 (6): 292–304.

Sarakomenos, D., 1930. The Olive Oil, Vol. 2, pp. 163–258. Pirsos, Athens (in Greek).

Sarikas, Z., 1982. Science and Domination over Nature. Sinchrona Themata 15 (September): 70 (in Greek).

Shnaiberg, A., 1980. The Role of Technology-Deus ex Machina or Social Creation? The Environment From Surpus To Scarcity, pp. 113–137. Oxford.

Sordinas, A., 1971. Preindustrial Olive Oil Workshops. Project on the Island of Corfu, Greece (Progress report, Memphis State University.

Sugar, P., 1980. Some Thoughts on the Modernization Preconditions. In: Modernization and Industrial Revolution in 19th Century Balcans. Athens. Themelio: 68–109 (in Greek).

Tanchen, S., Joncheua, N., Genon, N. & Kodounis, M., 1980. Detection of the Anthocyanins Contained in Olive Oil Fruits. Georgiki Erevna 4 (1): 5–13 (in Greek).

TEE Division of Southern Crete, 1981. Environmental Pollution Caused by the Wastewater of Olive Oil Mill. Technica Chronica (Athens) November: 5–22 (in Greek).

Tsoukalas, K., 1981. Social Development and the State. Themelio (Athens): 203 (in Greek).

Vernan, J.P., 1975. Mythe et Pensée chez les Grecs. Olcos: 286 (in Greek).

Working Group of the Ministries of Agriculture and Environment, 1983. A Study and Proposals for the Protection of the Environmental Pollution Caused by the Olive Oil Mills, Vol. 1. Athens (in Greek).

Xylogiannis, D., 1982. Annual Review of Evolution in the Olive Oil Sector 1981–82. Paseges, Athens (in Greek).

CHAPTER 25

Biomass: fuel versus food, a world problem?

D.O. HALL

Abstract

The overuse and undersupply of biomass is currently a serious problem and potentially a greater danger than ready lack of food. Today 14% of the world's primary energy is derived from biomass – equivalent to 20 m barrels oil/day. Predominant use is in the rural areas of developing countries where half the world's population lives, e.g. Nepal and Ethiopia derive nearly all, Kenya $^3/_4$, India $^1/_2$, China $^1/_3$, Brazil $^1/_4$ and Egypt and Morroco $^1/_5$ of their total energy from biomass. A number of developed countries also derive a considerable amount of energy from biomass, e.g. Sweden 15% and USA, Canada and Australia 3%. A number of European-wide studies have shown that about 5 to 10% of Europe's energy requirements could be met from biomass by the end of the century. An especially valuable contribution could be in the form of liquid fuels which have become so prone to fluctuating price and supply. Individual countries in Europe could provide up to half of their energy requirements from biomass if so desired. The resources available, the effect of agricultural surpluses and the factors which will influence biomass energy schemes are currently hotly debated.

Worldwide government expenditure on biomass energy systems is over $2bn. a year. However, biomass energy is not necessarily the panacea to any country's energy problems even though at present the process of photosynthesis produces an amount of stored energy in the form of biomass which is almost 10× the world's annual use of energy.

The world produces 10–20% more food than is required to feed its 4.5 billion people an adequate diet. In N. America and Europe the main problem with food is its easy over-production and general over-consumption. However, there are an estimated 450 m undernourished people, mostly in Asia and Africa. Simplistically, if available food production was increased by 1.5% (equivalent to about 25 mt of grains) and if this food was distributed equitably to those who need it, there would be no undernourished people in the world. The same argument applies if only 10% of the developed countries' grain production was diverted away from animals to humans. Health authorities have recommended lower meat consumption in the USA and UK; these and other diet and biotechnological changes will have long-term socio-economic consequences which are already underway.

The question is how to achieve both food *and* biomass fuel production *locally* on a sustainable basis. Both are required – thus planning and provision of the appropriate infrastructure and incentives must be provided. Increased support of R&D, training and firm establishment of top priority to agriculture and forestry are essential in many countries of the world – if necessary, with significant help from abroad.

D.O. Hall, N. Myers and N.S. Margaris (eds.), Economics of ecosystems management.
 ISBN-13: 978-94-010-8928-9

Introduction

At the start let me indicate what I am not going to advocate. I do not suggest that biomass will solve the energy problems of the world. I am not going to propose cutting down all the trees in the world or to reforest the world; and I am not advocating that we all become vegetarians. What I hope to clarify is that biomass already contributes a significant part of the world's energy and is likely to continue to do so in the forseeable future. But how much biomass will contribute in the long-term will depend very much on decisions that are made both at the local level and at the national level, in addition to international policy making, especially for energy and food.

'Food versus fuel'? This is the wrong question. It should rather be 'How can we distribute equitably the existing ample supply of food?' The problem of uneven distribution of the world's adequate supplies of food is well known and has been the subject of many reports over the last 50 years (at least!)

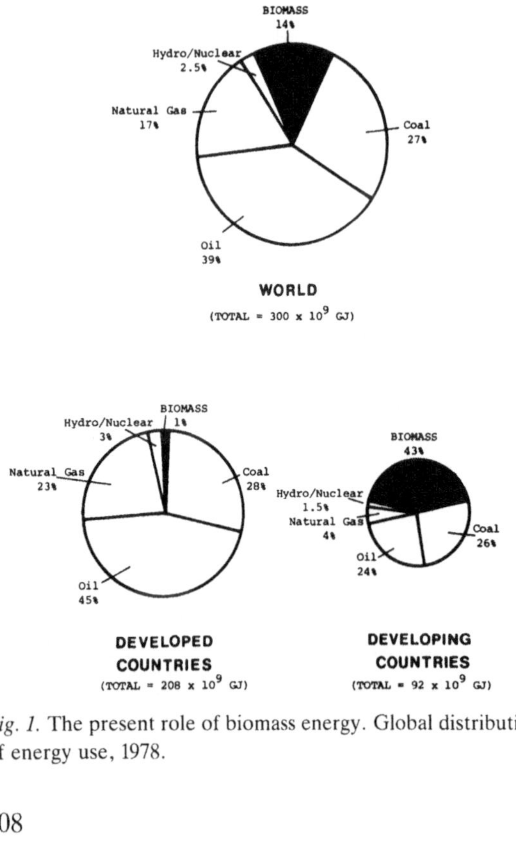

Fig. 1. The present role of biomass energy. Global distribution of energy use, 1978.

The author cannot improve on these studies and has little novel to say on the subject. What I do however wish to point out is that both food *and* fuel are crucial limiting factors in development and must be *locally* available on a *sustainable* basis. Appropriate infrastructure, incentives and planning related to agriculture and forestry enterprises certainly help alleviate shortages in order to allow socio-economic development to proceed. Nothing new, but unless real recognition of the importance of agriculture and forestry in many countries of the world is forthcoming very soon, they will continue to suffer national shortages of food and/or fuel which have debilitating effects on development.

About one tenth of the world's people are undernourished but the world food production is more than 10× greater than is required overall and per capita food production has been increasing at a compound rate of 0.5% in developing countries and 0.8% worldwide since 1950; even though the world's population has increased 76% in the last 30 years (increased from 2.5 to 4.5bn) the production of food has doubled. Biomass energy produced annually is about 10% greater than the total amount of energy used worldwide but many people in developing countries suffer chronic shortages of fuelwood and its overuse is having serious environmental consequences. The amount of grains needed to solve undernourishment in the developing countries is only about 25mt annually – this represents only 1.7% of the world's total production and only 15% of the world's trade in grains.

Today two thirds of the world's people depend on plants for nearly all their food and for the majority of their energy. The question should again be not 'food versus fuel?' but 'how can we increase the productivity of plant-based agriculture (and forestry) in order to provide the required levels of both food and fuel at the national and regional level?' Since only less than 10% of the world's food and hardly any of the world's biomass fuel enters transnational trade it is *local* production of plant products which is all important.

The average person in the rural areas of the developing world uses the equivalent of about one tonne of air dry wood. This is mainly for cooking and heating but also for small-scale industry, agri-

Table 1. Global distribution of energy use, population and food supply (1978).

	Developed countries	Developing countries	World
Population	1.1 b (26%)	3.1 b (74%)	4.2 b (100%)
Total energy use ($\times 10^9$)	208 GJ (69%)	92 GJ (31%)	300 GJ (100%)
per capita	189 GJ (6.3 tce)	30 GJ (1.0 tce)	71 GJ (2.4 tce)
Commercial energy use ($\times 10^9$)	206 GJ (80%)	52 GJ (20%)	258 GJ (100%)
per capita	187 GJ (6.2 tce)	17 GJ (0.6 tce)	61 GJ (2.0 tce)
Biomass energy use ($\times 10^9$)	2 GJ (5%)	40 GJ (95%)	42 GJ (100%)
per capita	1 GJ (0.03 tce)	13 GJ (0.4 tce)	10 GJ (0.3 tce)
Food supply – per capita	3353 Kcal	2203 Kcal	2571 Kcal
– % of requirement daily	129%	96%	106%
Commercial energy in total energy	98%	57%	86%
Biomass energy in total energy	2%	43%	14%

tce = tonnes coal equivalent.

culture, food processing and so on. The use of wood and charcoal in urban areas and for industry is also often much greater than is realized. I don't think many people recognised the importance of this because the statistics were not available to show this significance, and the consequences of biomass overuse were not readily evident. Until a few years ago world energy supply statistics listed biomass at 3–5%, if at all. It is now known that over half of all the trees cut down in the world today are used for cooking and heating. The problem of deforestation with its consequent flooding, desertification and agricultural problems is not solely due to over-cutting of trees for cooking and heating. There are also obviously other factors involved such as commercial and illegal cutting, absence of replanting, and so on.

Recently there have been a few good papers published on studies in Southern and East Africa which show that in an average family of six or seven, one person's sole job is to collect firewood and they will often have to walk great distances, which of course has other deleterious consequences. In urban environments, households can spend up to 40% of their income on fuelwood and charcoal. Another aspect which has been highlighted in Tanzania is the curing of tobacco; for each hectare of tobacco you need to burn the wood from one hectare of savannah woodland. There are many examples to show that it is not only domestic fuelwood use but also agricultural, urban and small-scale industrial uses which are having serious long-term consequences. Serious attempts are being made by a number of international and national groups to try and reverse this problem of deforestation by vigorously promoting reforestation, village fuelwood lots or community forestry, and agroforestry. The World Bank in 1980 concluded that if one was to reverse the deforestation problem one would need to spend 6.75 billion dollars over the next five years in order to start reforesting fifty million hectares. There is little hope that this will happen, but it was what was realistically thought to be needed. There are a number of reasons why this won't be possible, but I will mention only one here. It is the very low status that foresters have in developing (and developed) countries. Consequently it is all very well promulgating reforestation schemes to help solve the energy crisis in various parts of the world, but unless you have the people on the ground with the experience and knowledge, there is no way that these schemes can be implemented.

The realization that at least half the world's population are primarily dependent on biomass as their main source of energy has only widely occurred in the last few years – it has been called the 'fuelwood crisis' or the 'second energy crisis'. There is no doubt that half the people in the world have a much more serious energy problem than most of us who are primarily concerned about the price of petrol and whether we should turn the heat up or down. About 14% of the world's annual fuel supplies are currently derived from biomass.

Fourteen percent of the world's energy use represents the equivalent of twently million barrels of oil a day, which is slightly more than the total USA oil use.

The main issue in developing countries is that of scarcity and the problem of trying to maintain, or possibly even increase, the present level of use without harming agricultural or forestry and ecological systems. More efficient use of existing biomass and possible substitutes for biomass use, e.g. solar and wind-based technology, should be considered and implemented as quickly as possible to reverse the trend of excessive biomass use, as is already occurring in many countries. In the developed world the expertise exists and is already being used to implement biomass energy programs from the point of potential technology and economics. Biomass can provide a source of energy now and in the future; just how much it can contribute to the overall provision of energy will depend on existing local and national circumstances. It is imperative that each country establish its energy use patterns and the potential of biomass energy. This is not easy to accomplish quickly, but needs to be done as soon as possible.

How much photosynthetic production of biomass actually occurs on the earth? It has been shown that the world's total annual use of energy is only one tenth of the annual photosynthetic energy storage, i.e. photosynthesis already stores ten times as much energy as the world needs. The problem is getting it to the people who need it. Secondly, the energy content of stored biomass on the earth's surface today, which is about 90% in trees, is equivalent to our proven fossil fuel reserves. In other words, the energy content of the trees is equivalent to the commercially extractable oil, coal and gas. Thirdly, during the Carboniferous Era quite large quantities of photosynthetic products were stored, but in fact they only represent one hundred years of net photosynthesis. The overall photosynthetic efficiency during the Carboniferous Era was less than 0.0002%. Our total possible fossil fuel resources only represent a few number of years of net photosynthesis. Fourthly, the problem of CO_2 cycling in the atmosphere. Many people are rightly concerned about the problem of build up of CO_2 in the atmosphere if we continue to burn fossil fuels. It is a problem of cycling between two or three pools of carbon. The amount of carbon stored in the biomass is approximately the same as the atmospheric CO_2 and the same as the carbon as CO_2 in the ocean surface layers; there are three equivalent pools. The problem is how is the CO_2 distributed between these pools, and how fast does it equilibrate into the deep ocean layers. However, we should appreciate that increasing CO_2 concentrations in the atmosphere may be good for plants (since CO_2 is a limiting factor in photosynthesis and plants have better water use efficiency at higher CO_2 concentrations). Plants could also act as CO_2 sinks if photochemical means for fixing CO_2 were not available to alleviate the problem.

The oil/energy problem of the 1970's has had three clear effects on biomass energy and development. Firstly, in a number of developed countries large research and development programmes have been instituted which have sought to establish the potential, the costs and the methods of implementation of energy from biomass. The prospects look far more promising than was thought even three years ago. Demonstrations, commercial trials and industrial projects are being implemented. Estimated current expenditure is over a billion dollars per annum in North America and Europe. Secondly, in at least two countries, viz. Brazil (which currently spends over half of its foreign currency on oil imports) and China, with over 7 million biogas digesters, large scale biomass energy schemes are being implemented – the current investment is about 1.3 billion dollars per annum in Brazil. Thirdly, in the developing countries as a whole there has been an accelerating use of biomass as oil products have become too expensive and/or unavailable.

Biomass as a source of energy has problems and it has advantages. Like every other energy source one must realize that it is not the universal panacea. Some advantages and disadvantages are listed in Table 2.

I wish to emphasize one that is very interesting to me, namely the large biological and engineering development potential which is available for biomass. Presently we are using knowledge and experience which has been static for many years; the efficiency of production and the use of biomass as a source of energy has not progressed in the way agricultural yields for food have increased. Thus there is an undoubted potential to increase biomass yields.

Research in agriculture has paid off very well (Table 3) and this could also be the case for biomass R&D also shown from recent fuelwood schemes (Table 4). The most obvious problems that immediately come to mind are land use in competition with food production. Existing agricultural, forestry and social practices are also certainly a hindrance to promoting biomass as a source of energy, whether in a developing or a developed country.

In North America and Europe the problem in agriculture and nutrition is overproduction, excessive consumption of animal products via feeding of grains, and surpluses affecting world trade especially in relation to developing countries (Tables 5–7). Obviously this is too simplistic but the medium-term trends are important as they are likely to be aggravated by increased productivity, influence of new biotechnological processes and

Table 3. Studies of agricultural research productivity – direct cost–benefit type studies (World Bank, 1981, after Arndt et al. 1975).

Commodity	Country	Time period	Annual internal rate of return (%)
Hybrid corn	U.S.A.	1940–55	35–40
Hybrid sorghum	U.S.A.	1940–57	20
Poultry	U.S.A.	1915–60	21–25
Sugarcane	South Africa	1945–62	40
Wheat	Mexico	1943–63	90
Maize	Mexico	1943–63	35
Cotton	Brazil	1924–67	77+
Tomato harvester	U.S.A.	1958–69	37–46
Maize	Peru	1954–67	35–40
Rice	Japan	1915–50	25–27
Rice	Japan	1930–61	73–75
Rice	Colombia	1957–72	60–82
Soybeans	Colombia	1960–71	79–96
Wheat	Colombia	1953–73	11–12
Cotton	Colombia	1953–72	None
Aggregate	U.S.A.	1937–42	50
		1947–52	51
		1957–62	49

Table 2. Some advantages and problems foreseen in biomass for energy schemes.

	Advantages		Problems
1.	Stores energy	1.	Land and water use competition
2.	Renewable	2.	Land areas required
3.	Versatile conversion and products; some products with high energy content	3.	Supply uncertainty in initial phases
4.	Dependent on technology already available with minimum capital input; available to all income levels	4.	Costs often uncertain
5.	Can be developed with present manpower and material resources	5.	Fertilizer, soil, and water requirements
6.	Large biological and engineering development potential	6.	Existing agricultural, forestry, and social practices
7.	Creates employment and develops skills	7.	Bulky resource; transport and storage can be a problem
8.	Reasonably priced in many instances	8.	Subject to climatic variability
9.	Ecologically inoffensive and safe	9.	Low conversion efficiencies
10.	Does not increase atmospheric CO_2	10.	Seasonal (sometimes)

Table 4. Example of economic rates of return from fuelwood reforestation schemes financed by World Bank (Spears, 1981) (see reference for details of individual projects).

Country	Area (ha)	Invest-ment cost US$/ha	Economic rate of return (% annual)
Niger (1978)	400	5 000 (irrigated)	11
Philippines (1978)	28 000	300	22
Brazil (1978)	40 000	350	18
Burundi (1979)	8 500	388	17
Mali (1979)	3 400	1 900	11
Malawi (1980)	28 000	500	14
Nepal (1980)	11 000	900	16
Bangladesh (1980)	40 000	400	18

changes in diet. Thus, for very many countries the question should be how to achieve both food and biomass fuel production locally on a sustainable basis. Both are required – thus planning and provision of the appropriate infrastructure and incentives must be provided. Increased support of R&D, training and firm establishment of top priority to agriculture and forestry are essential in many countries of the world – if necessary, with significant help from abroad. Overall, biomass energy will not be a simple solution to the energy problems of the developing countries. Biomass systems will not necessarily be cheap, nor will they be implemented easily, without a major commitment from governments and a considerable amount of political will at the national and international level.

The gist of my paper which does not claim in any way to be novel, is that it is possible to produce sufficient food and some fuel on the land using good agricultural and forestry practices, today and in the future. How much a degree of self-sufficiency that a country or region decides to achieve in food and fuel depends very much on local conditions which are not easy to ascertain with accuracy from superficial surveys. However, an indication of what can be achieved towards providing both food and fuel is possible from past and present practises, and the geographical environment, in which the agriculture and forestry for both fuel and food will have to be practised.

If we really wish to solve the undernourishment/

Table 5. EEC production surpluses (OECD; EEC; 1982).

Sugar (past 3 year average)	2.6 Mt
Dairy cows	2.5–3.5 M
	= 1.2–1.7 Mha arable land
Fruit (withheld)	0.7 Mt
Fruit (losses)	1.6 Mt
Wine (distilled)	1.84 Gl
Potatoes (reject)	8.5 Mt
Cereals (1981/2)	8.6 Mt
(Wheat, barley, maize)	(7% of 122 Mt)

Table 6. Plant and animal protein production and consumption by man and livestock in 1975 (estimated, in million tonnes) (Pimentel & Pimentel, 1979).

	U.S.A		World	
Total cereal protein produced	17.0	100%	95	100%
Fed to livestock	15.5	91%	38	40%
Available to man	1.5	9%	57	60%
Total legume protein produced	9.3	100%	30	100%
Fed to livestock	9.0	97%	6	20%
Available to man	0.3	3%	24	80%
Total livestock protein produced	6.0	100%	33	100%
Fed to livestock	0.7	12%	3	9%
Available to man	5.3	88%	30	91%
Total protein produced[a]	34.1	100%	173	100%
Fed to livestock	26.1	77%	51	29%
Available to man	8.0	23%	122	71%

[a] Fish protein and other types of vegetable protein are included in this total.

Table 7. Total grain production and food gaps (excess of production to consumption projected to 1990 (t × 10^6 tonnes) (Barr, 1981).

	Population distribution	Production	Gap
Developed countries (excluding U.S.S.R. and E. Europe)	16%	640	+152
U.S.S.R. & E. Europe	8%	370	− 61
China	20%	290	− 10
Developing countries	8%	205	− 24
Developing countries (low income)	48%	260	− 47
World	100%	1765	+ 10

poverty problem at the peasant/village level in developing countries, where most of the world's poor and malnourished people live, we must understand what the limiting factors are in food (and biomass production) at this level. Is it lack of water, fertilizers, pesticides, credit facilities, extension services, transport, storage, etc, or factors such as lack of human and animal power at crucial times of the year, tax incentives, high energy costs, too low product costs and so on?

My contention is then, that with a reasonable degree of management, combined with the necessary incentives and infrastructures, many regions of the world could be self sufficient in their basic food requirements and in addition provide reasonable amounts of energy from natural or specifically grown biomass. Of course it is easy to make statements and much more difficult to put them into practice. This is realized, but to indicate what is feasible, with possibly not too many changes, is also important in worldwide context.

Biomass

Biomass is a jargon term used in the context of energy for a range of products which have been derived from photosynthesis; it can be recognized as the waste from urban areas and from forestry and agricultural processes, specifically-grown crops like trees, starch crops, sugar crops, hydrocarbon plants and oils, and also aquatic plants such as water weeds and algae. Thus everything which has been derived from the process of photosynthesis is a potential source of energy. We are talking essentially about a solar energy conversion system, since this is what photosynthesis is. The problem with solar radiation is that it is diffuse and it is intermittent, so that if we are going to use it we have to capture a diffuse source of energy and need to store it – this is what plants solved a long time ago. All our food and fuel is derived from this CO_2 fixation from the atmosphere. When looking at an energy process we need to have some understanding of what the efficiency of this process will be; one needs to look at the efficiency over the entire cycle of the system and in the process of photosynthesis we mean total incoming solar radiation converted to a stored end-product. Most people agree that the practical maximum efficiency of photosynthesis is between five and six per cent. It might not seem very good, but remember that this represents stored energy. Obviously if we can grow and adapt plants to increase the photosynthetic efficiency, the dry weight yields will increase and of course alter the economics of the crop. One of the very interesting areas of research is to try to understand what the limiting factors are in photosynthetic efficiency both for agriculture and for biomass energy.

Food production and use

World agriculture is already producing enough to feed everybody at a Western standard of diet – but inequalities and inefficiences in distribution seem to make it impossible for the poorest to ensure their share. The current supply of food is sufficient to feed our 1977 population of 4.2 billion at a level of 2,570 kilo calories per day, which is about 20% above requirements, and supply 69 g/day of protein. The disparity between developed ($^1/_4$ world population) and developing ($^3/_4$ population) countries is seen in Table 1 both in the amount of calorie intake and the contribution from animal sources (5× lower). Miller has estimated that of the total primary food production (62% are grains) equivalent to 4515 Kcal/person/day, only 2149 Kcal is directly consumed by man and 2365 Kcal is fed to animals, who in turn contribute 414 Kcal to man to give an average worldwide total daily intake of 2563 Kcal per capita. Over a third of the world's cereal production (about 1500 mt/year of which only 12% enters world trade) and over half of the soyabean production is fed directly to animals. 'Diverting only $^1/_4$ of the world soya bean harvest from feed use to direct human consumption would provide 5 kg per year of high protein food (750 Kcal energy/day) for everyone in the world'.

The problem in using the above facts to help 'solve' the world's food problem are well discussed by FAO in its recent 'Agriculture: Toward 2000'. These facts – which look so wrong morally – are often put forward to support the case for giving less cereals to animals and more to people. Unfortunately neither the facts nor the solutions are simple.

The great bulk of the feed use of cereals occurs in the developed countries, followed at a great distance by the more prosperous developing countries, whereas the low income developing countries use less than 2% of their total cereal supplies on animal feed. In the short run, when total cereal supplies cannot be changed, saving a part of current supplies from going to livestock feeding in the better-off countries would potentially make the quantities involved available as additional food. However, even if many people are desperately short of food, saving these cereals from being used as livestock feed would by no means automatically ensure that needy people would receive them. Supplies have to be financed, generally transported to other countries and then distributed.

The actual situation is thus very complex, but the issue is essentially one of insufficient incomes and demand on the part of the population groups which need to increase their cereal consumption. An 'imposed' reduction in the feed use of cereals (perhaps in another part of the world) can do little by itself to solve this basic problem. Cereals provide 59% of all food calories in the developing countries and no less than 83% of the calorie content of their food imports. The most vulnerable cereal importing countries which contain $^2/_3$ of the developing countries population had a net cereal deficiency of 22 mt in 1978/9. 'Thus it would be facile to conclude that only a small increase in production or a modest switch from feed to food use is called for. As emphasized repeatedly in this study a lasting solution to the problem of chronic undernourishment involves production and distribution, and employment/income. However, at present and in the future the quantities of additional food intakes required . . . are almost miniscule compared with what the world produces and consumes'.

Thus, although in all developing countries the per capita calorie consumption is 2203 kcal which is 96% of an 'average requirement' of 2300 kcal/day, the situation varies from country to country and within any given country can vary tremendously depending on individual income. In India for example the 38% of the population in the lowest 4 income classes (out of 9) have calorie deficits varying from 306 to 1108 kcal/day, with a national average in 1976 of 2217 kcal/person/day. In Kenya where in the country as a whole there is only a marginal calorie deficit, 40% of the rural population have a deficit of 640 calories and in urban areas 40% have a deficit of 340 calories/day.

The world's grain production since 1950 and trends until 1990 are well summarized in a recent article by Barr (Table 7). From 1950 to 1980 food production doubled while the population increased by $^2/_3$ – thus per capita food production worldwide increased at 0.8% compound annual rate. The compound annual rates of increased per capita food production have however been decreasing over the three decades from 1.6% to 0.6% to 0.3%. Developed countries have shown strong upward trends while most developing countries (with the exception of Africa) have experienced improved per capita output. Most of the increase in output both in developed and developing countries came 70–90% from gains in yields rather than from expansions in area. In developing countries population increases have absorbed more than $^4/_5$ of the gain in food production while in developed countries food production has increased twice as fast as population. A disquieting occurrence this last decade has been the fact that food production declined worldwide in 3 of the years while in the previous two decades (1950–1970) production did not decline in any one year. A trend analysis to 1990 from production and consumption patterns for the period 1960–80 suggests that the 'world can indeed produce sufficient grain to permit some limited improvement in per capita consumption'. (Table 7 shows a surplus of 10 mt in 1990).

The changing pattern of world grain trade shows the overwhelming reliance on N. America as a net exporter of 131 mt to over 100 countries. The great dependance of the USSR on imported grains (45 mt in 1981) is quite a recent and important development in the world food question. There is naturally concern over this reliance on one area but the productivity of N. America's farmers (with all their back-up support) is a reasuring factor in the world food equation. There is concern about the possibility of serious shortfalls in grain production but in the last 20 years the largest deficit in world production occurred in 1972 and resulted in a decrease in world stocks of 42 mt. However, during all this period the world's grain stocks have never fallen below 140 mt which represents about 40 days world consumption.

There is concern however as to whether increasing grain yields can be sustained to match and even improve on the world's increase in population (estimated at 6.2 billion in 2000 and hopefully levelling off at 10 billion). The aspects of concern are soil erosion, loss of agricultural land due to water logging, salinity, urbanization, shortages of water, environmental and pollution problems, political upheavals, and so on. There is however cause for optimism and one must not always believe bad news uncritically.

Increasing food production

Since the population will be increasing and high productivity land will be lost (an estimated net loss of 55 m ha by the year 2000) (see Table 8 for land use patterns according to Buringh), it seems imperative that crop productivity be increased on existing land and that new land is brought into production, and of course trying to stop the loss of good land. It is however considered that higher productivity will account for over 80% of the increased production by the year 2000. Thus rapid implementation of current knowledge on high-productivity agriculture is necessary if food consumption per capita is to hold its own, let alone increase. This does not imply a massive change in agricultural practice in the developing countries but it does require that the gains of the 'green revolution' be consolidated and be more widely applied even than at present. The rapid high-yielding varieties of

Table 8. World land use (mha) and productivity class – trends to 2000 (UNEP 1982 from Buringh, 1981).

	Land use 1975	Land losses	Land reclamation	Land use 2000	Net change 1975/2000
Cropland					
High	400	100	45	345	−55
Medium	500	80	325	745	+245
Low	600	40	150	710	+110
	1500	220	520	1800	+300
Grassland					
High	200	30	0	170	−30
Medium	300	20	30	320	+20
Low	500	90	40	510	+10
Zero	2000	0	0	2000	0
	3000	140	70	3000	0
Forest					
High	100	25	45	30	−70
Medium	300	90	180	100	−200
Low	400	75	205	230	−170
Zero	3300	0	0	3140	−160
	4100	190	590	3500	600
Non-agricultural	400	200	0	600	+200
Other land	4400	100	0	4500	+100

wheat and rice do indeed show that agriculture in developing countries can be 'modernised'. Production of rice in developing countries has grown by about 2.4% a year since the advent of the HYVs as compared with 0.9% in the early 1950s. Wheat output has grown by about 4% annually in recent years as compared with 2.4% in earlier years. During the same period production in developing countries of maize, sorghum millet, and other coarse grains fell. It augers well for the future, especially when suitable farm management practices are found and applied to medium and small farmers, and to the more marginal regions.

On a global basis Buringh has estimated the maximum food production capacity of the world and pointed out the advantages of 'modern agriculture' as opposed to 'labour-oriented' (no machinery of chemical fertilizers) agriculture. He calculates that the total potential agricultural land of the world is 2419 million hectares (equal to one quarter of the world's total land area) compared to the 1405 million hectates presently cultivated (11% of total) – at present two thirds of this land is used for cereal crop production. He makes the staggering claim that thirty times the present grain production could be achieved in the world compared to present production. He believes that it makes far more sense to increase the productivity of existing cultivated land by using modern techniques instead of opening up new land with its low productivity.

Only one fifth of the presently cultivated land is used for improved or 'modern' agriculture, while the remaining $^4/_5$ has low productivity levels. Reclaiming new land is seen as poor agricultural practice and also uneconomic. Productive agriculture on one third of the presently cultivated and grazing land can produce enough food for the present population and consequently even more land is available for forest and 'wildlife'. Of course, there are many problems in introducing 'modern' agriculture all over the world, e.g. water, fertilizers, energy, soil erosion, technical expertise, social practices, and so on. However, the advantages seem very great, especially since it would free, or leave untouched, large areas which could be reforested for other uses, such as fuelwood provision, maintaining ecological and climatological balances, energy farming operations, etc.

Looking at Europe and the USA it should be noted that the land areas devoted to arable (cultivated cropland) agriculture are really rather small – in the UK and the USA this is only about $^1/_5$ of their total land areas. In the USA about 60% of this cropland is used for growing animal feed, while in the UK 92% of all farm or 87% of arable and pasture land is devoted to feeding livestock – in addition we in the UK import 15 million tonnes of feedstuffs per year. A small change in the modern diet towards eating less animal products has been recently suggested in the UK and by the NAS in the US to alleviate heart and cancer problems and would release large amounts of plant material and areas for other uses. In the USA 91% of the vegetable protein and in the UK 70% of the primary products of agriculture are fed to animals. In both Europe and the USA there are surpluses of animal and plant products which can be difficult to store and economically subsidize. In the world as a whole 40% of the cereal production and 30% of the total protein production is fed to animals (Table 6).

Although livestock compete with people for food, and processing those foods through animals is a rather inefficient way of providing human food, livestock will likely be reared in large numbers for some time to come. Animals consume many high-fibre crops and crop residues, waste products, and by-products that are not useful to people and covert them into high-quality food. This is particularly evident for the ruminants, which can use both high-cellulose feed and non-protein nitrogen.

We can conclude that maintaining and even increasing agricultural productivity in the developed countries should not be an insurmountable problem if past history is any criterion. In fact the problem in the USA and the EEC these days is of increasing surpluses which are economically and politically troublesome, both to the countries themselves and to those involved in world food exports. The US agricultural productivity is now considered to be in the 'science' phase, with an annual growth rate of 1.6% since 1945 (total farm output per unit has doubled); between 1920 and 1945 was the 'mechanical power phase' with a 1.2% growth rate.

The world's main help should therefore be focussed on improving agriculture in the developing countries themselves – as the FAO says 'there is no other truly realistic option'. As Soedjamoko of the UN University says 'The only way out is for developing countries to grow more of their own food'; 'Additional food will have to be produced through higher yields per hectare – but without high-cost and energy intensive inputs'. As Caruthers says 'only by strengthening domestic agriculture will the impact of the food weapon be blunted. Furthermore it is not only good politics to seek food security but also good economics. However, such a strategy will require many governments to change their development philosophy, their investment and research priorities, and new levels and forms of aid to agriculture will need to be agreed'. 'The political and financial shocks of the 1970s may have some beneficial long-term effects for agriculture. It may be the final lesson to show developing nations that the vision of an industry-led, or export cash crop-led development policy is illusory. Perhaps urban elites who generally control development policy in developing countries, will recognize that it is now vitally important for national independence and integrity to commit scarce resources to agricultural investment on an unprecendentedly massive scale'. The problems associated with military expenditures in preempting funds and armed conflicts disrupting agriculture must also not be underestimated.

However, the problems of implementing new agricultural policies and R&D to stimulate agricultural productivity are well known. Not the least problem is the lack of money – of the $ 150 billion spent annually worldwide on R&D, only 3% is spent on agriculture with most of this being used in developed countries. Cost–benefit analysis suggests that 2% of the value of agricultural product should be spent on R&D – among developing countries, particularly in Asia, less than 0.5% is currently being allocated.

Research is successful only if the improved technologies are adopted by farmers. Where adoption is slow, the explanation may well be that price incentives to producers are inadequate. An attractive economic environment for agriculture, and in particular remunerative prices for producers, must be a part of the efforts made to strengthen the national agricultural research program. The potential benefits of some of the research have sometimes not been realized because of the failure of governments to provide a suitable economic environment that can encourage the adoption of a technology. Few developing countries can afford to by-pass the opportunity to maximize their agricultural (and forestry) productivity.

There are many factors involved in optimizing plant productivity. To mention a few, without producing a shopping list (they are just as applicable to biomass energy systems): Yield stability and resistance to stresses, water and nutrient use efficiences, resistance to diseases and pests, etc. These must be combined with farming systems which provide plant breeding and seed supply systems, adequate extension and credit services, and applied and adaptive research to back up the farmer whether he be farming on a large, medium or small scale. A decrease in post-harvest losses, which can be easily be 10 to 20% of the crop, is also imperative. Research on growing plants in marginal areas should not be neglected as they are already important environments for large numbers of subsistance farmers and are far more extensive than the fertile areas. The ability of plants to produce worthwhile yields in areas of water deficiency or flooding, and nutrient deficiency or toxicity, salinity, and high or low temperatures will be increasingly important.

Biomass production and increased productivity

The developed countries constitute about $^1/_4$ of the world's population but consume $^4/_5$ of the world's commercial energy. The average per capita consumption of energy (including biomass) in the developed countries is about 208 GJ (= 7.4 t coal) and about 28 GJ (= 1.0 t coal) in developing countries where non-commercial, mostly biomass, energy use is equivalent to 0.4 t coal or 1 t wood per person per year (Fig. 1). It has been calculated that the coal equivalent of 0.3 to 0.4 t per person per year is the energy requirement for minimum provision of food and habitat for *survival* in a rural environment.

Reddy has estimated that about 1.5 t coal equivalent is necessary to provide the minimum requirement for a *satisfactory* life including food, habitat and transportation.

In many countries, biomass fuel supplies can no longer meet demand and for a variety of reasons, severe shortages are developing. These problems, sometimes known as the 'second energy crisis', are superimposed on those caused by the rising price of oil. Despite their modest consumption levels, many of the developing countries have been hard hit by the oil crisis and are now facing acute balance of payments difficulties as a consequence. Palmedo in his study of 88 developing countries showed that over half of them depended on liquid fuels for at least 90% of their commercial energy; in only four of the oil importing countries did liquid fuels represent less than 50% of total commercial energy use. As a consequence many developing countries are obliged to devote over half of their export earnings to provide their liquid fuel requirements which so often represent relatively small percentages of their total energy use. Because of these dual energy problems, there is a clear need to develop indigenous energy sources in the developing countries – renewable as well as non-renewable. This is particularly urgent in view of the fact that expanding energy supplies will be necessary simply to maintain present levels of consumption for a growing population – let alone providing for the major increase in energy consumption that will be essential if economic development is to proceed. Among the various options there is significant scope for the development of biomass energy systems. Present biomass resources can be utilised more efficiently and a variety of approaches, new and old, are available for growing and using biomass for energy.

Developing low-cost methods of producing and using biomass energy is therefore of the highest priority if the future domestic energy requirements of the poor are to be met. Because of the heterogeneous nature of biomass energy systems it is difficult to make generalisations about their properties or their potential. One feature they do have in common, however, is the high degree to which they are interlinked with other factors – environmental, agricultural, socio-economic, and political. It is vital that these interactions are understood from the outset since they are of crucial importance in determining the limitations on biomass energy systems, as well as their potential impact, both good and bad. Limitations on the use of biomass energy systems, for example, can occur for a number of technical and economic reasons. One of the most important of these constraints is the competition for land between food and fuel production. In many cases, the introduction of new biomass systems will also be restricted by lack of financial and other resources, and the absence of the necessary infra-structures, extension services, manufacturing capabilities, and technical know-how.

Barriers to implementation of biomass systems can also arise for a variety of other reasons, more related to social and political factors. Conflicts of interest can be a major problem, for example, in cases where present ownership and control over resources is unequal, and where the introduction of new systems is likely to benefit some groups more than others. Other factors which slow down the rate of diffusion of new energy schemes include lack of perception of energy problems and general unwillingness to invest in systems which have not been fully proven.

One of the advantages of biomass energy systems, however, is that they often have the potential for yielding subsidiary benefits other than their simple effect on energy supplies. Some systems for example can help improve the environment whilst others can provide a stimulus for rural development through offering a source of employment, income and energy in rural areas. It is important to recognize, on the other hand, that none of these benefits is necessarily guaranteed. As with most other technologies, if they are applied without consideration of their possible environmental hazards, and without taking into account the needs of the local population, biomass systems can equally well have a negative impact – both in damaging the environment and in increasing the exploitation of rural areas and rural populations.

An examination of the presently available biomass energy and its current use on a regional or a country basis reveals a similar dichotomy to that seen in the case of food production and consumption. In N. America and Europe, the calculated energy potential from biomass is only half the present energy use; in Africa and Latin America biomass could provide many times more energy than required; while in Asia an excess is also calculated (Dunkerly; Revelle). However, the FAO has calculated that at present about 100 m people are suffering from 'acute scarcity' while about 1 billion people have a fuelwood 'deficit'. Also if present trends continue, these figures may reach 356 m and 2,770 m respectively in the year 2000 – a truly frightening proportion of the estimated population of 6.2 bn.

The use of the world's forest resources has increased significantly over the last few decades. For instance, the annual growth rate of fuelwood and charcoal consumption has been 2.1% in the developing countries over 1961/75 – this use represents nearly 60% of all the trees cut in the world. Simultaneously, besides this use as fuelwood, the use as sawnwood, paper and pulp has been increasing at the rate of 1.0 to 8.8% per annum; such industrial forest products are over 90% consumed in the developed countries.

If individual countries are studied with respect to their energy use and the possible sustainable yields of biomass (or wood only) quite widely varying estimates can be derived and have been published, depending on the assumptions of yields, land use, conversion routes, etc. A selection of these statistics is given by Dunkerly and Revelle. What they show is that in many developing countries of the world it is theoretically possible for biomass to be a significant and sustainable source of energy, if the appropriate actions could be and were implemented. As we realize this is not often possible in the prevailing circumstances.

Against this background a number of national and international agencies started aid programmes which stress reforestation, social forestry, agroforestry and other schemes aimed at increasing the production and supply of fuelwood. In many instances the effort required to actually reach sustainable yields in order to assure domestic requirements in the future is quite daunting – see Table 4 for a study by Spears of the World Bank. However, by 1980 the World Bank had rapidly expanded its reforestation projects, compared to 1975, and in 30 projects had shown good economic rates of return on the $ 100m's invested – above 10% in all cases. However, the World Bank did estimate that 50 m ha of land needs to be reforested in developing countries for fuelwood production at a cost estimated at $ 6.75 bn for 1980/5 alone – clearly impossible given the present limitations of finance, and the lack of qualified persons to implement such schemes. The previous neglect of forestry, even more so than agriculture, in many countries has resulted in a lack of personnel and adequate institutions and infrastructure.

Despite the well-known problems of overcutting for domestic, agricultural and industrial use in very many countries, there are some success stories of reforestation schemes. For example, China is reported to have increased its proportion of forests from 5% in 1949 to 13% in 1978 – an increase of 72 m ha. South Korea in the 1970's gave priority to rural community reforestation and planted 0.6 m ha of village woodlots (this is equal to half the area planted to rice). The social forestry scheme in Gujurat (India) is also well known for its successful implementation and take up by other regions in India. The scheme started with plantings along roads, railways and canals and then spread to village woodlots which were carefully planned in consultation with the villages in order to provide good maintenance and acceptable returns to the villagers. Other countries such as Indonesia, Philippines, Bangladesh, Thailand, Nepal, Kenya, Malawi, Mali and Jamaica, to name just a few, are now known to have embarked on fuelwood projects.

Of course not all 'biomass' schemes are forestry-based and not all of them occur in developing countries. Here follow some examples of biomass programmes around the world.

Brazil. The largest programme is in Brazil, which is currently spending about $ 1.3 bn annually of government money on subsidizing the production of alcohol, primarily from sugar-cane. The new programme started in 1975 (even though they have been blending alcohol in gasoline since the 1930's) due to the fact that currently Brazil spends about \$ 11 billion, over half of its total foreign income, on importing oil. What Brazil is trying to do is to reverse this great dependence on imported petroleum by the production of alcohol; currently they are producing nearly 5 bn litres, and this is proposed to be expanded up to about 11 billion litres by 1987, which is 10% of forecast oil consumption. At present all the petrol sold in Brazil is blended to 20% with alcohol. About $^3/_4$ million cars now run on hydrated alcohol which is 95% by volume. Everything is not perfect with this programme. They have recently published an assessment of the programme which shows their great awareness of the problems themselves. One of the problems which was immediately evident is pollution in rivers. Most of the alcohol currently being produced comes from sugar-cane plantations in the South West of Brazil where often the stillage, which is a byproduct from the distillation of the alcohol, has been put into the rivers. For every litre of alcohol produced about 8–10 litres of stillage is produced and this has a very high chemical and biological oxygen demand. There are certainly ways of countering this pollution problem.

Zimbabwe. Nearly 3 years ago they opened an alcohol distillery which saves them \$ 10–12 million of foreign currency per annum. They produce 40 million litres of ethanol a year from sugar-cane which is blended into petrol by the oil companies at 15% blend; it is planned to increase this up to 25% blend when a new refinery/distillery is opened. The yields of sugar-cane are high at about 120 t per hectare per annum. Using the stillage from the fermentation as a fertilizer on the sugar plantations increases the yields by about 6%, and this increase allows for the extra installations to be paid for within one year.

USA. The biomass resources of the US are indeed quite large. Currently they derive about 3% of their total energy requirements from biomass which is equivalent to over one million barrels of oil a day and is greater than the nuclear energy contribution. The energy content of the standing forests of the US are at least 50% greater than the oil reserves and about equivalent to the gas reserves. In 1980 the OTA estimated that 'Depending on a variety of factors, including the availability of cropland, improved crop yields, the development of efficient conversion processes, proper resource management, and the level of policy support, bioenergy could supply as few as 4–6 Quads/yr, or as many as 12 to 17 Quads/yr by 2000 (or up to 15 to 20 per cent of current U.S. energy consumption (about 74 Quads))' ($Q = 10^{18}J$). Both the quantity of biomass that can be obtained on a renewable basis, and the economic, environmental and other consequences of obtaining it will depend critically on the behaviour of growers and harvesters. The U.S. has a gasohol programme which blends 10% alcohol, primarily derived from corn (maize), with unleaded gasoline to provide a high octane fuel. This gasohol programme is relevant to the discussion that we are having about lead in petrol in Europe. The use of alcohol in gasoline has been known for a long time; cars were sold in 1902 which ran on ethanol! Since then cars have been run on blends of alcohol with anywhere from 5 to 35% in various parts of the world. For each 1% ethanol that is added to gasoline the octane rating is raised by an average of 0.24 for the motor octane number and 0.36 in research octane number. 10% alcohol added to gasoline increases the octane number by 3–4 points. One of the serious discussions about using biological material for producing alcohol is whether you obtain more energy out than you have put into the system, and secondly whether food is diverted from the food market and from world trade.

The use of agricultural and forestry residues is also proposed as a renewable source of energy as liquid fuels and/or heat. There is considerable discussion as to the net energy realisable and also the costs and environmental effects of such collection. Costs of using such generally low density material reflect its collection, transportation and processing. However, adverse effects on the soil (erosion, water and nutrient loss, etc.) are the major concern. Nevertheless there are definite opportunities for use of residues especially with changed agricultural practices such as minimal tillage.

Europe. Nearly all the countries in Europe have energy-from-biomass schemes. Why in Europe should we look seriously at biomass as a source of energy? Currently the EEC derives over half its total energy requirements from imported oil. For Europe as a whole some substitution of imported liquid fuels is what makes biomass look so interesting. The EEC programme examines the use of agricultural resources, the use of forestry resources, the use of algae, the digestion of biological materials to produce methane, and thermochemical routes such as gasification to produce methanol. We in Europe are fairly crowded, so how much land do we have to produce energy from biomass? A recent study indicated that the EEC could produce over 85 million tonnes of oil equivalent, which is nearly 2 million barrels of oil a day, providing about 1% of our estimated 1985 energy demand (Table 9). This is twice what we use for agriculture. The study showed that if there was a maximum disturbance to agriculture and forestry, where a crash programme was required, we could possibly achieve a 20% provision of our energy requirement in the EEC. It is highly unlikely that this will happen. Thus a 7% energy provision is there for the taking, with minimal disturbance to current agricultural and forestry practices, with costs which look quite interesting (Table 10).

The EEC currently spends nearly $^3/_4$ (equal to $14 bn) of its budget on the Common Agricultural Policy and most of this goes to subsidizing animals. To provide the food for these animals we devote a very large percentage of our land area for growing grains and we import from the US and other countries. We produce milk, butter and cheese – it is a problem getting rid of these mountains, so very often they are fed back to animals. Since 90% of the energy in the food is lost every time you go through the animal, it seems rather an unusual behaviour. Even two years ago it was unthinkable to mention the possibility that we could use some of our agricultural land and our agricultural surpluses to produce something different from them, such as energy.

Sweden currently derives about 15% of its total energy requirements from biomass. They have a very interesting series of advanced biomass trials with fast-growing willows. They now have nine clones which have twice the yield (about 15 dry tonnes/ha/yr) of those previously used. If so desired the Swedes could derive half their energy requirements from biomass.

Table 9. Total net energy potential (after conversion) of biomass in EEC by year 2000.

	mtoe
a) *Wastes and residues*	
1. Animal wastes	11.5
2. Crop residues	12.5
3. Forestry/wood residues	7.9
4. Municipal solid wastes	8.9
5. Sewage	1.8
b) *Firewood*	6.6
c) *Energy Forestry and crops*	
Dry (forestry and short rotation plantations)	28.1
Wet (Catch crops, algae and aquatic plants)	8.5
Total	85.8

In addition *agricultural surpluses* could be available, for example:

	mt 1980/81
Grains	7[a]
Wheat only	11[b]
Sugar	4

[a] Difference between exports and imports (66% of imports are maize).
[b] Increasing to 17 by 1985.

Table 10. Costs of production of biomass fuels (useful energy estimated for EEC).

	Relative cost	Actual cost (1982)
1. Heat from dry biomass, gasification, etc.	80–100	
2. Methanol from wood	120–150	¢ 350–700/t
3. Anaerobic digestion of wastes[a] and catch crops		
a) gas	150–300	
b) electricity	more than 300	
4. Ethanol		
a) sugar crops	150–300	¢ 700–1000/t
b) grains	100[b]–200	

Relative to oil at 100 units in 1980
[a] No credits for disposal.
[b] Assuming by-product credits.

Vegetable oils

It has been known for a long time that you can put all types of vegetable oils into diesel engines. Numerous studies show that in the sunny countries, if a maize farmer devoted 10% of his land area to growing sunflowers, he could run all the diesel-powered machines that he uses in his farming operation. Probably a blend of from 10–30% is preferable. There is work on the esterification of sunflower oil as methyl or ethyl esters; the esterified oil has fuel properties very close to those of diesel and the esterification can be done on the farm. The Brazilians are devoting considerable effort to extraction of oil from peanuts and carefully ascertaining the use of palm oil (which has high yields) as a 6% blend into diesel. Oils from soyabean, castor seeds, and indigenous plants such as marmeleiro and babassu nut are also being investigated. The recent success of a UK/Malaysian R&D programme in selecting high yielding clones of oil palm (over 12t/ha/yr – over 2 to 3× current average yields) and their rapid propagation using tissue culture technique is a dramatic example of what can be done to exploit oil producing (and other) plants once such a programme is initiated and fully supported.

Hydrocarbon plants

Proposals to use plants directly to produce petroleum have been frequent, with the main recent proponent being Calvin who advocates growing *Euphorbia lathyrus* for the extraction of hydrocarbons which have molecular weights close to those of petroleum. There are trials, mostly in Arizona, to establish whether this is economically viable. Inital hopes of high yields were not substantiated, but recent studies show yields of about 10 barrels (1.5 tonnes) of oil per hectre per annum under irrigation. The question is whether such yields are sustainable in arid environments. There are at least 5 other trials in various parts of the world to see if this is economically viable. There is another requirement from oil, and that is for the manufacture of synthetic rubber. Guayule, *Parthenium argentatum*, which grows naturally in Northern Mexico and the southern USA, can be used as a source of rubber with properties which are similar to that of the rubber tree. There is now a pilot plant of one tonne a day in Saltillo, Mexico, and a 50 tonne a day plant is now proposed there. Australian research has shown that guayule is the most promising of all the hydrocarbon plants for their conditions.

An alga *Botryococcus braunii* has been shown to yield 70% of its extract as a hydrocarbon liquid closely resembling crude oil. This has led to work on immobilizing these algae in solid matrices such as alginates and polyurethane and using a flow-through system to produce hydrocarbons. A green alga called *Dunaliella* discovered in the Dead Sea produces glycerol, β-carotene, and also protein. This alga does not have a cell wall and grows in these very high salt concentrations; thus to compensate for the high salt externally it produces glycerol internally.

Conclusion

The miseries of hunger, cold and excessive labour are too commonplace for anyone to talk without compassion about the problems of food and fuel shortages for so many of the world's population. There is excessive food and energy in the world as seen from the produce 'mountains' in many developed countries and the surplus oil and coal presently available. However, it still is very expensive to import fuel and it can consume a large percentage of a country's external income; the same argument is often not the case for food imports since these usually only provide the topping up – of course there are exceptions, but remember that only about $^1/_{10}$ of the world's grain is traded internationally while well over half of the world's energy is traded.

It is essential that countries with food shortages develop their agriculture to the utmost. As has been said many times this will not occur until agriculture is given its due recognition financially, administratively and politically. It is easy to provide shopping lists of what is required (R&D, institutions, extension services, incentives, etc.) but sustained food production on a local basis, which can tolerate weather fluctuations, needs commitments with long term planning – recognised nationally and internationally. This type of recognition is starting to happen, but it will be some time before the neglect of the last 2–3 decades can be reversed. An overreliance on the world agricultural commodity markets for export earnings can be disasterous in terms of fluctuation in earnings potential and diversion of scarce resources away from local food production; obviously a correct balance between local and export oriented production is very necessary.

There is optimism that food production can keep ahead of population growth, within limits of course, since crop productivity can be increased from quite a good base of understanding which we now have in agriculture. What is needed most is a pragmatic application of all this knowledge with a substantial input of funds rather than the meagre percentage (usually much less than 10%) of most countries' investment in R&D, infrastructure, etc., which now occurs. There is quite sufficient evidence that such investment pays off in economic terms if the programmes are reasonably well managed. Actually, what impresses me more often than not is how much is achieved with such low levels of investment, both in production and delivery systems, which usually occur at present – with substantial and well directed programmes crop production could increase dramatically in very many parts of the world.

Of greater concern, in my opinion, are the more serious consequences of overuse of biomass as a source of energy (but also for commercial purposes) which is occurring in so many countries. Excessive costs and shortages of fossil fuels have forced many people back to a reliance on biomass as the main source of energy. This is not only a problem for rural people as urban dwellers in many parts of the world use wood, charcoal and dung for cooking and heating; increasing denudation of vegetation for large distances around cities and towns are already manifestations of this increasingly serious problem. The quality of planning and control required to manage and/or change biomass uses is generally lacking where it is most needed. It is also likely to become even more serious as urbanization increases and fossil fuel availability and price become problematical.

There is also now a widespread recognition of the biomass energy/deforestation problem by national and international agencies. For example, foresters and their research efforts are now being encouraged from previously abyssmally low levels of incentives and finance by nearly all the national and international agencies. The reawakened realization that forestry can also involve a contribution of trees to agriculture, energy, rural welfare, environmental protection, etc., augers well for the future. The greatest reason for optimism, in my opinion, is that the potential yield increases which are possible for energy production (and other products) are very great and are only now being more widely recognised. Techniques applied to agriculture such as advanced breeding, genetics, plant tissue culture and propagation, root mychorriza and N_2 fixation, physiological control, etc., (besides advanced techniques of general forestry management) can be, and are being, applied in the more modern approaches to tree production as forests, energy plantations, agroforestry cropping, etc. These ideas and techniques need to be much more widely spread. The greatest impediment now to the implementation of such techniques is the lack of trained personnel due to the previous near-total neglect of 'modern' forestry by planning authorities.

Both food and biomass energy production in all its various forms must be interlinked with the problems of rural development and poverty. A knowledge of such linkages at village and regional levels is very important. However, because the relationships of agricultural systems are often difficult to comprehend, the sustained development of adequate food and fuel production must not be neglected as being too difficult to solve singly or together. Again, not novel thoughts, but probably worth repeating, if only to encourage further serious work in trying to understand the limiting factors in food (and biomass?) production in any given area.

Given the right commitments, I believe that the

evidence shows that food production and distribution can be easily improved so as to provide adequate diets to the world's current and foreseeable population into the beginning of the next century. The surplus food producing countries of the developed world could feed the world, if the politics and economics were so decreed. However, the developing countries need to improve their own food producing and distribution capabilities – which I believe is feasible, given the proper incentives. The more serious problem is how to stimulate biomass production and its more efficient use and management. This problem can be overcome given time, but because of the longer time lags in biomass production compared to agriculture and the very inadequate base from which biomass energy systems start, the overuse of biomass is the more serious problem of the two. However, it is most encouraging to note that in the recent agreement for cooperation between India and the USA that improvement of food production and biomass production were the top two priorities in the programme. This is the right direction!

References

Arnon, I., 1981. Modernization of Agriculture in Developing Countries: Resources, Potentials and Problems. John Wiley & Sons, Chichester.

Barnard, G.W. & Hall, D.O., 1983. Energy from renewable sources ethanol fermentation and anaerobic digestion. In: H. Dellweg (ed.), Biotechnology, Vol III, pp. 595–625. Verlag Chemie, Weinheim.

Barr, T.N., 1981. The world food situation and global grain prospects. Science 214: 1087–1095.

Berg, A., 1981. Malnourished People: a Policy Review. Poverty and Basic Needs Series. World Bank, Washington DC 20433.

Bioenergy Council, 1981. International Bio Energy Directory. The Bioenergy Council, Washington DC 20006, U.S.A.

Biomass Abstracts, IEA, 1976. Biomass conversion technical information service, Natl. Board. Sci. Technol., Dublin 4, Ireland.

Biomass Panel of Energy Research Advisory Board U.S.A., 1983. Biomass energy. Solar Energy 30: 1–31.

Biotech 83. Proc. Intl. Commercial applications and implications of biotechnology, Online Conf. Ltd., Northwood HA6 1TS, U.K.

Blaxter, K. (ed.), 1980. Food Chains and Human Nutrition. Applied Science Publ., London.

Blaxter, K. & Fowden, L. (eds.), 1982. Food, Nutrition and Climate. Applied Science Publishers, London.

Boyer, J.S., 1982. Plant productivity and environment. Science 218: 443–448.

Brady, N.C., 1982. Chemistry and world food supplies. Science 218: 847–853.

Brown, L.R., 1980. Food or fuel – new competition for the world's cropland. Paper No. 35, Worldwatch Institute, Washington DC 20036.

Brown, L.R., 1981. World population growth, soil erosion and food security. Science 214: 995–1002.

Buringh, P., van Heemst, H.D.J. & Staring, G.J., 1975. Computation of the Absolute Maximum Food Production of the World. Agricultural University Wageningen, Dept. of Tropical Soil Science, The Netherlands; also 1977, An estimation of World Food Production. Based on Labour-oriented Agriculture.

Burley, J., 1983. Obstacles to tree planting in arid and semiarid lands. U.N. Universtity, Tokyo 150.

Bungay, H.R., 1981. Energy, the biomass options. Wiley-Interscience, N.Y.; also Science 218: 643–646.

Bunting, A.H., 1981. Changing Perspectives in Agriculture in Developing Countries: An Agronomist's View. J. Agric. Econ. 32: 287–302.

Carruthers, S.P. & Jones, M.R., 1983. Biofuel production – strategies for UK agriculture. Centre for Agric. U.K. Strategy, Univ. Reading, U.K.

CIBA Foundation Symposium 97, 1983. Better crops for food. Pitman, London.

Coale, A.J., 1983. Recent trends in fertility in less developed countries. Science 221: 828–832.

Coombs, J., 1980. Renewable sources of energy (Carbohydrates). Outlook Agric. 10: 235–245.

Coombs, J., Hall, D.O. & Chartier, P., 1983. Plants as solar collectors: optimizing productivity for energy. D. Reidel, Dordrecht, Netherlands.

Deudney, D. & Falvin, C., 1983. Renewable energy: the power to choose. W.W. Norton & Co., N.Y.

Dritschilo, W., 1983. Energy as food resources ratios for alternative technologies. Energy 8: 255–265.

Dunkerley, J., Ramsay, W., Gordon, L. & Cecelski E., 1981. Energy strategies for developing Countries. Publ. by Johns Hopkins Univ. Press, for Resources for the Future, 1955 Massachusetts Ave. N.W., Washington DC.

Earl, D.E., 1975. Forest energy and economic development. Clarendon Press, Oxford.

Eckholm, E., 1982. Down to Earth, Environment and Human Needs. Pluto Press, London.

Experientia (Special Issue), 1982. New trends in research and utilization of solar energy through biological systems. Vol. 38: 1–88; 145–222.

FAO, 1980. Agriculture Toward 2000. FAO, Rome, Italy.

FAO, 1981. Energy Cropping versus Food Production. FAO, Rome, Italy.

FAO, 1983. World food report; and Food works – 20 years of food aid for development. FAO, Rome, Italy.

Flaim, S. & Hertzmark, D., 1981. Agricultural Policies and Biomass Fuels. Ann. Rev. Energy 6: 89–121.
Foley, G. & Barnard, G., 1983. Biomass gasification in developing countries. Earthscan Technical Reports, London, U.K.
Fritz, M., 1981. Future Energy Comsumption of the Third World, with Special Reference to Nuclear Power. Pergamon Press, Oxford.
Gray, C.W., 1982. Food consumption parameters for Brazil and their application to food policy. Int. Food Policy Res. Inst., Washington DC 20036, U.S.A.
Hall, D.O., 1979. Solar energy use through biology – past present and future. Solar Energy 22: 307–318. 1981, Solar energy through biology – fuel for the future. In: J.T. Massassah & E.J. Briskey (eds.), Advances in food producing systems for arid and semi-arid lands, pp. 105–137. Academic Press, N.Y. 1982, J. Roy. Soc. Arts 130: 457–471.
Hall, D.O., 1980. Renewable Resources (hydrocarbons). Outlook Agri. 10: 246–254.
Hall, D.O., 1983. Energy from biomass in Europe. In: Strub et al., pp. 51–62.
Hall, D.O., Barnard, G.W. & Moss, P.A., 1982. Biomass for energy in developing countries. Pergamon Press, Oxford.
Holdgate, M.W., Kassas, M. & White, G.F. (eds.), 1982. The World Environment 1972–1982. A Report by the United Nations Environment Programme, Natural Resources and the Environment Series, Vol. 8, Tycooly International publ. Co., Dublin.
Hulse, J.H., 1982. Food science and nutrition. The gulf between rich and poor. Science 216: 1291–1294.
Huxley, P.A., 1982. Agroforestry – a range of new opportunities. Biologist 29: 141–143.
Idso, S.B. & Hansen, J. et al., 1983. Climate effects of atmosphere carbon dioxide. Science 220: 873–875.
Jablonski, D.M. (ed.), 1982. Future energy sources; national development strategies, Vol I–III. Mc Graw-Hill, N.Y.
Kovarik, B., 1982. Fuel alcohol – energy and environment in a hungry world. Earthscan, London, U.K.
Lanly, J.P., 1983. Assessment of the forest resources of the tropics. CAB Forestry Abst. 44: 287–318.
Moss, R.P. & Morgan, W.B., 1981. Fuelwood and Rural Energy Production and Supply in the Humid Tropics. Natural Resources and Environment Series, Vol. 4, UN University Press, Tokyo.
Nair, P.K.R., 1980. Agroforestry species – a crop sheets manual. Intl. Council Research Agroforestry, P.O. Box 30677, Nairobi, Kenya.
NAS, 1981. Firewood crops; Shrub and Tree Species for Energy Production. Natl. Acad. Sci. USA, Washington DC 20418.
NAS, 1982. Diffusion of biomass energy technologies in developing countries. Natl. Acad. Sci. USA, Washington DC 20418.
NAS, 1982. Priorities in biotechnology research for international development. Natl. Acad. Sci. USA, Washington DC 20418.
OTA, 1980. Energy from Biological Processes. Office of Technology Assessment, U.S. Congress, Washington DC 20510, USA.
Palmedo, P.F., Nathans, R., Beardsworth, E. & Hale, S., 1978. Energy Needs Uses, and Resources in Developing Countries. Policy Analysis Div., Natl. Centre for Analysis of Energy Systems, Brookhaven Nat'l Laboratory, Upton, N.Y. 11973.
Palz, W., Chartier, P. & Hall, D.O. (eds.), 1981. Energy from Biomass. Applied Sci. Publ., London.
Parikh, J.T., 1980. Energy Systems and Development; Constraints, Demand and Supply of Energy for Developing Regions. Macmillan, New Delhi.
Parikh, K. & Rabar, F. (eds.), 1981. Food for All in a Sustainable World: the IIASA food and Agriculture Program, IIASA, Laxenburg, Austria.
Pellizzi, G. (ed.), 1981. New and renewable energy in agriculture. FAO, Rome, Italy.
Pimentel, D. et al., 1978. Biological Solar Energy Conversion and U.S. Policy. Bio Science 28: 376–382.
Pimentel, D. & Pimentel, M., 1979. Food, Energy and Society. Edward Arnold Publ., London.
Plunkett, D.L. & Smith, N.J.H., 1982. Agricultural research and third world food production. Science 217: 215–220.
Rabson, R. & Roberts, P., 1981. The role of fundamental biological research in developing future biomass technologies. Biomass 1: 17–38.
Reddy, A.K.N., 1981. An Indian village agricultural ecosystem – case study of Ungra village. II Discussion. Biomass 1: 77–78.
Revelle, R., 1980. Energy dilemmas in Asia: the needs for research and development. Science 209: 164–168.
Sanchez, P.A. et al., 1982. Amazon Basin Soils: Management for Continuous Crop Production. Science 216: 164–168.
Sen, A., 1982. Poverty and Famines: An Essay on Entitlement and Deprivation. Clarendon Press, Oxford; also Third World Quart. 4: 447–466.
Shacklady, C.A. (ed.), 1983. The use of organic residues in rural communities. U.N. University Press, Tokyo 150.
Skutsch, M.M., 1983. Why people don't plant tress. Resources for the Future, Washington DC 20036.
Smil, V., 1979. Energy Flows in the Developing World. Am. Sci. 67: 522–531.
Smil, V., 1983. Biomass energies. Plenum Press, N.Y.
Smil, V. & Knowland, W. (eds.), 1980. Energy in the developing world. Oxford University Press, Oxford.
Spears, J.S., 1980. Overcoming constraints to Increased Investment in Forestry. World Bank, 1818, H. St. NW., Washington, DC 20433, U.S.A.
Stewart, G.A. et al., 1982. Potential for Production of hydrocarbon fuels from crops in Australia. CISRO Div. Chem. Tech., Canberra, Australia.
Stout, B.A., 1979. Energy for world agriculture. FAO, Rome, Italy.
Strub, A., Chartier, P. & Schleser, G. (eds.), 1983. Energy from biomass: 2nd E.C. conference. Appl. Sci. Publ., London.
Swaminathan, M.S., 1982. Biotechnology research and third world agriculture. Science 218: 967–972.
TRW, 1982. Energy Balances in the reproduction and end-use of alcohol derived from biomass. Report to US Dept. En-

ergy, Washington, DC 20545.

Unasylva, 1981. Vol 33, Issue No. 131. Wood Energy Special Edition 1. FAO, Rome, Italy.

Vergara, W. & Pimentel, D., 1978. Fuels from Biomass Comparative Study of the Potential in Five Countries. The United States, Brazil, India, Sudan and Sweden. Advances in Energy Systems and Technology 1: 125–173.

World Bank, 1978. Forestry Sector Policy paper. World Bank, Washington DC 20433, U.S.A.

World Bank, 1980. Energy in the Developing Countries. World Bank, Washington DC 20433, U.S.A.

World Bank, 1981. Agricultural Research: Sector Policy Paper. World Bank, Washington DC 20433, U.S.A.

World Bank, 1981. Mobilizing Renewable Energy Technology in Developing Countries, strengthening local-capabilities and Research. World Bank, Washington DC 20433, U.S.A.

World Bank and FAO, 1981. Forestry Research Needs in Developing Countries – Time for a Reappraisal. World Bank, Washington DC 20433, U.S.A.

World Bank, 1982. World development report – agriculture and economic development. World Bank, Washington DC 20433, U.S.A.

World Bank/IBRD, 1983. The energy transition in developing countries. World Bank, Washington DC 20433, U.S.A.

Wortman, S., 1980. World food and nutrition; The scientific and technological base. Science 209: 157–164.

Zoborsky, O.R. (ed.), 1981. Handbook of biosolar resources, Vol I and II. CRC Press, Boca Raton, Florida, U.S.A.

CHAPTER 26

Wood as a source of fuel: a simulation of supply and demand in New England

ELIZABETH TITUS

Over the past few decades both the industrial and residential demand for firewood have grown steadily in the United States. Although at the national level this trend is expected to level off, in certain regions demand continues to increase (EIA, 1981). Associated with the increasing demand for firewood are concerns about suburban air pollution, wood poaching, and potential overexploitation of the forest resource (Lipfert & Dungan, 1983; West et al., 1980).

Although forests are a renewable resource, they are renewable only within limits. Both biological and economic factors influence these limits; such factors include a forest's intrinsic rate of growth, ambient environmental quality, market conditions, and harvesting practices. In addition, one cannot discount the possibility that the supply will decline because available areas of forest land may decrease; this could result from expanding suburban development, from new conservation policies, from monopolist behavior by forest landowners, or even potentially from competition from other energy crops, if biomass fuel technologies enter the market.

This paper presents a computer simulation model which was developed to evaluate the long-term consequences of the use of wood as a heating fuel. Results of a sensitivity analysis are also presented, indicating the responsiveness of the model to changes in each parameter. With this model and sensitivity analysis both biological and economic factors affecting the demand and supply of wood can be explored. The model was calibrated with data from the New England region, which includes six states, Maine, New Hampshire, Vermont, Massachusetts, Connecticut, and Rhode Island. The model has been used to identify when and under what conditions biological overexploitation of the forest resource will occur. It is furthermore possible to estimate the consequences of chronic air pollution, large-scale land development, or alternative harvest practices on wood supply and demand in the region.

New England is an appropriate case study for several reasons. Within the 6.5 million acre area, there is a mixture of urban and rural populations. Over fifty percent of the region is forested and the region supports commercial forestry and forest-products industries. In addition, this region is experiencing rapid growth in peak demands for energy; furthermore, prices of conventional fuels are higher than the national average in New England; there is demonstrable interest in energy independence and the region's energy needs have been well-studied (EIA, 1983; Blair, 1979).

Also significant with respect to regional heating requirements is the number of households. The housing census indicates gradual growth in the number of households, which is explained in part by increasing numbers of single parents and by the greater spending power of working couples (Jackson et al., 1981). The housing stock itself is older than average, and oil is the predominant heating fuel. The census reveals a very small percentage of

D.O. Hall, N. Myers and N.S. Margaris (eds.), Economics of ecosystems management.

 ISBN-13: 978-94-010-8928-9

homes heated by wood; since wood heat may be an unrecorded supplementary source in many homes, it is possible that wood use is underestimated by this statistic. Analysis of census data of the past decades characterizes New England as a region with an older than average population, and with higher than average incomes. These features help to explain the observation that household use of energy for space heating in particular, electricity, is expected to increase over time, despite the fact that the region's energy prices are high. Finally, New England is a region of historical interest, in that prior to the industrial revolution, dependence on wood as a fuel in New England led to overexploitation and utter depletion of the region's forests. One can safely assume that most of the forests which exist today are 'second growth forests', generally 80 to 120 years old. New England's history demonstrates that the limits of exploitation are indeed approachable. It underscores the importance of recognizing limits of exploitation.

The simulation model is composed of two major sections, one which forecasts regional demand for wood and one which simulates net forest growth in the region. The general structure of the model is illustrated in Figure 1.[1] Simulation of the demand for wood involves the separate estimation of commercial and residential demands, using econometric models. In estimating commercial demand, it was assumed that no recycling or sales of firewood to other sectors occurs; these assumptions help to account for some inefficiency associated with the commercial harvest effort. Residential demand for firewood is estimated by a two-stage process. First household heating requirements are projected, then the share of heating demand to be met by wood is derived from a logit model of fuel choice. The logit algorithms presented below demonstrate the importance of non-economic as well as economic variables.

$$\ln (\frac{S_i}{S_k})_t = B_0 - B_1 \ln (\frac{P_i}{P_k})_t - B_2 LT_t + B_4 Y_t + B_5 \ln (\frac{S_i}{S_k})_{t-1} \quad (1)$$

$$\ln (\frac{S_i}{S_k})_t = A_0 - B_1 \ln (\frac{P_i}{P_k})_t - A_2 MT_t - A_3 LT_t + A_4 V_t + A_5 \ln (\frac{S_i}{S_k})_t -1 \quad (2)$$

$$\sum_{n=i}^{k} S_n = 1 \quad (3)$$

where for each year t:

S_n = share of fuel n

MT = mean seasonal high temperature

[1] Those who desire a detailed description of the model should contact the author.

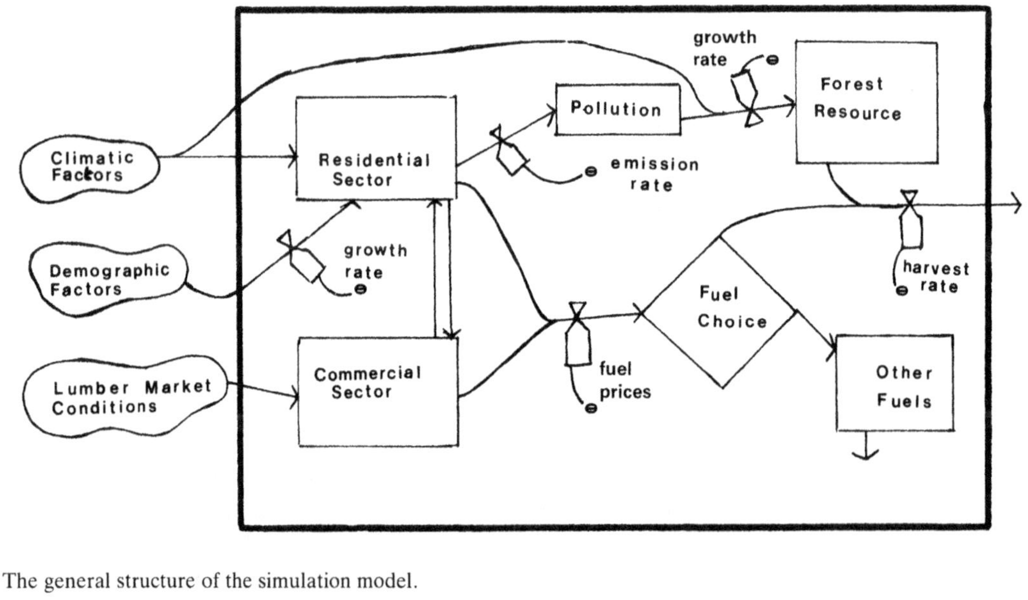

Fig. 1. The general structure of the simulation model.

LT = mean seasonal low temperature
P_n = price of fuel n
Y = personal income

The model was designed so that aggregate demand can be increased by an inefficiency factor, in order to explore variable amounts of energy expended in the harvest effort. Simulation of net forest growth is achieved by means of a modified version of the JABOWA model of forest succession. 'JABOWA', a widely used forest simulation model was developed as recently as the 1970's for the Hubbard Brook Watershed Study in Dartmouth, New Hampshire (Botkin, 1977). It has since appeared in modified form, to include the occurrence of random disturbance and to include cumulative effects of sulfur dioxide pollution. In this formulation, a logistic growth equation is approximated, with both harvesting and pollution effects incorporated. Presently, pollution effects are treated as a hypothetical influence on the growth of selected forest species, in the absence of empirical or experimental observations.

The model is run incrementally over relatively small time steps to approximate continuous growth. Both the annual growth and the entire stock are recorded in tons per acre. Harvesting is simply a subtraction of tons per acre of demand from the entire stock. The net supply is then reassigned as existing forest stock which will undergo growth in the next iteration. The model can be run either until biological exploitation is reached or until a target year is reached.

To test the responsiveness of the entire model, a sensitivity analysis was performed. The sensitivity analysis is useful in various ways. If a particular parameter is found to be sensitive, then that parameter must be accurately known to produce accurate model results. Secondly, the sensitivity of a parameter is indicative of whether the most important processes of the model are governed by that parameter. By evaluating sensitivity for more than one time stream, one can observe that certain processes exert greater influence over the short run, while others influence the long-run results more heavily. Finally, the sensitivity analysis is useful in revealing processes and data that warrant further study.

The sensitivity analysis was accomplished by varying each parameter individually by some percentage of its original value. The same percentage was not used for each parameter, after experimentation revealed that a very wide range of sensitivities existed; in many cases small (10%) changes would produce small but undetectable changes.

Three output variables were chosen as indicators; these are annual growth, net stock, and annual harvest. The absolute value of the ratio of the fractional change of the indicator to the fractional change in the model parameter was calculated, and this was defined as the sensitivity of the indicator variable. Absolute values were chosen because changes in harvest are inversely related to changes in growth or net stock, so the directional aspect of sensitivity has relatively little value for these indicators. As described below,

$$S = \left| \frac{(V_{new} - V_{old})}{V_{old}} \div \frac{(P_{new} - P_{old})}{P_{old}} \right| \qquad (4)$$

where S = sensitivity
V = output variable
P = input parameter

Results of the sensitivity analysis are presented in Table 1. Note that if $S = 1$, then a change in the parameter produce a change of the same magnitude in the appropriate variable. S = values above .5 may be considered very sensitive; those from .1 to .5 may be considered sensitive, while values below .1 indicate that a variable is insensitive to parameter changes.

From the table of results, it is clear that sensitivities change with time. The harvest variable, for example, becomes increasingly sensitive with time, but the relative influence of parameters changes. Annual growth, on the other hand, diminishes in sensitivity with time. As may be expected, the ecological parameters exert greater influence on the forest variables, while demographic parameters influence harvest more heavily. The acreage parameter and efficiency parameter are noteworthy in that they are non-ecological, yet they exert a substantial influence on stock, one which increases with time. With respect to harvest, the analysis reveals that this variable is insensitive to changes in

Table 1. Sensitivity analysis: fraction change in indicators: fraction change in parameter. Values ranked (where blanks appear, S = 0).

% Δ in Par:		2010						2050					
		Ann growth		Harvest		Net stock		Ann growth		Harvest		Net stock	
K	25	*K*	1.98	DAYS	0.9989	*K*	.697	R_2	2.65	*MT*	6.36	*Q*	1.15
Q	57	*Q*	0.907	HH	0.743	*Q*	.575	*K*	2.28	INC	5.35	AC	1.11
R_1	80	R_1	0.869	INC	2.59E-5	*P*	.441	*Q*	1.11	AC	2.72	*K*	.95
R_2	40	R_2	0.476	*LT*	2.75E-6	R_1	.436	R_1	0.56	*K*	2.62	R_1	.45
P	256	*P*	0.431	PE	2.41E-6	R_2	.390	*P*	0.003	*LT*	1.89	R_2	.25
AC	90	AC	1.57E-4	PW	2.28E-6	AC	6.55E-5	PW	1.15E-7	R_2	1.66	EFF	.004
HH	50	HH	3.49E-5	*MT*	1.19E-6	EFF	6.55E-6			EFF	1.62	*P*	.0002
EFF	33	EFF	1.74E-5	LXO	3.39E-7	DAYS	6.54E-6			HH	1.31	MT	7.38E-6
DAYS	78	DAYS	1.73E-5	LXW	0.6E-7	HH	3.35E-6			DAYS	.99	INC	6.2E-6
LXO	160	LXO	1.21E-6			LXO	4.26E-7			R_1	.78	LT	2.2E-6
INC	13	INC	3.76E-7			INC	2.87E-7			LXO	.24	HH	1.5E-6
PW	50	PW	2.54E-7			LXW	.96E-7			LXW	.24	DAYS	1.16E-6
LXW	270	LXW	2.42E-7			PW	.9E-7			*Q*	.13	LXO	2.82E-7
LT	34	*LT*	1.77E-7			*LT*	6.1E-8			*P*	.003	LXW	2.82E-7
MT	10	*MT*	8.75E-8			*MT*	2.92E-8					PW	1.4E-8
PE	50	PE	3.0E-8			PE	1.2E-8					PE	2.0E-9

Where the parameters are:
K = carrying capacity
Q = initial forest stock
R_1 = intrinsic growth rate of forest type *i*
P = percent growth reduction caused by air pollution
AC = acres of forest
HH = number of households
EFF = harvest efficiency
DAYS = number of days per year space heating is required
LXO, LXW, LXG = shares of fuel types in region
PPO, PW, PG = fuel prices
LT, MT = mean seasonal low and high temperatures (F°)

the logit fuel-choice parameters, yet is sensitive to available land area and household growth rate.

In short, accurate estimation of the ecological parameters, available acreage, and household growth rates are most crucial to accurate outputs from the model. The economic parameters and the pollution parameter, which are most likely to fluctuate, are, conveniently, less influential. In addition, it is significant to note that with this model, changes in economic conditions do register small, but detectable influence on overall forest stock. If, as is suspected, the available stock is less than was estimated here, the economic influences would be larger, more measurable and perhaps meaningful for energy policy or harvesting decisions.

Of all the parameters included in the model, the efficiency parameter appears to be one which merits further study and more accurate estimation. Although the model is not highly sensitive to this parameter, both stock and harvest are more sensitive to efficiency than the other economic parameters. In addition, efficiency is more controllable at a regional level than are demographic parameters. The model results suggest that resource management practices which determine efficiency clearly influence supply over the long run. The self-regulating power of market forces is not fully reflected in the model. Two assumptions are especially noteworthy. First, it is assumed that wood in New England is perceived as an abundant resource and accordingly, it is assumed that the price of firewood is independent of supply; wood is simply 'priced to sell', its price will be no greater than that of the cheapest alternative heating fuel. Secondly, there is an assumption of 0 inventory. In short, the harvest in any given year exactly satisfies demand.

Although these assumptions are a departure from reality and from the standard bioeconomic models, there is some justification for them given that the market for firewood is informal. The assumptions which were made define conditions befitting a sort of worst case scenario in fact, less wood would satisfy demand because with inventory, wood could dry to a condition where it possessed a higher energy content. And, by specifying that the price of wood range between 0 and the price of its cheapest competitor, one is restricting the model to an examination of the maximum likely demand for wood.

In summary, results generated by the model under various values of the parameters suggest that wood fuel can serve as an additional energy source for this region. Barring drastic cutbacks in available acreages of forest land and drastic alterations of growth rates, demographic or ecological, the supply will prove more than adequate in the foreseeable future. Most fluctuations in energy price, in air quality or other pollution, in income, choice of heating technology, or length of heating season will not jeopardize or alter the relative abundance of wood.

References

Blair, P., 1979. Load Management Strategies for Electric Utilities: A Production Cost Simulation. Simulation: 93–102.

Botkin, D.B., 1977. Life and Death in a Forest. In: C. Hall & J. Day (eds.), Ecosystem Modeling in Theory and Practice. John Wiley and Sons, Inc., New York.

Energy Information Administration, 1981. Estimates of U.S. Wood Energy Consumption from 1949 to 1981 and Residential Energy Consumption Survey: Housing Characteristics. Reviewed in Science 217: 1017 (September, 1982).

Energy Information Administration, 1983. Energy Prices, 1970–1980. U.S. Department of Energy, Pacific Northwest Laboratory, Richland, Washington, PNL-4256 Vol. 1.

Jackson, G. & Bolton R. et al., 1981. Regional Diversity. Audobon House Publishing Co., Boston.

Lipfert, F. & Dungan, J.L., 1983. Residential Firewood Use in the United States. Science 219: 1425–1427.

West, D., McLaughlin, S. & Shugart, H., 1980. Simulated Forest Response to Chronic Air Pollution Stress. J. Environ. Qual. 9 (1): 43–49.

CHAPTER 27

Biomass refining: a non-waste approach

EMMANUEL G. KOUKIOS

Abstract

A biomass refinery is an integrated complex of industrial processes that separate the major components of bioresources and convert them to fuels, food, feed, fiber, fertilizer and chemicals by physical, chemical, and biochemical methods. To optimize biomass refining one needs to maximize (a) the efficiency and (b) the selectivity of the fractionation, and (c) the quality of its products (nutritional, energetic, mechanical properties etc.), and at the same time minimize (d) the amount of wastes and the environmental problems caused thereof. These criteria are applied here in several pilot-cases that represent the most promising routes for refining the main structural units of plant-biomass, namely leaves, seeds and lignocellulosic stalks by nonwaste technologies. Experimental data from biomass refining in the cases of various agricultural residues are used to illustrate the principles of the proposed approach.

Introduction and conceptual framework

The production of energy and chemicals from plant-biomass resources by whole-crop utilization has been quite a popular idea during this last decade, with a gradual shift of interest from fuels to organics (Szego & Kemp, 1975; Wilke, 1975; Hall, 1978; Alexander, 1981; Bungay, 1983). Despite the obvious appeal of an oil-free petrochemical industry, biomass refining should contribute to more than that. The present critical situation in the areas of food and feed resources, especially proteins (Pirie, 1976; FAO, 1977; Präve et al., 1980), and their relationships with energy (Pimentel & Pimentel, 1979; Spedding, 1979), calls for the development of novel related products, by extractive and/or biosynthetic processes. Soil-plant relationships constitute another troubled area, where biomass-derived organic fertilizers have certainly a vital role to play in order to establish viable production systems (Friend & Maher, 1982; Tunney, 1980). Finally, in the pursuit of new fiber sources to meet the world's expanding needs for paper, board and synthetics, non-wood fibers from various annual plants are reasonably considered as a presently underutilized and very promising element (Archison, 1974; Jahn, 1979).

Solid, liquid and gaseous fuels, basic and speciality chemicals, new polymers, proteins of food and feed quality, various composts and humic fertilizers, as well as cellulosic pulps for the paper and the textile industry will be among the main products of refining plant-biomass (Koukios & Economides, 1977) as shown in Figure 1. Such a multipurpose scheme can, at the same time, take advantage of the highly heterogeneous structure and chemical composition of plant matter, which otherwise would be a problem. In fact, the engineering of such complicated systems can be greatly facilitated by distinguishing a few major structural units that are typical in plant-biomass, and can be easily

D.O. Hall, N. Myers and N.S. Margaris (eds.), Economics of ecosystems management.
 ISBN-13: 978-94-010-8928-9

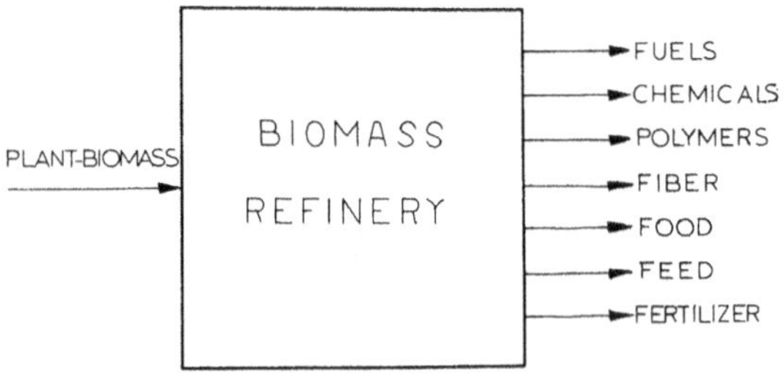

Fig. 1. Simplified view of plant-biomass refinery.

separated from each other by the use of techniques well-known to conventional agriculture and agro-industry. A tentative classification follows:

Plant-Biomass I : leaves, tops, and young stems;
Plant-Biomass II : seeds and grains (endosperms);
Plant-Biomass III : stalks and mature stems, hulls and brans, and the lignocellulosic part of roots;
Plant-Biomass IV : carbohydrate juices and other fruit components.

An analysis of the annual agricultural production in this country according to the above classification is given in Table 1 (Koukios & Hatziyannakis, 1982). A further distinction between products and residues is introduced to give a clear picture of the actual situation. Lignocellulosics are by far the most significant fraction, 99% of which is presently a residue, followed by leafy materials, 50% of which are today consumed by humans (vegetables) and animals (grass), with seeds and grains in the third place of total production but first in present consumption, due to the importance of cereal grains and pulses; the 30% of this fraction classified as residue consists mainly of oilseeds and cakes. The remaining fraction (IV), of relatively limited importance, will not be examined in this work.

The original fractionation of plant matter in to these major structural units, usually mechanical in essence, is followed by a complete fractionation of each type of material to its most significant components and conversion of the latter to valuable end-products. Physical, chemical, and biochemical agents can be employed to achieve these transformations, as indicated in Table 2 based on all the available literature listed in this paper. Research needs in this area tend to focus on the problem of developing viable fractionation and conversion processes in terms of economy, energy and environmental impact (Moreira et al., 1981; Lipinsky, 1979). In such an effort, criteria like (a) fractionation efficiency, (b) fractionation selectivity, (c) quality of products, byproducts and intermediates, and (d) amount and pollution load of wastes are of the outmost importance. The object of this paper is the application of these criteria in the analysis of several promising fractionation processes, of a physical and chemical nature, in the perspective of an integrated biomass refinery.

Leaf protein fractionation

The green, leafy material or plant-biomass I is especially rich in proteins, but is also characterized by high amounts of fibrous components, and often

Table 1. Annual production of major plant – biomass fractions from Greek agriculture (1980, dry t).

Major fraction	Product		Residue		Total	
	Mt/yr	%	Mt/yr	%	Mt/yr	%
I	2.4	11.7	2.4	11.5	4.8	23.2
II	2.7	13.1	1.1	5.0	3.8	18.1
III	0.1	0.6	11.0[a]	52.6	11.1	53.2
IV	0.9	4.3	0.3	1.2	1.2	5.5
Totals	6.1	29.7	14.8	70.3	20.9	100.0

[a] Stalks 5.0 (23.9%), hulls 1.0 (4.8), roots 5.0 (23.9%).

Table 2. Potential products from plant-biomass fractionation.

Fraction	Component	End-Products[a]						
		Fuel	Food	Feed	Fiber	Fertilizer	Polymer	Chemical
I	LPC/LPI		P	P			P, C	
	Fiber	P, C		P	P, C	P, B		
	NPN[b], solubles	B		P, B		P, B		B
	Lipid fraction		P					C, B
II	Lipids	P, C	P				C	C
	SPC/SPI		P	P			P, C	
	Residue			P	P	P, B		
	NPN[b], solubles	B		P, B		P, B		B
III	Cellulose	P, C, B	C, B	C, B	P, C	P, B	P, C	C, B
	Lignin	P, C					C	C
	Hemicellulose	B	B	B			C, B	C, B

[a] Conversion by P = physical, C = chemical, and B = biochemical agents.
[b] Non-protein-nitrogeneous compounds.

contains toxic and other harmful substances. As a result, only a few such materials can be used as animal feed, and even fewer can be consumed directly as foods, although these proteins are of excellent quality. Leaf protein recovery by mechanical extraction is a relatively simple process leading to the production of leaf protein concentrates (LPC) or isolates (LPI) (Fig. 2). This process has been extensively examined in the past from the points of view of agronomy and nutrition; however, considerably less attention has been given to the engineering problems (Pirie, 1971, 1982). Nevertheless, the idea has found industrial application in a few countries, mainly for the production of LPC as a high-value by-product in forage processing, to be used as animal feed (Pirie, 1971; Edwards et al., 1978).

According to a different approach, supported by Pirie, in his numerous publications on the subject, and other researchers (Pirie, 1976; Spedding, 1979), the best use of LPC, in terms of both total protein production per ha and energy required per kg of protein, is in its direct consumption by humans. Such a way to the optimum passes through the maximization of protein extraction, e.g. at levels above 70–80%, an objective systematically pursued only in a small number of cases (Crook, 1946). Overall, this process should be easily optimized in any given situation by taking into account the quantity and quality of all fractions, provided that the level of protein and dry matter (DM) fractionation could be adjusted to maximize the economic efficiency, while requiring the lowest possible energy consumption per unit of product. For that purpose,

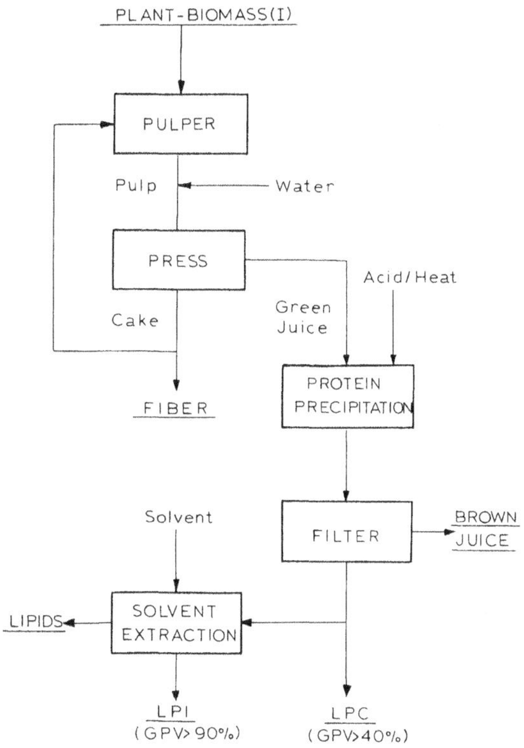

Fig. 2. Flow-sheet for the fractionation of leafy biomass.

a good knowledge of the relationships between the operating conditions in leafy biomass factionation and the efficiency and selectivity of the process is necessary.

Two sets of factorial experiments were designed to investigate the response surfaces of those relationships in the cases of leaf pulping and juice-fiber separation, respectively. The experimental regions of the factor space are given in Table 3, and the results in Figures 3 and 4. The pulping experiments were carried out in a modified domestic mincer, where the leaves of a green vegetable (*Cichorium divaricatum*) were forced by screw expeller to pass through a set of revolving blades and a perforated cylindrical die. The resulting pulps, appropriately diluted with water, were pressed under constant coditions. As it can be seen in Figure 3, 33–67% of DM and 54–82% of N-compounds were extracted from the original biomass, at selectivities between 1.2 and 2.5. Low cutting and high extrusion degrees are in favor of DM extraction, while low cutting and low extrusion degrees maximize protein extraction. Repulping increases both efficiencies, however the loss of DM is dominant, leading to lower selectivities, unless more water is added to facilate the separation of proteins during pressing, in which case, best results were obtained (trial No. 6). Simple equations can be derived from those

Table 3. Experimental regions investigated factors affecting plant protein fractionation.

Factor	Units	Codel levels		
		−(1)	0	+(1)
A. *Pulping* (Fig. 3)				
(x_1) Cutting	No. of blades in pulper	2	3	4
(x_2) Extrusion	Diameter (mm) of holes in output die	8	6	4
(x_3) Repulping	No. of passes from pulper	1	2	3
(x_4) Water	Water added in pulpafter each pulping (W/W)	0	1/1	3/1
B. *Separation* (Fig. 4)				
(x_5) Water added to pulp	W/W	1/1	2/1	3/1
(x_6) Pressure	Kg/cm^2	7.1	35.4	63.7
(x_7) Amount of pulp pressed	g of original pulp/cm^2 of pressed surface	0.88	1.77	2.65
C. *Extraction* (Fig. 4)				
(x_8) Temperature	°C	65	80	95
(x_9) pH	–	10	11	12
(x_{10}) Time	min	60	90	120

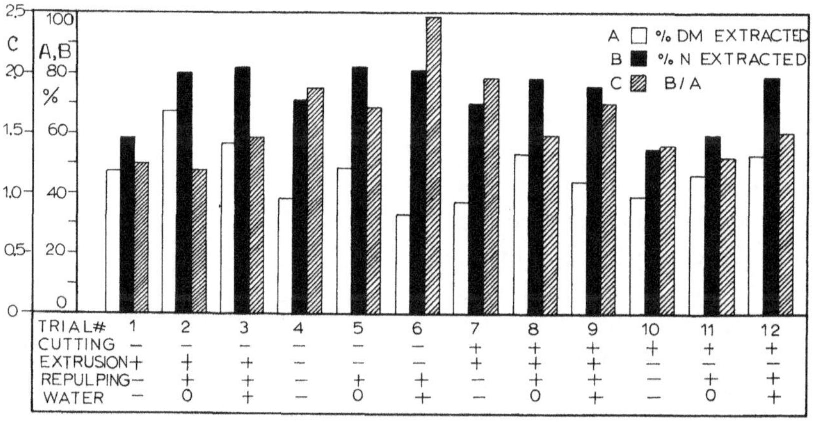

Fig. 3. Effects of major parameters in pulping green leaves of *Cichorium divaricatum*.

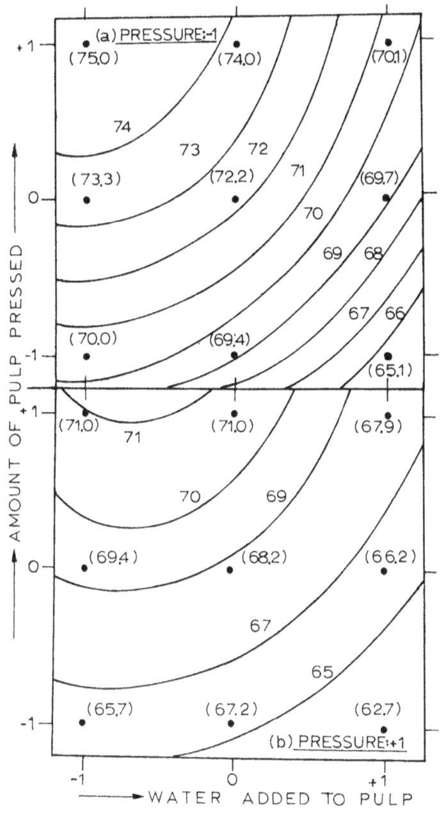

Fig. 4. Constant response plots for % DM remaining in fiber after pressing pulp of sugar beet (*Beta vulgaris*). Curves drawn according to equation (2) (see Table 4).

results making possible the exploration of the response surfaces in extraction and, in particular, the analysis of the interaction effects between factors,

e.g. for pulping:

$$\text{DM in fiber (\% of original)} = 53.30 + 3.33x_1 - 4.0x_2 - 6.73x_3 + 3.08x_1x_2 + 2.53x_2x_3. \quad (1)$$

Coded values are used for the three factors in (1) (Table 3).

The especially strong effect from the addition of water is a strong indication for the critical role of juice expression in the fractionation of leaf components. The separation experiments were done in a cylindrical expression cell, equipped with a compression machine. The pulp was prepared from sugar beet (*Beta vulgaris*) leaves, at the optimal conditions determined according to the previous set of experiments. Figure 4 verifies the significance of this step for the whole process and permits an investigation of the corresponding response surfaces. The efficiency of the fractionation is increased by high dilution rates, high pressure applications, and low amounts of solids to be pressed, as described by the equation (in coded values, as in Table 3):

$$\text{DM in fiber (\% of original)} = 70.46 - 1.95x_5 - 1.72x_6 + 2.49x_7 - 1.06x_5^2 - 0.83x_7^2 + 0.32x_5x_6 \quad (2)$$

Seed protein fractionation

The endosperm of seeds and grains or biomass II contains considerable amounts of proteins, usually 10–40% in Gross Protein Value (GPV), and even higher amounts, usually 20–50%, of fatty components. As in the case of leaf protein, the presence of toxic and other unacceptable substances has limited human and animal consumption to a few cereal grains, pulses and oil-seeds. Seed protein recovery by alkaline extraction of the defatted meals, followed by acid precipitation of the proteins, has been proposed as an alternative to seed detoxification and other treatments aimed to increase feeding value (van den Berg, 1972; Pirie, 1975) (Fig. 5). The production of seed protein concentrates (SPC) and isolates (SPI) from various sources by this process has been experimentally examined in a great number of cases, and although it has reached industrial application a few times, the problems of process engineering are far from being solved (FAO, 1977, 1978; van den Berg, 1972; Pirie, 1975; Iammartino, 1974). This is especially true for seeds and cakes presently available as residues from an agro-industry of local or regional importance, e.g. olive oil kernels, tomato-processing wastes and fruit-canning residues (Fernández Diez, 1960; Miladi & Hegsted, 1974).

A series of factorial experiments was designed to investigate the relationships between operating conditions at the extraction stage and the efficiency and selectivity of seed protein fractionation. The endosperm fraction of the pits of apricot (*Armenica vulgaris*) was used as raw material. The original meal, containing 50.9% lipids and 5.4% ash, had a GPV of 33.5%. After extraction with hexane the de-fatted meal contained 3.8% lipids, 5.5% ash, and had a GPV of 52.2%. N-compounds were extracted from this material by a 10/1 (w/w) aqueous solution of NaOH under the conditions shown in Table 3. The solid residue was separated and

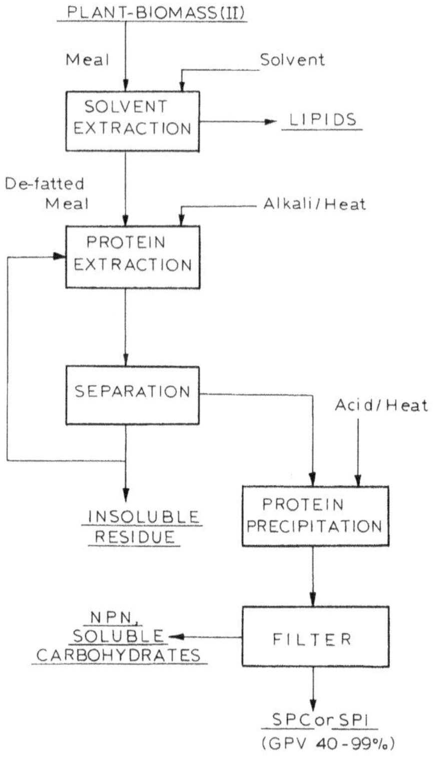

Fig. 5. Flow sheet for the fractionation of oil seeds.

washed, then proteins were precipitated from the liquid fraction by addition of 1N HCl to pH 4 and mild heating. The effects and interactions of the three factors examined here can be seen in Figure 6. The recovery of N-containing compounds was in the area of 21–46%, compared to 13–40% for DM, while SPC/SPI separated had a GPV between 60 and 98%. The selectivity of the fractionation varied in the area of 1.2–1.9. High pH values at low temperature favor both protein and DM recovery in SPC, resulting in low selectivities. Best selectivities were obtained at low pH values. The various response surfaces can be investigated by equations like the one following (in coded values as in Table 3):

$$\%\text{N recovery in SPC} = 33.00 - 2.75x_8 + 3.25x_9 + 1.25x_{10} - 4.00\,x_8x_9 + 1.50x_9x_{10} + 5.25x_8x_9x_{10} \quad (3)$$

Fractionation of lignocellulosic biomass

The search for an optimal fractionation of the three major components of lignocellulosic biomass has been one of the most fascinating technological problems of our time. This is mainly due to the fact that, because of the relative importance of lignocellulosics (Table 1), the future of biomass refining depends heavily upon the efficiency, selectivity, and product quality of this type of fractionation.

Numerous processes have been proposed and

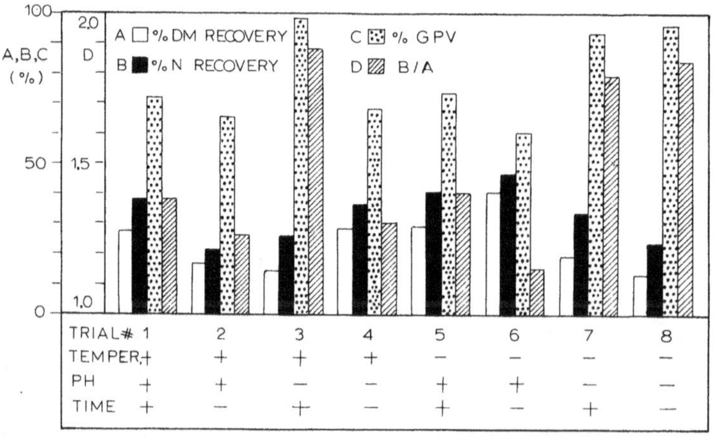

Fig. 6. Effects of major parameters on SPC production from the de-fatted meal of apricot (*Armenica vulgaris*) endosperms.

tested in the past; however, only two-stage acid hydrolysis has found an industrial application, although presently limited to a series of plants in the USSR. Most novel processes give emphasis to the physical and/or chemical pretreatments of this complicated structure to enhance its accessibility to chemical and biochemical agents (Millet et al., 1975). In an effort to understand the potential role and interrelations of some typical fractionation stages in the optimization of the whole process, we have attempted to analyze the scheme presented in Figure 7.

Prehydrolysis is by far the best method to separate hemicellulose as a sugar solution. In almost all proposed processes there is a step, where a contact with an acidic aqueous phase is employed for that purpose (Aronovsky & Gortner, 1936; Dietrichs et al., 1978; Lora & Wayman, 1978; Lee et al., 1978; Marchessault & St. Pierre, 1978;Ladisch et al., 1979; Sitton et al., 1979; Katzen et al., 1980; Wilke et al., 1981). Higher yields in the enzymatic conversion of the residue were observed, when prehydrolysis was carried out at high temperature (Knappert et al., 1981). Previous projects in this

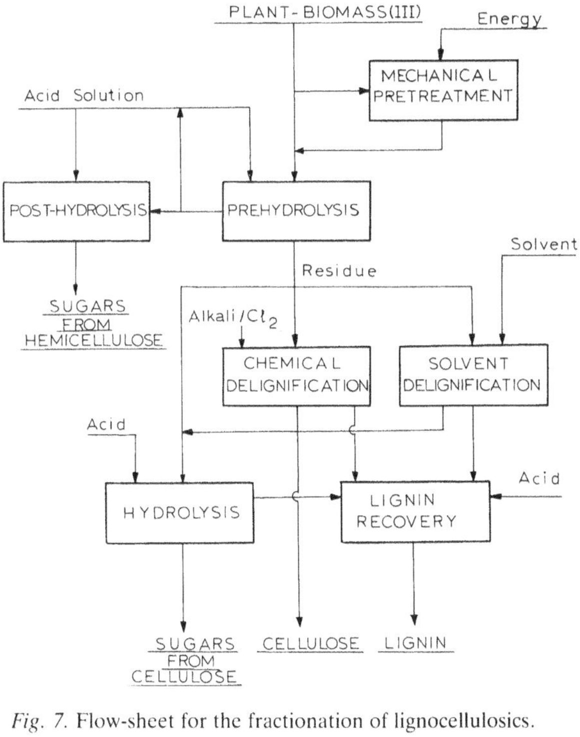

Fig. 7. Flow-sheet for the fractionation of lignocellulosics.

laboratory helped to investigate the effects of this prehydrolytic pretreatment (Koukios & Nicolacopoulos, 1979; Koukios, 1981; Koukios & Valkanas, 1982). Experiments on wheat (*Triticum sativum*) straw have shown that pentosan hydrolysis under strong conditions follows pseudo-first order kinetics with a rate constant given by the equation (Koukios, 1981)

$$k_p = 1.95 \cdot 10^{13} \cdot C_{\text{acid}}^{0.93} \cdot e^{-\frac{27.2}{RT}} \qquad (4)$$

Under those conditions, 90% of the original hemicellulose can be recovered as reducing sugars (Koukios & Valkanas, 1982). At the same time, the structure of the residue is significantly modified: a considerable fraction of cellulose, as high as 30% of the original value, is depolymerized also following pseudo-first order kinetics, and more than 50% of the original lignin becomes soluble in organic solvents, possibly after depolymerization, according to the same apparent kinetic pattern (Koukios, 1981).

Such modeling of prehydrolysis makes possible a preliminary investigation of its three major effects, especially regarding further treatment of the residue. The results are summarized in Table 4 and Figure 8, where for simplicity the yield of the residue is considered as an overall characteristic parameter (Koukios & Valkanas, 1982).

Another area of interest in this laboratory is that of the possible interactions between prehydrolysis and chemical delignification. Removal of lignin leads to a considerable improvement in the accessibility to enzymes (Millet et al., 1975; Fan et al., 1981), but not to acids, unless the delignified pulps are first mercerized or regenerated (Saeman, 1981). Overall, a highly selective and efficient delignification prcedure would be extremely useful as a fractionation tool. Coventional, alkaline pulping of prehydrolyzed biomass was proposed for that purpose as early as in 1952 (Sweeny et al., 1952). On the other hand, prehydrolysis has been found to have beneficial effects on the pulping of annual plants (Rudel, 1984; Richter, 1956).The potential of this combination can be investigated with the aid of Table 5 and a set of factorial experiments is

Table 4. The intensity of prehydrolysis as measured by the amount of the residue.

% Residue	Separation of hemicelluloses	Polysaccharides in residue	Lignin in residue
80	Limited (30%)	High DP[a] (800–900)	Unaltered
70	Extensive (65%)	Moderate DP (700–800)	Altered
60	Optimal (95%)	Low DP (550–650)	Degradation
50	Sugar degradation	Extensive attack on a-cellulose	Condensation

[a] Average degree of polymerization.

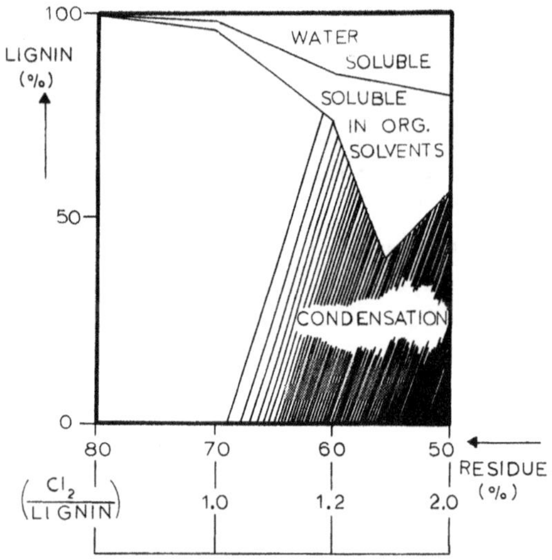

Fig. 8. The modifications of wheat straw lignin as a function of prehydrolysis residue (schematic). Condensation measured by chlorine consumption in delignification.

described in Table 6. Even at mild conditions of prehydrolysis (Table 5), high selectivities cannot be combined with the production of extensively delignified pulps. No substantial improvement is observed with the kraft method. Figure 9 makes possible an exploration of the peculiar response surface of selectivity, and particularly of the interactions between the two stages, according to the equation:

$$\text{Selectivity} = 1.49 + 0.55x_{11} - 0.09x_{12} - 0.06x_{14} + 0.01x_{16} + 0.30x_{13} + 0.19x_{11}x_{13} - 0.16x_{15} - 0.25x_{13}x_{15} \quad (5)$$

(the coded values of the factors are given in Table 6).

According to Figure 9, the selectivity of delignification is strongly affected by the intensity of

Table 5. Delignification of prehydrolyzed Wheat Straw[a] by alkaline agents.

Conditions				Pulp (% on straw)	Delignification (%)	Selectivity (lignin/sugars dissolved)
Temperature (°C)	Time (hr)	NaOH (%)	Liquid/solid			
98	2	0.5	7/1	59.4	41.0	2.94
98	2	1.0	7/1	53.9	52.5	1.85
98	2	2.0	7/1	43.9	74.0	1.19
98	2	3.0	15/1	38.7	84.7	0.92
98	4	3.0	15/1	39.5	87.4	1.03
98	6	3.0	15/1	36.4	89.6	0.88
160	3	1.6[b]	7/1	39.0	77.5	0.61
160	3	2.1[b]	7/1	33.5	94.5	0.65
160	3	2.7[b]	7/1	32.5	97.5	0.65

[a] 70% residue.
[b] Kraft, sulfidity 32%.

Table 6. Factors affecting the fractionation of lignocellulosics experimental regions investigated (biomass: corn stalk).

Factor Units	Code levels		
	−1	0	+1
A. *Prehydrolysis* (liquid/solid 10/1)			
(x_{11}) Temperature °C	130	150	170
(x_{12}) Time min	10	35	60
(x_{13}) Concentration of HCl % in liquid	0.5	1.0	1.5
B. *Delignification* (kraft, sulfidity 32%, liquid/solid 7/1)			
(x_{14}) Temperature °C	140	155	170
(x_{15}) Time min	30	45	60
(x_{16}) Active alkali % in liquid	2	2.5	3

prehydrolysis; however, high selectivities are observed at prehydrolytic conditions leading to extensive sugar degradation and, consequently, limited efficiency in the fractionation of corn (*Zea mays*) stalk hemicellulose.

Unconventional delignification of prehydrolyzed biomass by high-consistency chlorination followed by alkaline extraction of chlorolignins has been recently proposed as a high-efficiency, high selectivity procedure (Rudel, 1984). This is verified by the results shown in Table 7 and Figure 10. Moreover, an optimal point can be selected for each lignocellulosic material. Recovery of waste chlorolignins from the low-concentration pulping liquors of these experiments was quite satisfactory, although significantly species-selective (Table 7).

When lignocellulosics are converted to pulp by chemical delignification, less than half of the original biomass is actually utilized (cellulose), while the rest (hemicellulose, lignin) finds its way to the effluent (black liquor). As a result, this waste is characterized by (See among others Richter, 1956; Koukios, 1975; Gehm, 1965; Lamb, 1972; Ramalho, 1977)

(a) high levels of total solids (TS) per kg of pulp produced, usually in the range of 2:1;
(b) high oxygen demands, both chemical and biochemical, for the degradation of such an organic load;
(c) high values of viscosity, resulting in flow problems and difficulties in handling the waste stream; and
(d) high amounts of inorganics, especially silica, causing also a series of problems in the operation and maintenance of the conventional recovery and recycle plants.

The effects from the introduction of prehydrolysis before alkaline pulping can be seen in Table 8, while Table 9 shows the effects of the delignification method in combination to prehydrolysis. The

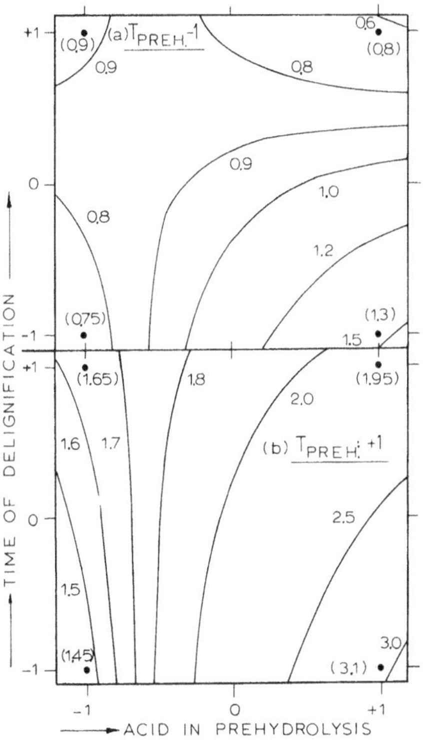

Fig. 9. Constant response plots for the selectivity of corn stalk delignification by prehydrolysis – pulping. Curves drawn according to equation (5) (see Table 5).

Table 7. Fractionation of lignocellulosic biomass by prehydrolysis – chlorination – lignin precipitation.

Biomass	Prehydrolysis[a]		Chlorination[c]				Precipitation[d]	
	Sugar (%)	Residue (%)	Pulp (%)	Delignific. (%)	Cl_2 Lignin[e]	Selectivity	Lignin (g/L)	Recovery (%)
Wheat straw	16.8	76.9	51.8	92.4	0.94	5.31	1.5	32.3
Rice straw	18.8	68.8	41.0	88.4	0.79	3.55	1.1	57.9
Corn stalk	35.4	57.1	36.5	89.9	1.21	7.35	1.7	67.4
Reed stalk	20.4	65.6	39.5	89.7	1.29	2.77	1.7	33.6
Tobacco stalk[b]	14.7	58.2	34.2	84.5	1.10	1.79	1.5	53.2
Grapevine stalk	18.9	69.1	45.8	72.7	1.41	12.93	1.7	68.8

[a] 98° C, 0.5% HCl, to maximum; post-hydrolyzed with 1 g/L HCl for 30 min at 98° C.
[b] 0.75% HCl.
[c] High – consistency, followed by alkaline extraction of chorolignins.
[d] Acidified by H_2SO_4 to pH 1 at room temperature.
[e] Dissolved.

pulping methods examined here are: soda, prehydrolysis-soda, kraft, prehydrolysis-kraft, alkaline sulfite, prehydrolysis-sulfite, and prehydrolysis-chlorination.

Tables 8 and 9 summarize the major results of our investigation. As a rule, the introduction of prehydrolysis leads to a substantially modified black liquor, and, in almost all cases, to significantly reduced loads. In particular, TS is found reduced by 18–25% (Table 8), its organics containing relatively more lignin than sugars (Table 9). The oxygen demand, both chemical and biochemical (of 5 and 7 days), is also reduced compared to the delignification of untreated straw in all cases, with the exception of kraft, where the pulp yield is the lowest. According to Table 9 prehydrolysis

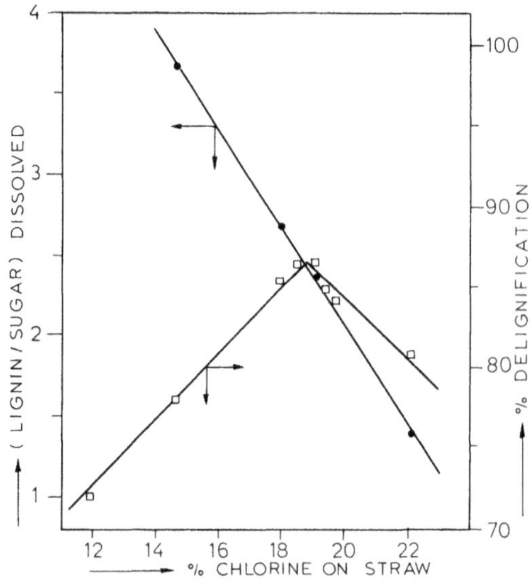

Fig. 10. Chlorination of prehydrolyzed (70% residue) wheat straw. Effects of Cl_2 supply.

Table 8. Effects of wheat straw prehydrolysis and delignification on the black liquor.

Method	Pulp yield (%)	TS (g/g pulp)	pH	BOD_7 (g/g pulp)	Viscosity (cst)	
					Initial	After precipitation[a]
Soda[b]	50.8	2.14	12.3	0.22	0.81	0.73
Prehydrolysis[c]-soda[b]	38.8	1.76	12.2	0.16	0.61	0.59
Soda[d]	48.8	2.46	12.3	0.27	0.84	0.77
Prehydrolysis[c]-Soda[d]	36.5	1.85	12.2	0.22	0.65	0.61

[a] Precipitation of ca. 90% of lignin from black liquor.
[b] Treatment for 2 hr at 98° C with 3% NaOH (15:1).
[c] Treatment for 6 hr at 98° C with 3% NaOH (15:1).
[d] Treatment for 3 hr at 98° C with 0.55% HCl (10:1). Residue ca. 70%.

Table 9. Effects of wheat straw pulping method on the organic load of black liquor.

Method	Pulp yield (%)	pH	Ash dissolved[a] (g/g pulp)	Organics (%)		COD (g/gpulp)	BOD_5 (g/gpulp)
				Lignin	Sugars		
Kraft[b]	51.5	10.6	0.12	28	72	1.99	0.31
Prehydrolysis[c]-kraft[b]	33.0	10.0	0.08	36	64	2.43	0.31
Alkaline sulfite[d]	55.5	9.0	0.11	32	68	1.73	0.23
Prehydrolysis[c]-Alk.sulfite[d]	37.5	8.5	0.06	41	59	1.50	0.12
Prehydrolysis[c]-chlorination[e]	50.1	6.5	0.03	66	34	0.56	0.03

[a] From original straw (i.e. in both pretreatment and treatment).
[b] Treatment for 120 min at 145° C with 18% chemicals (sulfidity 30%, liquid/solid 7:1).
[c] Treatment as in Note c of Table 8.
[d] Treatment for 180 min at 130° C with 24% chemicals (1/3 NaOH, liquid/solid 7:1).
[e] High-consistency treatment for 90 min at room temperature up to a Cl_2 consumption of 19.7% on untreated straw, followed by alkaline extraction (60° C, 2% NaOH).

decreases by 35–45% the amount of inorganics dissolved from straw under alkaline conditions (mostly SiO_2). This effect, combined with the reduction in the viscosity of the black liquor by 25%, obviously because of the change of its chemical composition, results in much improved possibilities for applying conventional recovery methods. Moreover, a comparison with the change of viscosity because of lignin precipitation (reduction by 10% or less) indicates that the major factor here is sugar content.

Chlorination appears by far to be the best method for the delignification of prehydrolyzed lignocellulose under the perspectives of this work. The utilization of the three major components is maximized while the pollution load is minimized. There also exists a considerable potential for the design of an efficient recovery system for HC1, $C1_2$, and NaOH.

Conclusions

Prehydrolysis of hemicellulose prior to chemical delignification can significantly contribute to the solution of the pollution and recovery problems in the pulp industry. The quality and load of the black liquor can be modified according to a given objective by simply adjusting a few process parameters.

The preceding analysis can serve as an outline of a systematic approach to the synthesis and optimization of plant-biomass fractionation sequences, in the prospective of a biomass refinery (Fig. 1). First, the biomass forms available will be considered, and their primary fractionation to typical units envisaged (Table 1). Then, based on the particular needs for specific products (Table 2) at the regional/national/international level, the most promising flow-sheets will be selected (Figs. 2, 5, and 7). The various stages of these sequences can be modeled by appropriate statistical and engineering techniques (Figs. 3, 4, 6, 8–10, Table 3). Besides, special attention should be given to modeling major interactions between stages (Figs. 8–10, Tables 4–7). Consequently, the proposed scheme will be described by a set of simple equations, like (1–5) presented above. Technical optima could be easily observed in this phase (Figs. 3, 6, 8–10). Finally, since all costs and energy needs can be expressed as functions of the examined process parameters (x_1 – x_{16} in this work), the optimization of the system, or its various subsystems, is greatly facilitated. Heuristic and feedback techniques should then employed to further improve the tested scheme.

References

Alexander, A.G., 1981. ACS symposium Series, 144: 49.
Archison, J.E., 1974. Tappi, 57 (11): 5.

Aronovsky, S.T. & Gortner, R.A., 1936. Ind. Eng. Chem., 28: 1270.
van den Berg, I.C., 1972. Plant protein Concentrates. In: Dechema-Monographien Band 70, Nr. 1327–1350 Weinheim.
Bungay III, H.R., 1983. Environ. Sci. Technol., 17 (1): 25A.
Crook, E.M., 1946. Biochem. J., 40: 197.
Dietrichs et al., H.H., 1978. Holzforchung 32 (6): 193.
Edwards et al., R.H., 1978. AICHE Symposium Servies 74: 158.
Fan et al., L.T., 1981. Biotechnol. Bioeng. Symp. 11: 29.
FAO, 1977. New Feed Resources. Animal Production and Health Paper 4, Rome.
FAO, 1978. Agricultural Residues: Compendium of Technologies. Agric. Serv. Bull. 33, Rome.
Fernández Diez, M.J., 1960. Grasas Y Aceites 11: 19.
Friend, G. & Maher D., 1982. Biofuels Development and Soil Productivity. Office of Appropriate Technology. State of California.
Gehm, H.W., 1965. Gurnham C.F. (ed.), Industrial Wastewater Control 2, p. 357. Academic Press, New York.
Hall, D.O., 1978. Fuel 57: 322.
Iammartino, N.R., 1974. Chem. Eng., August 5, p. 50.
Jahn, E.C., 1979. Tappi 62 (2): p. 31.
Katzen et al., R., 1980. Chem. Eng. Progress. February, p. 62.
Knappert et al., D., 1981. Biotechnol. Bioeng. Symp. 11: 67.
Koukios, E.G., 1981. Lignocellulose Hydrolysis Reactions. Proc. AICHE National Meeting, New Orleans.
Koukios, E.G., 1975. Dissertation. NTU of Athens (in Greek).
Koukios, E.G. & Economides, D.G., 1977. Bull. Greek Assoc. Chem. Eng. 17: 22 (in Greek).
Koukios, E.G. & Hatziyannakis, C., 1982. Organic Wastes in Greece. Proc. 7th Nath, Conf. Chem. Assoc., Ioanina (in Greek).
Koukios, E.G. & Nicolacopoulos, P., 1979. Optimization of Sugar, Cellulose and Lignin from cornstalks. Proc. 27th IUPAC Congress, Helsinki.
Koukios, E.G. & Valkanas, G.N., 1982. Ind. Eng. Chem. Prod. Res. Dev. 21: 309.
Ladisch et al., M.R., 1979. Energy 4: 263.
Lamb, J.C., 1972. Pulp and Paper Manufacturing Wastes, Reprint C-85, World Health Organization.
Lee et al., Y.Y., 1978. Biotechnol. Bioeng. Symp. 8: 75.
Lipinsky, E.S., 1979. ACS Adv. Chem. Series 181: 1.
Lora, J.H. & Wayman, N., 1978. Tappi 61 (6): 47.
Marchessault, R.H. & St-Pierre, J., 1978. A New Understanding of the Carbohydrate System. Proc. CHEMRAWN I Conf., Toronto.
Miladi, S. & Hegsted, D.M., 1974. Nutrition Reports International 9 (2): 117.
Millet et al., M.A., 1975. Biotechnol. Bioeng. Symp. 5: 193.
Moreira et al., A.R., 1981. Overview of Research Needs for Unconventional Food Production. FPBE/AICHE 2 (5): 1.
Pimentel, D. & M., 1979. Food, Energy and Society. E. Arnold, London.
Pirie, N.W. (ed.), 1971. Leaf Protein. Blackwell, Oxford.
Pirie, N.W., 1975. Food Protein Sources. Cambridge Univ. Press, Cambridge.
Pirie, N.W., 1976. Food Resources. Pelican.
Pirie, N.W., 1982. Ind. Eng. Chem. Prod. Res. Dev. 21: 299.
Präve et al., P., 1980. Interdisciplinary Science Reviews 5 (1): 6.
Ramalho, R.S., 1977. Introduction to Wastewater Treatment Processes. Academic Press, New York.
Richter, G.A., 1956. Tappi 39 (4): 193.
Rudel, E., 1894. D.P. 77: 598.
Saeman, J.F., 1981. ACS Symposium Series 144: 185.
Sitton et al., O.C., 1979. Chem. Eng. Progress. December, 52.
Spedding, C.R.W., 1979. An Introduction to Agricultural Systems. Applied Science Publishers, London.
Sweeny et al., O.R., 1952. U.S.P. 2 615: 883.
Szego, G.C. & Kemp, C.C., 1973. CHEMTECH 3: 275.
Tunney, H., 1980. Agricultural Wastes as Fertilizers. In: N.W.M. Bewick (ed.), Handbook of Organic Waste Conversion. Van Nostrand Reinhold, New York.
Wilke, C.R., (ed.), 1975. Cellulose as a Chemical and Energy Resource. Interscience, New York.
Wilke et al., C.R., 1981. Biotechnol. Bioeng. 23: 163.

GPSR Compliance

The European Union's (EU) General Product Safety Regulation (GPSR) is a set of rules that requires consumer products to be safe and our obligations to ensure this.

If you have any concerns about our products, you can contact us on ProductSafety@springernature.com

In case Publisher is established outside the EU, the EU authorized representative is:

Springer Nature Customer Service Center GmbH
Europaplatz 3
69115 Heidelberg, Germany

Batch number: 10762209

Printed by Printforce, the Netherlands